STATISTICAL TECHNIQUES
FOR NEUROSCIENTISTS

FRONTIERS IN NEUROSCIENCE

Series Editor
Sidney A. Simon, PhD

Published Titles

Apoptosis in Neurobiology
Yusuf A. Hannun, MD, Professor of Biomedical Research and Chairman, Department of Biochemistry and Molecular Biology, Medical University of South Carolina, Charleston, South Carolina
Rose-Mary Boustany, MD, tenured Associate Professor of Pediatrics and Neurobiology, Duke University Medical Center, Durham, North Carolina

Neural Prostheses for Restoration of Sensory and Motor Function
John K. Chapin, PhD, Professor of Physiology and Pharmacology, State University of New York Health Science Center, Brooklyn, New York
Karen A. Moxon, PhD, Assistant Professor, School of Biomedical Engineering, Science, and Health Systems, Drexel University, Philadelphia, Pennsylvania

Computational Neuroscience: Realistic Modeling for Experimentalists
Eric DeSchutter, MD, PhD, Professor, Department of Medicine, University of Antwerp, Antwerp, Belgium

Methods in Pain Research
Lawrence Kruger, PhD, Professor of Neurobiology (Emeritus), UCLA School of Medicine and Brain Research Institute, Los Angeles, California

Motor Neurobiology of the Spinal Cord
Timothy C. Cope, PhD, Professor of Physiology, Wright State University, Dayton, Ohio

Nicotinic Receptors in the Nervous System
Edward D. Levin, PhD, Associate Professor, Department of Psychiatry and Pharmacology and Molecular Cancer Biology and Department of Psychiatry and Behavioral Sciences, Duke University School of Medicine, Durham, North Carolina

Methods in Genomic Neuroscience
Helmin R. Chin, PhD, Genetics Research Branch, NIMH, NIH, Bethesda, Maryland
Steven O. Moldin, PhD, University of Southern California, Washington, D.C.

Methods in Chemosensory Research
Sidney A. Simon, PhD, Professor of Neurobiology, Biomedical Engineering, and Anesthesiology, Duke University, Durham, North Carolina
Miguel A.L. Nicolelis, MD, PhD, Professor of Neurobiology and Biomedical Engineering, Duke University, Durham, North Carolina

The Somatosensory System: Deciphering the Brain's Own Body Image
Randall J. Nelson, PhD, Professor of Anatomy and Neurobiology, University of Tennessee Health Sciences Center, Memphis, Tennessee

The Superior Colliculus: New Approaches for Studying Sensorimotor Integration
William C. Hall, PhD, Department of Neuroscience, Duke University, Durham, North Carolina
Adonis Moschovakis, PhD, Department of Basic Sciences, University of Crete, Heraklion, Greece

New Concepts in Cerebral Ischemia
Rick C.S. Lin, PhD, Professor of Anatomy, University of Mississippi Medical Center, Jackson, Mississippi

DNA Arrays: Technologies and Experimental Strategies
Elena Grigorenko, PhD, Technology Development Group, Millennium Pharmaceuticals, Cambridge, Massachusetts

Methods for Alcohol-Related Neuroscience Research
Yuan Liu, PhD, National Institute of Neurological Disorders and Stroke, National Institutes of Health, Bethesda, Maryland
David M. Lovinger, PhD, Laboratory of Integrative Neuroscience, NIAAA, Nashville, Tennessee

Primate Audition: Behavior and Neurobiology
Asif A. Ghazanfar, PhD, Princeton University, Princeton, New Jersey

Methods in Drug Abuse Research: Cellular and Circuit Level Analyses
Barry D. Waterhouse, PhD, MCP-Hahnemann University, Philadelphia, Pennsylvania

Functional and Neural Mechanisms of Interval Timing
Warren H. Meck, PhD, Professor of Psychology, Duke University, Durham, North Carolina

Biomedical Imaging in Experimental Neuroscience
Nick Van Bruggen, PhD, Department of Neuroscience Genentech, Inc.
Timothy P.L. Roberts, PhD, Associate Professor, University of Toronto, Canada

The Primate Visual System
John H. Kaas, Department of Psychology, Vanderbilt University, Nashville, Tennessee
Christine Collins, Department of Psychology, Vanderbilt University, Nashville, Tennessee

Neurosteroid Effects in the Central Nervous System
Sheryl S. Smith, PhD, Department of Physiology, SUNY Health Science Center, Brooklyn, New York

Modern Neurosurgery: Clinical Translation of Neuroscience Advances
Dennis A. Turner, Department of Surgery, Division of Neurosurgery, Duke University Medical Center, Durham, North Carolina

Sleep: Circuits and Functions
Pierre-Hervé Luppi, Université Claude Bernard, Lyon, France

Methods in Insect Sensory Neuroscience
Thomas A. Christensen, Arizona Research Laboratories, Division of Neurobiology, University of Arizona, Tuscon, Arizona

Motor Cortex in Voluntary Movements
Alexa Riehle, INCM-CNRS, Marseille, France
Eilon Vaadia, The Hebrew University, Jerusalem, Israel

Neural Plasticity in Adult Somatic Sensory-Motor Systems
Ford F. Ebner, Vanderbilt University, Nashville, Tennessee

Advances in Vagal Afferent Neurobiology
Bradley J. Undem, Johns Hopkins Asthma Center, Baltimore, Maryland
Daniel Weinreich, University of Maryland, Baltimore, Maryland

The Dynamic Synapse: Molecular Methods in Ionotropic Receptor Biology
Josef T. Kittler, University College, London, England
Stephen J. Moss, University College, London, England

Animal Models of Cognitive Impairment
Edward D. Levin, Duke University Medical Center, Durham, North Carolina
Jerry J. Buccafusco, Medical College of Georgia, Augusta, Georgia

The Role of the Nucleus of the Solitary Tract in Gustatory Processing
Robert M. Bradley, University of Michigan, Ann Arbor, Michigan

Brain Aging: Models, Methods, and Mechanisms
David R. Riddle, Wake Forest University, Winston-Salem, North Carolina

Neural Plasticity and Memory: From Genes to Brain Imaging
Frederico Bermudez-Rattoni, National University of Mexico, Mexico City, Mexico

Serotonin Receptors in Neurobiology
Amitabha Chattopadhyay, Center for Cellular and Molecular Biology, Hyderabad, India

TRP Ion Channel Function in Sensory Transduction and Cellular Signaling Cascades
Wolfgang B. Liedtke, MD, PhD, Duke University Medical Center, Durham, North Carolina
Stefan Heller, PhD, Stanford University School of Medicine, Stanford, California

Methods for Neural Ensemble Recordings, Second Edition
Miguel A.L. Nicolelis, MD, PhD, Professor of Neurobiology and Biomedical Engineering,
    Duke University Medical Center, Durham, North Carolina

Biology of the NMDA Receptor
Antonius M. VanDongen, Duke University Medical Center, Durham, North Carolina

Methods of Behavioral Analysis in Neuroscience
Jerry J. Buccafusco, PhD, Alzheimer's Research Center, Professor of Pharmacology and Toxicology,
    Professor of Psychiatry and Health Behavior, Medical College of Georgia, Augusta, Georgia

In Vivo Optical Imaging of Brain Function, Second Edition
Ron Frostig, PhD, Professor, Department of Neurobiology, University of California,
Irvine, California

Fat Detection: Taste, Texture, and Post Ingestive Effects
Jean-Pierre Montmayeur, PhD, Centre National de la Recherche Scientifique, Dijon, France
Johannes le Coutre, PhD, Nestlé Research Center, Lausanne, Switzerland

The Neurobiology of Olfaction
Anna Menini, PhD, Neurobiology Sector International School for Advanced Studies, (S.I.S.S.A.),
    Trieste, Italy

Neuroproteomics
Oscar Alzate, PhD, Department of Cell and Developmental Biology, University of North Carolina,
    Chapel Hill, North Carolina

Translational Pain Research: From Mouse to Man
Lawrence Kruger, PhD, Department of Neurobiology, UCLA School of Medicine, Los Angeles,
    California
Alan R. Light, PhD, Department of Anesthesiology, University of Utah, Salt Lake City, Utah

Advances in the Neuroscience of Addiction
Cynthia M. Kuhn, Duke University Medical Center, Durham, North Carolina
George F. Koob, The Scripps Research Institute, La Jolla, California

Neurobiology of Huntington's Disease: Applications to Drug Discovery
Donald C. Lo, Duke University Medical Center, Durham, North Carolina
Robert E. Hughes, Buck Institute for Age Research, Novato, California

Neurobiology of Sensation and Reward
Jay A. Gottfried, Northwestern University, Chicago, Illinois

The Neural Bases of Multisensory Processes
Micah M. Murray, CIBM, Lausanne, Switzerland
Mark T. Wallace, Vanderbilt Brain Institute, Nashville, Tennessee

Neurobiology of Depression
Francisco López-Muñoz, University of Alcalá, Madrid, Spain
Cecilio Álamo, University of Alcalá, Madrid, Spain

Astrocytes: Wiring the Brain
Eliana Scemes, Albert Einstein College of Medicine, Bronx, New York
David C. Spray, Albert Einstein College of Medicine, Bronx, New York

Dopamine–Glutamate Interactions in the Basal Ganglia
Susan Jones, University of Cambridge, United Kingdom

Alzheimer's Disease: Targets for New Clinical Diagnostic and Therapeutic Strategies
Renee D. Wegrzyn, Booz Allen Hamilton, Arlington, Virginia
Alan S. Rudolph, Duke Center for Neuroengineering, Potomac, Maryland

The Neurobiological Basis of Suicide
Yogesh Dwivedi, University of Illinois at Chicago

Transcranial Brain Stimulation
Carlo Miniussi, University of Brescia, Italy
Walter Paulus, Georg-August University Medical Center, Göttingen, Germany
Paolo M. Rossini, Institute of Neurology, Catholic University of Rome, Italy

Spike Timing: Mechanisms and Function
Patricia M. Di Lorenzo, Binghamton University, Binghamton, New York
Jonathan D. Victor, Weill Cornell Medical College, New York City, New York

Neurobiology of Body Fluid Homeostasis: Transduction and Integration
Laurival Antonio De Luca Jr., São Paulo State University–UNESP, Araraquara, Brazil
Jose Vanderlei Menani, São Paulo State University–UNESP, Araraquara, Brazil
Alan Kim Johnson, The University of Iowa, Iowa City, Iowa

Neurobiology of Chemical Communication
Carla Mucignat-Caretta, University of Padova, Padova, Italy

Itch: Mechanisms and Treatment
E. Carstens, University of California, Davis, California
Tasuku Akiyama, University of California, Davis, California

Translational Research in Traumatic Brain Injury
Daniel Laskowitz, Duke University, Durham, North Carolina
Gerald Grant, Duke University, Durham, North Carolina

Statistical Techniques for Neuroscientists
Young K. Truong, University of North Carolina, Chapel Hill, North Carolina
Mechelle M. Lewis, Pennsylvania State University, Hershey, Pennsylvania

STATISTICAL TECHNIQUES
FOR NEUROSCIENTISTS

Edited by

Young K. Truong
The University of North Carolina at Chapel Hill, USA

Mechelle M. Lewis
Pennsylvania State University, Hershey, Pennsylvania, USA

CRC Press
Taylor & Francis Group
Boca Raton London New York

CRC Press is an imprint of the
Taylor & Francis Group, an **informa** business

First published in paperback 2024

First published 2016 by CRC Press
2385 NW Executive Center Drive, Suite 320, Boca Raton FL 33431

and by CRC Press
4 Park Square, Milton Park, Abingdon, Oxon, OX14 4RN

CRC Press is an imprint of Taylor & Francis Group, LLC

© 2016, 2024 Taylor & Francis Group, LLC

Library of Congress Cataloging-in-Publication Data

Names: Truong, Young K., editor.
Title: Statistical techniques for neuroscientists / editor, Young K. Truong.
Other titles: Frontiers in neuroscience (Boca Raton, Fla.)
Description: Boca Raton : Taylor & Francis, 2016. | Series: Frontiers in neuroscience | Includes bibliographical references and index.
Identifiers: LCCN 2016003627 | ISBN 9781466566149 (alk. paper)
Subjects: | MESH: Statistics as Topic | Neurosciences
Classification: LCC RC337 | NLM WL 16 | DDC 616.80072/7--dc23
LC record available at http://lccn.loc.gov/2016003627

ISBN: 978-1-4665-6614-9 (hbk)
ISBN: 978-1-03-292027-6 (pbk)
ISBN: 978-1-315-37434-5 (ebk)

DOI: 10.1201/9781315374345

Visit the Taylor & Francis Web site at
http://www.taylorandfrancis.com

and the CRC Press Web site at
http://www.crcpress.com

Dedication

To David R. Brillinger
and
Chuck J. Stone
and
my lovely Psyche Lee

Contents

PART I   Statistical Analysis of Neural Spike Train Data

PART II   *Statistical Analysis of fMRI Data*

Chapter 3   A Hypothesis Testing Approach for Brain Activation 103

Wenjie Chen, Haipeng Shen, and Young K. Truong

Chapter 4   An Efficient Estimate of HRF 137

Wenjie Chen, Haipeng Shen, and Young K. Truong

Chapter 5   Independent Component Analysis: An Overview 215

Dong Wang, Seonjoo Lee, Haipeng Shen, and Young K. Truong

Chapter 6   Polynomial Spline Independent Component Analysis with
                    Application to fMRI Data 227

Atsushi Kawaguchi and Young K. Truong

Chapter 9   Diagnostic Probability Modeling for Longitudinal Structural Brain MRI Data Analysis 361

Atsushi Kawaguchi

Chapter 10   Supervised SVD of fMRI Data with Time-Varying Frequency Components 375

Avner Halevy and Young K. Truong

Preface

Why Data Analysis in Neuroscience

The brain contains about ten billion neurons (nerve cells) and they exchange signals with each other. Some form a group of neurons, each of which receives signals from a large number of other neurons, then sends off its own signals to the other neurons. One of the central problems for neural scientists is to relate signals (electrical activity) observed from the brain to the underlying physiological processes. A variety of recording methods are being used to observe this activity in the active brain. Intracellular recordings of the membrane potentials of individual neurons, extra-cellular spike recordings from one or more individual neurons, and recordings of signals that measure the activity of an ensemble of neurons either locally as the local field potential, or from larger brain regions via electroencephalography (EEG), magnetoencephalography (MEG) or functional magnetic resonance imaging (fMRI). Some recordings are very high in temporal resolution and are only possible for a single or small ensemble of neurons. Others tend to explore the spatial characteristics for the whole brain but lack the ability to capture the real dynamic of the activity. Thus any particular choice of recording method will be closely related to the research aims of the investigator about the mechanisms of neuronal processing.

Inspired by Hebb's influential work, and motivated by more recent physiological and anatomical findings, it is now very common to observe the activity of multiple single neurons simultaneously in order to understand the principles of coordinated neuronal activity and its spatio-temporal scales. While the coordinated dynamics are apparent in time-resolved multiple-channel measurements among neurons and groups of neurons, methods based on EEG, MEG or fMRI for brain activation focus more on the brain regions of interest. Thus, the analysis of data from simultaneously recorded spike trains or larger brain regions (EEG or fMRI) will allow us to relate concerted activity of ensembles of neurons or brain function to behavior and cognition. Different statistical data analyses are thereby relevant to distinguish different or even complementary spatio-temporal scales. Statistical analysis of brain data is the logical follow-up to improve our understanding of the neuronal activity or network underlying information processing in the brain.

Purpose of the Book

The book aims at introducing new and useful methods for data analysis involving simultaneous recording of a neuron or large cluster (brain region) of neuron activity. The statistical estimation and tests of hypotheses are based on the likelihood principle derived from stationary point processes and time series. Algorithms and software development are given in each chapter; the main objective is to reproduce the computer simulated results described therein.

Intended Audience

This book is intended for neural scientists who are interested in analyzing multi-channel time series data arising from brain functions or activities. It is also for statisticians who are familiar with traditional multivariate time series analysis and spatial modeling. It is especially for readers who wish to employ intense computing methods to extract important features or information directly from the data rather than relying heavily on models built upon leading cases such as linear regression or Gaussian processes. It is also for practitioners or scientists who are interested in reproducible research. Basic knowledge of R and MATLAB® is essential for understanding and reproducing the numerical results.

Organization of the Book

The first part of the book deals with the traditional multivariate time series analysis applied in the context of multichannel spike trains and fMRI using, respectively, the probability structures or likelihood associated with time to fire and discrete Fourier transforms (DFT) of point processes. The second part introduces a relatively new way of statistical spatio-temporal modeling for fMRI and EEG data analysis.

Some useful facts and computing tools are collected in the appendices for quick reviews or tutorials in using the software described in each chapter.

How to Read This Book

For multichannel spike train or fMRI voxel time series data analysis, the reader should start with Part I of the book. For spatio-temporal modeling, start with Part II. Instructions or tutorials for setting up the computing environment and our software packages are given in the appendices at the end of the book.

Software Download

Software packages can be acquired by contacting the first editor (YT) and will be made available online at http://sph.unc.edu/adv_profile/kinh-truong-phd/.

Acknowledgments

We wish to thank Professor David R. Brillinger for his vision of applications of time series to neural science. He was one of the pioneers in the field long before we started to analyze neural spike train and fMRI data. We are very grateful for his teaching and research contributions to statistical computing and neural science. Chuck Stone has played a very critical role in my (YT) academic career. As teacher and PhD advisor, Chuck has defined the way I view statistics and its applications. This book is dedicated to both of them.

We also wish to acknowledge the immense contributions of Dr. X. Huang for her continuing support of this project; her insights, energy, and the human brain data from her lab have been instrumental in shaping our statistical approaches to fMRI data analysis. Some of the procedures in Chapters 3 and 4 were developed based on the time series package written by Professor J. Newton. The book was initiated by Dr. Sidney Simon, the editor of this series, his encouragement and patience have led to countless improvements in preparing this book. We wish to thank Barbara Norwitz, editor of the book series at Taylor & Francis, for her patience and understanding of my family issues during the course of preparing the book. It has been a pleasure to work with her. We also thank Karen Simon for her superb proofreading of the manuscript, and Jill Jurgensen for lending her technical support on the LATEX package and template for this book series. Part of this project has been supported by National Science Foundation NSF DMS-0707090 and DMS-1106962, National Institute of Child Health and Human Development NIH U54HD079124 (PI: Dr. Piven), and UNC Clinical Translation Science Award, Biostatistics Core TR000083-05.

Finally, YT wishes to dedicate this book to his lovely wife Psyche for her unconditional support of this project and he is grateful to his parents Cung Binh, Hong, Arthur and Jenny.

MATLAB® is registered trademark of the MathWorks, Inc. For product information please contact

The MathWorks, Inc.
3 Apple Hill Drive
Natick, MA 01760-2098 USA
Tel: 508-647-7000
Fax: 508-647-7001
E-mail: info@mathworks.com
Web: www.mathworks.com

Editors

Young K. Truong

Young K. Truong, PhD, is professor of biostatistics at the University of North Carolina-Chapel Hill. His research expertise includes statistical learning, functional modeling, time series, spatio-temporal data analysis, and event-history analysis. He has contributed significantly in the areas of statistical time series/longitudinal modeling using splines, window or kernel-based smoothing methods, and wavelets. His current research focuses mainly on spatio-temporal data analysis with the aim to spatially localize dynamic processes in the functional magnetic resonance imaging (fMRI) or electroencephalography (EEG) human brain data. He is also interested in developing statistical inference for group comparison based on human brain imaging data, and methods for analyzing neuro spike train data. His teaching and research experience in linear mixed modeling has provided new insights and approaches to many statistical analyses for handling between- and within-subject variability.

Mechelle M. Lewis

Professor Mechelle M. Lewis, PhD, has been with Departments of Neurology and Pharmacology at the Pennsylvania State College of Medicine in Hershey since 2008. She has a broad background in neurobiology, with specific training in key research areas (Parkinson's disease, MRI techniques, data acquisition and analysis). She has worked on all aspects of several clinical translational projects from design, subject recruitment, imaging data collection, management, and analysis to interpretation, manuscript writing, and grant preparation. Dr. Lewis has a keen interest in both normal and dysfunctional brain processes, as well as the development of tools and markers to help better understand brain function. Currently, she is a co-investigator on NIH-funded R01 and U01 projects, one focused on development of imaging biomarkers in Parkinson's disease and the other on understanding the role of environmental neurotoxicants on the development of neurobehavioral disorders. Dr. Lewis also is a co-investigator on a grant funded by the Michael J. Fox Foundation that aims to develop a tool for discriminating Parkinson's from parkinsonism patients. She is a member of the Society for Neuroscience, American Association for the Advancement of Science, the Movement Disorders Society, and the American Neurological Association, and actively reviews manuscript for several journals.

Contributors

Wenjie Chen
American International Group
New York, New York

Avner Halevy
University of North Carolina
Chapel Hill, North Carolina

Atsushi Kawaguchi
Kyoto University
Kyoto, Japan

Seonjoo Lee
Columbia University
New York, New York

Shih-Chieh Lin
National Institutes of Health
Bethesda, Maryland

Haipeng Shen
University of Hong Kong
Hong Kong, China

Dong Wang
University of North Carolina
Chapel Hill, North Carolina

Ruiwen Zhang
Statistical Analysis System (SAS)
    Institute Inc.
Cary, North Carolina

List of Figures

List of Tables

Part I

Statistical Analysis of Neural Spike Train Data

1 Statistical Modeling of Neural Spike Train Data

Ruiwen Zhang
SAS Institute

Shih-Chieh Lin
NIH

Haipeng Shen
University of Hong Kong, China

Young K. Truong
University of North Carolina at Chapel Hill

CONTENTS

The advance of the multi-electrode has made the field of neural science feasible to record spike trains simultaneously from an ensemble of neurons. However, the statistical techniques for analyzing large-scale simultaneously recorded spike train data have not developed as satisfactorily as the experimental techniques for obtaining these data. This chapter describes a very flexible statistical procedure for modeling an ensemble of neural spike trains, followed with the associated estimation method for making an inference for the functional connectivity based on the statistical results. To approach this problem in a more concrete manner, we use stochastic point processes to develop a statistical model for analyzing an ensemble of spiking activities from noncholinergic basal forebrain neurons [11]. The formulation is equipped with the likelihood (or loosely, the probability) of the occurrence of the neural spike train data, based on which the statistical estimation and inference will be carried out. The model can assess the association or correlation between a target neuron and its peers.

1.1   INTRODUCTION

It is known that neurons, even when they are apart in the brain, often exhibit correlated firing patterns [22]. For instance, coordinated interaction among cortical neurons is known to play an indispensable role in mediating many complex brain functions with highly intricate network structures [23]. A procedure to examine the underlying connectivity between neurons can be stated in the following way. For a target neuron i in a population of N observed neurons, we need to identify a subset of neurons that affect the firing of the target in some statistical sense.

In the study of neural plasticity and network structure, it is desirable to infer the underlying functional connectivity between the recorded neurons. In the analysis of neural spike trains, functional connectivity is defined in terms of the statistical dependence observed between the spike trains from distributed and often spatially remote neuronal units [6]. This can result from the presence of a synaptic link between neurons, or it can be observed when two unlinked neurons respond to a common driving input.

Characterization of the functional network requires simultaneous monitoring of neural constituents while subjects carry out certain functions. Technology developments in multi-electrode recording enable us to easily obtain the activities of ensembles of spiking neurons simultaneously. Easier access to data has underlined the need for developing analysis methods that can process these data quickly and efficiently [10, 14, 9, 21, 2]. In this chapter, we apply and develop statistical approaches to analyze simultaneously recorded neural spike trains and infer functional connectivity between neurons that act in concert in a given brain region or across different regions.

Early methods of inferring connectivity usually focus on analyzing pairs of neurons using time series techniques (e.g., cross-correlograms or joint peri-stimulus time histograms) or frequency-domain methods (e.g., cross-coherence) [2], but pairwise methods generally provide only an incomplete picture of the connections among several neurons. More recent methods, such as gravitational clustering and spike-pattern classification methods [20], have made steps toward estimating functional connectivity that is greater than pairwise, but they are still mostly suitable for pairs or triplets. When large-scale recording of multiple single units became more common, using these methods to infer a complex dynamic network structure became nontrivial.

1.2   POINT PROCESS AND CONDITIONAL INTENSITY FUNCTION

Likelihood methods under the generalized linear model (GLM) framework are increasingly popular for analyzing neural ensembles [10, 9, 21, 17, 12, 15]. We ap-

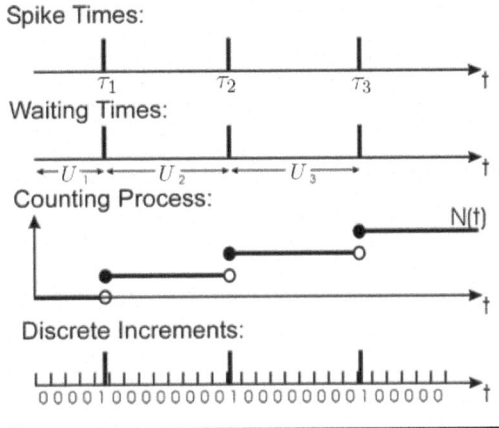

Figure 1.1: Multiple specifications for point process data.

plied a point process likelihood model under the GLM framework [9] for some real applications. The spike train data come from noncholinergic basal forebrain neurons [11], which have never been analyzed under the point process likelihood framework. Harris's method approaches the problem similarly in the spirit of [1], but it models the intensity function as the sum of weighted Gaussian kernels and adds a penalty term in the log likelihood. We will discuss more details in Section 1.4.

A point process may be specified in terms of spike times, spike counts, and inter-spike intervals. Let $(0, T]$ denote the observation interval, and suppose that a neuron fires at times τ_i, $i = 1, 2, ..., n$, where $0 < \tau_1 < \tau_2 < \cdots < \tau_{n-1} < \tau_n \leq T$ is a set of n spike times. For $t \in (0, T]$, a spike train is defined as

$$S(t) = \sum_i \delta(t - \tau_i),$$

where τ_i is the time of the ith spike and $\delta(\cdot)$ is known as Dirac delta function which follows

$$\delta(x) = \begin{cases} 1 & \text{if } x = 0 \\ 0 & \text{otherwise.} \end{cases}$$

When the τ_i are random, one has a stochastic point process $\{\tau_i\}$.

For spikes $\{\tau_i\}$ randomly scattered along a line, the counting process $N(t)$ gives the number of points observed in the interval $(0, t]$

$$N(t) = \#\{\tau_i \text{ with } 0 < \tau_i \leq t\} \tag{1.1}$$

where # stands for the number of events (which have occurred up to time t). The counting process satisfies

(i) $N(t) \geq 0$;

(*ii*) $N(t)$ is an integer-valued function;
(*iii*) if $s < t$, then $N(s) \leq N(t)$;
(*iv*) for $s < t$, $N(t) - N(s)$ is the number of events in (s,t).

A counting process is said to have *independent increments* if the number of events that occur in disjoint time intervals is independent. A counting process is said to have *stationary increments* if the number of events that occur in any time interval depends only on the length of the interval. Independent increment and stationary increment are strong conditions and rarely hold for neuronal data since spike trains depend on history, on which we shall elaborate in the next section.

As we know, a Poisson process is a special case of a point process. A counting process $N(t)$ $t \geq 0$ is a homogeneous (stationary) Poisson process with rate λ, $\lambda > 0$, if

(*i*) $N(0) = 0$;
(*ii*) the process has independent increments;
(*iii*) $\mathrm{P}(N(t+\Delta) - N(t) = 1) = \lambda\Delta + o(\Delta)$; and
(*iv*) $\mathrm{P}(N(t+\Delta) - N(t) \geq 2) = o(\Delta)$;

where Δ is a very small interval and $o(\Delta)$ refers to all events of order smaller than Δ, such as two or more events that occur in an arbitrarily small interval.

The probability mass function is

$$\mathrm{P}(N(t) = k) = \frac{e^{-\lambda t}(\lambda t)^k}{k!}, \tag{1.2}$$

for $k = 0, 1, 2, \ldots$. The mean and the variance of the Poisson process are λt.

To construct the *interspike interval (ISI) probability* for the Poisson process, we denote U as the *interspike interval* between two conjoint events (also called the waiting time), and

$$\mathrm{P}(N(u) = 0) = \mathrm{P}(U > u) = e^{-\lambda ut} \tag{1.3}$$

or

$$\mathrm{P}(u \geq U) = 1 - e^{-\lambda u}. \tag{1.4}$$

Differentiating (1.4), we have the probability density of the interspike interval, which is the exponential density

$$p(u) = \lambda e^{-\lambda u}. \tag{1.5}$$

We can show that, given the assumption that if the interspike interval probability density of a counting process is exponential, the counting process is a Poisson process.

Once we understand the interevent probability density, we can consider the spike train as segments of interspike intervals, $\{u_i\}$, and the *waiting time density* function is the probability density of the time until the occurrence of the kth event. Denote $s_k = \sum_{i=1}^{k} u_i$ as the waiting time up to k^{th} spike. Then using the properties of a Poisson process, we obtain

$$
\begin{aligned}
P(t < s_k < t+\Delta) &= P(N(t) = k-1 \text{ and one event in } (t,t+\Delta]) + o(\Delta) \\
&= P(N(t) = k-1)\lambda\Delta + o(\Delta) \\
&= \frac{(\lambda t)^{k-1}}{(k-1)!}e^{-\lambda t}\lambda\Delta.
\end{aligned} \tag{1.6}
$$

Hence, the waiting time density function until the kth event is

$$
p_k(t) = \lim_{\Delta \to 0} \frac{P(N(t) = k-1)\lambda\Delta + o(\Delta)}{\Delta} = \frac{(\lambda t)^{k-1}}{(k-1)!}e^{-\lambda t}\lambda, \tag{1.7}
$$

which also gives

$$
p_k(t) = \frac{\lambda^k t^{k-1}}{\Gamma(k)}e^{-\lambda t}. \tag{1.8}
$$

The waiting time density is a gamma distribution with parameters k and λ.

The Poisson process is one of the simplest and most commonly used classes of neural spiking models. Poisson processes are characterized by lack of memory, meaning that the probability distribution of spikes at any point is independent of all previous activities. In some cases, especially when spikes are rare compared with the time scale of the intrinsic membrane dynamics or the effect of history has been averaged out by combining multiple spike trains, Poisson processes can accurately describe spiking activity.

However, Poisson processes are rarely realistic for various neural spike train data. In particular, the biophysical properties of ion channels limit how fast a neuron can recover immediately following an action potential, leading to a refractory period during which the probability of firing another spike is zero immediately afterward and then significantly decreased further after the previous spike. This is perhaps the most basic illustration of history dependence in neural spike trains.

Since most neural systems have a history-dependent structure, it is necessary to define a firing rate function that depends on history. Any point process can be completely characterized by the *conditional intensity function*, $\lambda(t|H_t)$ [5], which is defined as follows,

$$
\lambda(t|H_t) = \lim_{\Delta \to 0} \frac{P(N(t+\Delta) - N(t) = 1|H_t)}{\Delta}, \tag{1.9}
$$

where H_t is the history of the spike train up to time t, including all the information of the track and number of spikes in $(0,t]$ and any covariances up to time t as well. Therefore, $\lambda(t|H_t)\Delta$ is the probability of a spike in $(t,t+\Delta]$ when there is history dependence in the spike train. In survival analysis, the conditional intensity function $\lambda(t|H_t)$ is called the *hazard function*, the probability of an event in the interval $[t,t+\Delta)$ given that there has not been an event up to t [4].

The conditional intensity function generalizes the definition of the Poisson rate [3,4]. If the point process is an inhomogeneous Poisson process, then $\lambda(t|H_t) = \lambda(t)$.

It follows that $\lambda(t|H_t)\Delta$ is the probability of a spike in $[t,t+\Delta)$ when there is a history-dependent spike train.

When Δ is small, (1.9) can be rewritten as

$$P(N(t+\Delta) - N(t) = 1|H_t) \approx \lambda(t|H_t)\Delta, \qquad (1.10)$$

which approximates the probability of a spike firing in a small time interval $(t,t+\Delta]$ given the spike history up to time t. Thus the intensity function $\lambda(\cdot)$ is proportional to the probability of neuron firing. The estimation of the probability distribution of the next firing time for arbitrary time-dependent input current is one of the major goals in the theory of noisy spiking neurons [7].

There are two primary ways to characterize a point process. The first is in terms of the interevent probability model or interspike interval probability model specifically, and the second is the conditional intensity function. Actually, defining one defines the other and vice versa.

Under the framework of survival analysis, $\lambda(t|H_t)$ can be defined in terms of the interspike interval density at time t, $p(t|H_t)$, as

$$\lambda(t|H_t) = \frac{p(t|H_t)}{1 - \int_0^t p(t|H_t)dt}, \qquad t \geq 0. \qquad (1.11)$$

The conditional intensity function completely characterizes the stochastic structure of the spike train. In any time interval $(t,t+\Delta]$, $\lambda(t|H_t)\Delta$ defines the probability of a spike given the history up to time t. If the spike train is an inhomogeneous Poisson process, then $\lambda(t|H_t) = \lambda(t)$ becomes the generalized definition of the Poisson rate. We can write

$$\lambda(t|H_t) = -\frac{d\left[\log[1 - \int_0^t p(s|H_s)ds]\right]}{dt}. \qquad (1.12)$$

Integration gives

$$-\int_0^t \lambda(s|H_s)ds = \log\left[1 - \int_0^t p(s|H_s)ds\right]. \qquad (1.13)$$

Finally, exponentiating yields

$$\exp\left\{-\int_0^t \lambda(s|H_s)ds\right\} = 1 - \int_0^t p(s|H_s)ds. \qquad (1.14)$$

Therefore, by (1.11–1.14), we have

$$p(t|H_t) = \lambda(t|H_t)\exp\left\{-\int_0^t \lambda(s|H_s)ds\right\}. \qquad (1.15)$$

So, given the conditional intensity function, the interspike interval density function is specified and vice versa. Hence, defining one completely defines the other.

1.3 THE LIKELIHOOD FUNCTION OF A POINT PROCESS MODEL

The likelihood function is formulated by deriving the joint distribution of the data and then, viewing this joint distribution as a function of the model parameters with the data fixed. The likelihood approach has many optimality properties that makes it a central tool in statistical theory and modeling.

With the interspike interval density function for any time t, the likelihood of a neural spike train is defined by the joint probability density function of all the data, or in another word, it is the joint probability density function of exactly n events happening at times τ_i, for $i = 1, 2, \ldots n$. This is given by

$$
f(\tau_1, \tau_2, \ldots, \tau_n, N(T) = n) = \prod_{k=1}^{n} \lambda(\tau_k | H_{\tau_k}) \exp \left\{ -\int_0^T \lambda(s|H_s) ds \right\}
$$

$$
= \exp \left\{ \int_0^T \log \lambda(s|H_s) dN(s) - \int_0^T \lambda(s|H_s) ds \right\}.
$$

$$(1.16)$$

Also, the joint probability density of a spike train can be written in a canonical form in terms of the condition intensity function [24]. That is, when formulated in terms of the conditional intensity function, all point process likelihoods have the form given by (1.16).

If the density function in (1.16) depends on an unknown q-dimensional parameter θ to be estimated, then we write the likelihood equation as a function of θ:

$$
L(\theta) = \exp \left\{ \int_0^T \log \lambda(s|H_s, \theta) dN(s) - \int_0^T \lambda(s|H_s, \theta) ds \right\}. \qquad (1.17)
$$

One of the most compelling reasons to use maximum likelihood estimation (MLE) in neural spike train data analyses is that for a broad range of models, these estimates have other important optimality properties in addition to being asymptotically Gaussian. First, there is consistency, which states that the sequence of maximum likelihood estimates converges in probability (or more strongly almost surely) to the true value as the sample size increases. Second, the convergence in probability of the estimates means that they are asymptotically unbiased. That is, the expected value of the estimate $\hat{\theta}$ is θ as the sample size increases. For some models and some parameters, unbiasedness is a finite sample property. The third property is invariance. That is, if $\psi(\theta)$ is any transformation of θ and $\hat{\theta}$ is the maximum likelihood estimate of θ, then $\psi(\hat{\theta})$ is the maximum likelihood estimate of $\psi(\theta)$. Finally, the maximum likelihood estimates are asymptotically efficient in that as the sample size increases, the variance of the maximum likelihood estimate achieves the Cramer–Rao lower bound. This lower bound defines the smallest variance that an unbiased or asymptotically unbiased estimate can achieve. Like unbiasedness, efficiency for some models and some parameters is achieved in a finite sample. Detailed discussions of these properties are given in the statistical inference book [3].

1.4   CONTINUOUS STATE-SPACE MODEL

Synchronized firing of related neurons is an interesting fact which motivates us to develop statistical analysis using the valuable information of concurrent spiking activities (also called *peer activities*). To develop statistical approaches for the analysis of simultaneously recorded neural spike trains and infer functional connectivity between interactive neurons in the brain, we employ the likelihood of a neuron (the target) and model its firing probability as a function of the other neurons (the peers) in the network or ensemble. In this chapter, we introduce two models for this type of analysis. A more flexible one will be described in the next chapter.

We have shown that the point process likelihood function can be analyzed under the GLM framework and that the conditional log-intensity function can be expressed as a general function consisting of extrinsic covariates, its own past activity, and concurrent spiking activities. A "peer prediction" method was introduced for the situations when external factors cannot explain some unpredictable spike times [9]. We also showed that the firing rate can be predicted better by incorporating the spike times of a simultaneously recorded assembly population than by using only external variables. There are three major steps in the method. Initially, all the peer spike trains are smoothed with a Gaussian filter in the time domain.

$$s(t)^\alpha = \frac{1}{\sqrt{2\pi\sigma^2}} \sum_{j=1}^{n^\alpha} \exp\left(-\frac{(t - \tau_j^\alpha)^2}{2\sigma^2}\right), \qquad (1.18)$$

where α is the index of peer cells, $\{\tau_j^\alpha\}$ is the spike train of neuron α, and n^α is the number of spikes of peer α. σ is the smoothing bandwidth of the Gaussian kernel, termed *peer prediction timescale* by the author.

Then, under the generalized linear model, the predicted intensity function at time t is given by

$$\lambda(t) = g\left(\sum_\alpha s(t)^\alpha w^\alpha\right), \qquad (1.19)$$

where w^α is the prediction weight of peer α. The sign of the weight represents the positive or negative correlation between a certain member neuron and the target neuron. Here $g(\cdot)$ is the link function and has the following form

$$g(x) = \begin{cases} \exp(x) & x < 0 \\ x + 1 & \text{otherwise.} \end{cases} \qquad (1.20)$$

The advantage of this link function is that it will not lead to excessively high intensity when many positively correlated peer cells fire simultaneously compared with a simple exponential link function.

The final step is to estimate the weight by maximizing the penalized log-likelihood on the training set given by

$$\log(L_f) = \sum_t [-\lambda(t)\Delta + \Delta N_t \log(\lambda(t)\Delta)] - \frac{1}{4} \sum_\alpha (w^\alpha)^2. \qquad (1.21)$$

The maximum is carried out by Newton's method with an analytically calculated Hessian matrix.

A 10-fold cross-validation procedure is used to repeatedly divide the recorded data into a training set and a test set. For each training set, the mean firing rate is calculated, f_0, as the number of spikes during the training period divided by the length of the training period. The prediction quality of the test set, termed *predictability*, is defined as the difference, $\log(L_f) - \log(L_{f_0})$, over the base of $\log(2)$. Then, the predictability of the entire data set is defined by a cross-validation procedure, where the data are divided into 10 segments, each segment is used in turn as test set, and the log likelihood ratios for each segment are summed and divided by the total time.

As the spike train is smoothed by the Gaussian kernel, then the choice of the bandwidth, σ in Equation (1.18), is critical. The optimal bandwidth, σ, is chosen by optimizing the predictability; therefore, it maximizes the log-likelihood function $\log(L_f)$ among all the values of σ [9].

1.4.1   KERNEL SMOOTHING

The firing rate is a fundamental concept for the description of a spiking neuron (a point process in general). The underlying firing rate $\lambda(t)$ is a non-negative deterministic function of time, such that the integral $\int_t^{t+\Delta} \lambda(u)du$ represents the expected number of spikes encountered in an observation of the neuron during the observation interval $(t, t+\Delta]$. The rate function underlying the spiking of a real neuron, however, cannot be observed directly; it must be reconstructed from the recorded spike trains.

The convolution method with a fixed kernel function can be described to estimate the neuronal firing rate from single-trial spike trains [13]. Consider a single spike train, comprised of a finite number of discrete spike events at times $\tau_1, ..., \tau_n$. We define the estimation of the time-varying rate function as

$$\lambda(t) \doteq \sum_{i=1}^{n} K(t - \tau_i), \tag{1.22}$$

where $K(t)$ is called the *kernel function*. Thus, the desired underlying rate function $\rho(t)$ is estimated from a single-trial spike train by taking the sum over kernel functions $K(t - \tau_i)$, centered at spike occurrence times τ_i. Figure 1.2 illustrates this concept with a normalized triangular kernel function. See Table 1.1.

The kernel $K(t)$ is required to be non-negative to avoid negative rates. Moreover, the kernel should be normalized such that each spike contributes with unit area to the rate function; this also guarantees that the integral of $\lambda(t)$ is equal to the total number of spikes n recorded during the interval $(0, T]$. Finally, the first moment of $K(t)$ is required to be zero to preserve the center of mass of the spike train.

There are two important aspects to specify a kernel function: the shape of the kernel function and its width. The kernel shape determines the visual appearance of the estimated rate function. The kernel width or bandwidth is defined as

$$\sigma = \sqrt{\int_{-\infty}^{\infty} t^2 K(t)dt}, \tag{1.23}$$

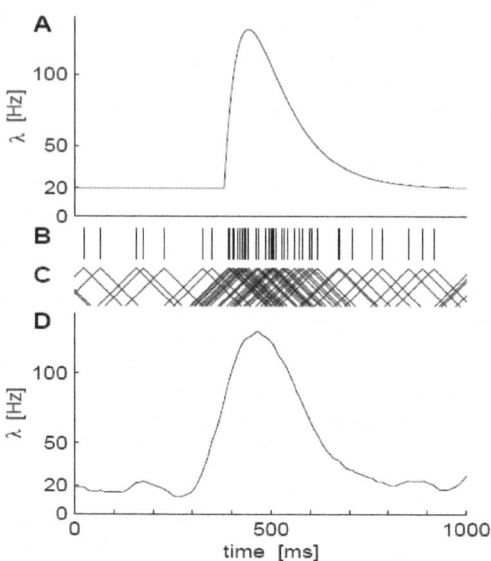

Figure 1.2: Concept of single-trial rate estimation by means of the kernel approach. (A) The true underlying rate function. (B) One single-trial spike train. (C) Kernel functions centered at spike occurrence times. (D) The empirical rate function. In this particular example, a triangular kernel with a standard width of 40 ms was used.

which can be viewed as a smoothing parameter. Table 1.1 lists four kernel functions of different shapes, parameterized by their standard width (deviation).

We evaluated the rate estimators depending on shape and width of kernel functions, and the integrated square error (ISE) is defined as

$$ISE \doteq \int_0^T (\lambda(t) - \rho(t))^2 \, dt. \tag{1.24}$$

The conclusion was drawn that the choice of a specific kernel shape is not critical for the estimate; while the performance of the estimate in terms of minimizing the mean ISE improves when the choice of bandwidth is close to the optimal [13]. Among all the kernel functions, a Gaussian shaped kernel, i.e., a Gaussian kernel density estimator, is most often used. Thus, (1.22) is given by

$$\lambda(t) = \frac{1}{\sqrt{2\pi\sigma^2}} \sum_{i=1}^{n} \exp\left(-\frac{(t - \tau_i)^2}{2\sigma^2}\right). \tag{1.25}$$

When $\lambda(t)$ varies slowly, Gaussian kernels do a good job of estimating the rate and filtering out the noise. Nevertheless, when the firing rate varies quickly, Gaussian filters are not able to capture the variation without introducing artificial high-frequency fluctuations. In other words, to filter high-frequency noise, the Gaussian kernel den-

Table 1.1

Kernel functions. The kernel functions are defined to be zero outside the support. All kernels have been normalized to have unit area (probability density function) and have standard deviation σ.

Kernel	$K(t, \sigma)$	Support
Boxcar	$\dfrac{1}{2\sigma\sqrt{3}}$	$[-\sigma\sqrt{3}, \sigma\sqrt{3}]$
Triangle	$\dfrac{1}{6\sigma^2}\left(\sigma\sqrt{6} - |t|\right)$	$[-\sigma\sqrt{6}, \sigma\sqrt{6}]$
Epanechnikov	$\dfrac{3}{4\sigma\sqrt{5}}\left(1 - \dfrac{t^2}{5\sigma^2}\right)$	$[-\sigma\sqrt{5}, \sigma\sqrt{5}]$
Gaussian	$\dfrac{1}{\sigma\sqrt{2\pi}}\exp\left(-\dfrac{t^2}{2\sigma^2}\right)$	$[-\infty, \infty]$

sity estimator must remove the high-frequency firing rate. In general, the most obvious advantage of kernel smoothing is its simplicity. Kernel smoothing methods are extremely fast and simple to implement. However, as the results depend critically on the choice of the smoothing parameter σ, the lack of a global choice of the bandwidth is typically considered a major shortcoming of kernel smoothing methods.

1.4.2 ADAPTIVE KERNEL SMOOTHING

There are two concerns about the standard fixed-bandwidth kernel estimation. One is that it requires the investigator to choose the parameter σ arbitrarily, which produces significantly different estimates of firing rates. The other problem is that the bandwidth of the standard kernel estimator is constant throughout the time interval of the neuronal response. Some reported to solve those two problems by varying the width of the estimation kernel throughout the trial, letting the data themselves determine how much to vary the width of the kernel. This process is called *adaptive-kernel estimation*, because the width of the kernel adapts to the local density of the data points [16]. The procedure is basically as follows:

1. Form a fixed-bandwidth kernel estimate (called a pilot estimate) from the data, the bandwidth is σ_p.
2. This pilot estimate of the density is used as a measure of the activity over small time periods throughout the entire response interval. The definition of "small" here depends on the choice of the fixed kernel bandwidth, σ_p.
3. These local kernels are then used to produce a smoothed firing rate that changes more rapidly in regions of high firing, and less in regions of less firing.

The pilot estimate is used to define a set of local bandwidth factors, λ_i,

$$\lambda_i = \sqrt{\frac{f(i)}{\mu}} \tag{1.26}$$

where $f(i)$ is the pilot estimate at the ith point, and μ is the geometric mean of the pilot estimates,

$$\mu = \exp\left[\frac{1}{n}\sum_{i=0}^{n-1} \ln f(i)\right]. \tag{1.27}$$

Finally, the adaptive kernel estimate is computed by convolving each point with a kernel density function having a width that is the product of the fixed bandwidth σ_p and the factor for each point

$$m(k) = \frac{1}{n}\sum_{0}^{n-1} K\left(t - \tau_i | \sigma = \frac{\sigma_p}{\lambda_i}\right). \tag{1.28}$$

Adaptive kernel smoothing benefits from the simplicity of kernel smoothing methods, and the additional complexity of the local kernel widths increases the computational effort only very slightly. Further, this approach lifts the strict stationarity requirement of many models. A possible shortcoming is that, even though it adapts the kernel width, the adaptive kernel smoothing still requires a specific choice of kernel width for the pilot estimate.

1.4.3 KERNEL BANDWIDTH OPTIMIZATION

The kernel smoother and a time histogram are classical tools for estimating an instantaneous firing rate. The optimization method was initially proposed for the joint *peristimulus time histogram* (PSTH) of spike counts over multiple neurons [18]. The method can select the bin width of the time histogram automatically based on the principal of minimizing the mean integrated squared error (MISE), defined as follows, without knowing the underlying rate.

$$MISE = \int_0^T E(\lambda(t) - \hat{\lambda}(t))^2\, dt, \tag{1.29}$$

where $\lambda(t)$ is the underlying rate and $\hat{\lambda}(t)$ is the estimation, and E refers to the expectation with respect to the spike generation process under a given time-dependent rate $\lambda(t)$.

Similarly, we consider a kernel rate estimator as $\hat{\lambda}(t)$ and select the width of a kernel under the MISE criterion. Suppose independently and identically obtained m spike trains which contain M spikes as a whole, then a superposition of the spike trains can be regarded as being drawn from an inhomogeneous Poisson point process, due to the general limit theorem of the sum of independent point process [19]. Define $\bar{Y}_t = \frac{1}{m}\sum_{i=1}^{M}\delta(t - \tau_i')$, where τ_i' is the ith spike of the superimposed spike sequence and $\delta(\cdot)$ is still the Dirac delta function. The estimator $\hat{\lambda}(t)$ can be constructed by a kernel function $K_w(\cdot)$ as $\hat{\lambda}(t) = \int \bar{Y}_s K_w(t-s) ds$, where w refers to the bandwidth.

Following Equation (1.29), the integrand can be decomposed into three terms: $\lambda(t)^2 - 2\lambda(t)E(\hat{\lambda}(t)) + E(\hat{\lambda}(t)^2)$. Since the first component does not depend on the choice of the kernel, we subtract it from the MISE and define a cost function as a function of the bandwidth w given by

$$C_m(w) \doteq MISE - \int_0^T \lambda(t)^2 dt = -2\int_0^T \lambda(t)E(\hat{\lambda}(t)) dt + \int_0^T E(\hat{\lambda}(t)^2) dt. \quad (1.30)$$

From a general decomposition rule of a covariance of the two random variables, we obtain

$$\int_0^T \lambda(t)E(\hat{\lambda}(t)) dt = \int_0^T E(\bar{Y}_t \hat{\lambda}(t)) dt - \int_0^T E(\bar{Y}_t - E(\bar{Y}_t))(\hat{\lambda}(t) - E(\hat{\lambda}(t))) dt$$

$$= E\left(\int_0^T \bar{Y}_t \hat{\lambda}(t) dt\right) - \frac{K_w(0)}{n} E\left(\int_0^T \bar{Y}_t dt\right). \quad (1.31)$$

To obtain the next equality, we used the assumption that the spike sequence is a Poisson process so that the spikes are independent of each other. Hence, the cost function is estimated as

$$\hat{C}_m(w) = \frac{2K_w(0)}{m} \int_0^T \bar{Y}_t dt - 2\int_0^T \bar{Y}_t \hat{\lambda}(t) dt + \int_0^T \hat{\lambda}(t)^2 dt$$

$$= \frac{2K_w(0)}{m^2} M - \frac{2}{m^2} \sum_{i=1}^M \sum_{j=1}^M K_w(\tau_i' - \tau_j') + \frac{1}{m^2} \sum_{i=1}^M \sum_{j=1}^M \psi_w(\tau_i' - \tau_j'), \quad (1.32)$$

where $\psi_w(t)$ is given by $\psi_w(t) \doteq \int_0^T K_w(s)K_w(s+t) ds$.

The optimal bandwidth w^* minimizes the score function $\hat{C}_m(w)$.

1.4.4 SMOOTHING SPLINES

The penalty smoothing estimates the rate function by maximized the penalized likelihood function [8]. We first consider a nonparametric regression problem $Y_i = \eta(x_i) + \varepsilon_i$, $i = 1, 2, ..., n$, where n is the total number of points of a spike train, $x_i \in [0, T]$, $\varepsilon_i \sim N(0, \sigma^2)$ and $\eta(\cdot)$ is the smoothing spline function. Then we may estimate $\eta(\cdot)$ via the penalized least square

$$\frac{1}{n} \sum_{i=1}^n (Y_i - \eta(x_i))^2 + \kappa \int_0^1 (\eta''(x))^2 dx \quad (1.33)$$

where the first term, $\sum_{i=1}^n (Y_i - \eta(x_i))^2$, measures the goodness-of-fit of the smoothing function η to the data; the second term, $J(\eta) = \int_0^1 (\eta''(x))^2 dx$, penalizes the roughness of $\eta(x)$; and the smoothing parameter κ controls the trade-off between the two conflicting goals. The solution of (1.33) is known as the *natural cubic spline*. As $\kappa \to \infty$, the estimate "shrinks" into the null space of the roughness penalty, $\{\eta : \int_0^1 \eta''(x)dx\}$. When $\kappa \to 0$, it converges to the minimum curvature interpolation, i.e., $Y_i = \eta(x_i)$ for $i = 1, 2, ...n$.

For the point process data, log-likelihood is used as the goodness-of-fit measure, instead of the least square. As we know, when the error terms of the regression are assumed to be independent and identically distributed normal random variables, the ordinary least square (OLS) turns out to be equivalent to the maximum likelihood estimation.

Consider the so-called exponential family distributions with densities of the form

$$f(y|\eta, \phi) = \exp\{(y\eta - b(\eta))/a(\phi) + c(y, \phi)\}, \tag{1.34}$$

where $a > 0$, b and c are known functions, η is the parameter of interest and ϕ is either known or considered as a nuisance parameter. Observing $Y_i\ f(y|\eta(x_i), \phi)$, one may estimate $\eta(x)$ via the general penalized likelihood,

$$-\frac{1}{n}\sum_{i=1}^{n}\{Y_i\eta(x_i) - b(\eta(x_i))\} + \kappa J(\eta), \tag{1.35}$$

where $J(\eta) = \int_0^1 (\eta''(x))^2\,dx$ and $c(y, \phi)$ is dropped as it does not depend on η.

A special case of the penalized likelihood method is Poisson regression. For the density function $f(y|\eta) = \exp\{y\eta - e^\eta - \log y!\}$, one may minimize the penalized likelihood

$$-\frac{1}{n}\sum_{i=1}^{n}\left\{Y_i\eta(x_i) - e^{\eta(x_i)}\right\} + \kappa J(\eta), \tag{1.36}$$

where $\eta(x_i) = \log\lambda(x_i)$ is the log intensity and the $\lambda(x_i) > 0$ but $\eta(x_i)$ is free of constraint.

1.4.5 REAL DATA ANALYSIS

The 16 spike trains for the analysis were recorded simultaneously from noncholinergic basal forebrain neurons [11]. Let us first explore neuron 1, our target spike train, in Figure 1.3.

Another spike train from neuron 5 that is positively related to the target spike train was selected as the first peer to predict the firing rate of neuron 1. Figure 1.4 is the cross-correlogram (CCG) which peaks at lags close to 0. An intuitive interpretation of the CCG is that the target neuron is more likely to fire immediately before or after the peer. We can expect a positive weight estimated between this target neuron and its peer due to the synchronous firing.

Predictability based on the log-likelihood ratio was first introduced in predicting earthquake occurrence. One unit of predictability (information bit) would mean that uncertainty of earthquake occurrence is reduced on average by a factor of 2 by using a particular model. We quantify the predictability of a spike train in a similar way, so that we can reduce the uncertainty as much as possible. Comparisons were made between predictability from one peer (neuron 5) or two peers (neuron 4 and neuron

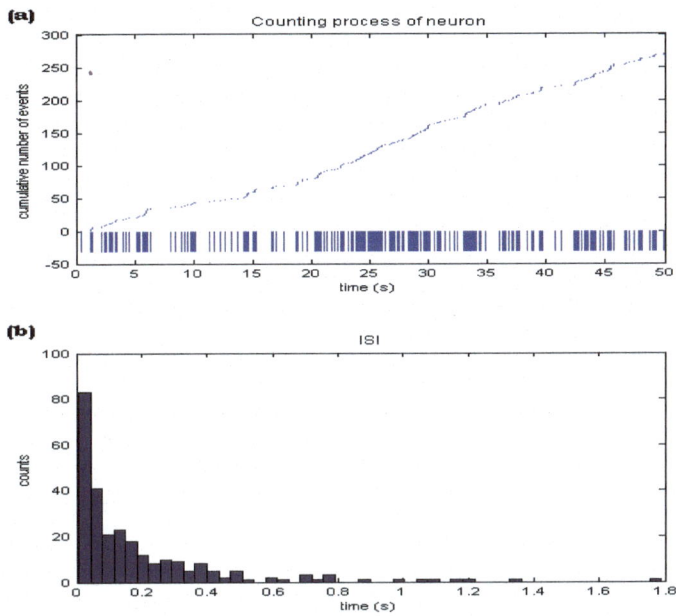

Figure 1.3: (a): The non-uniform distribution of the ticks on the raster plot is shown at the bottom of the figure, which indicates a model based on an inhomogeneous Poisson process for this spike train. (b): Interspike interval (ISI) histogram for the neural spike train of neuron 1.

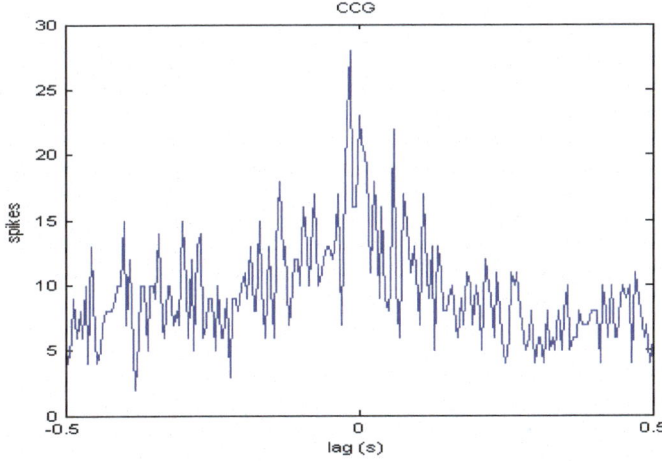

Figure 1.4: Cross-correlogram of neuron 1 (the target neuron) and neuron 5 shows a peak for lags that are less than 0.1 second.

5). Also, predictability from two positively correlated peers (neuron 4 and neuron 5) or one of them (neuron 15) has no significant correlation with the target neuron.

Neuron 4 has an excitatory impact on the target neuron as well. When we estimate the rate function based on the two peers (neuron 4 and neuron 5), we can see in Figure 1.5 that the green curve overall is above the blue one, which is the predictability from one peer (neuron 5) only.

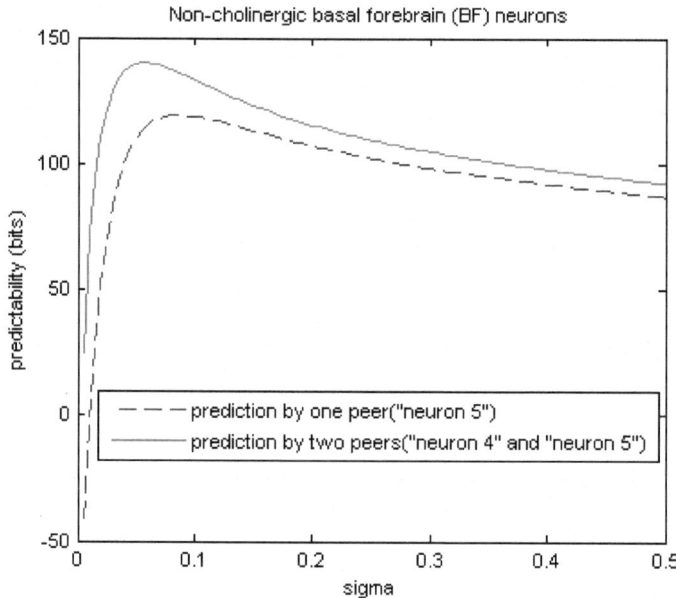

Figure 1.5: Both neuron 4 and neuron 5 are positively correlated to the target neuron. Predictabilities were estimated by one peer (neuron 4) or two peers (neuron 4 and neuron 5). Prediction by those two peers save more information bits overall.

What if we involve another spike train that has no strong evidence of the synchronization with the target spike train? We can see in Figure 1.6(b) that involving neuron 15, which has no correlation with the target neuron, does not improve the predictability.

In the peer prediction framework, it is possible to estimate the timescale with which neurons are coordinated into synchronization by varying the temporal bandwidth, σ. Figure 1.7 shows the predictability as a function of timescale for peer neuron 5.

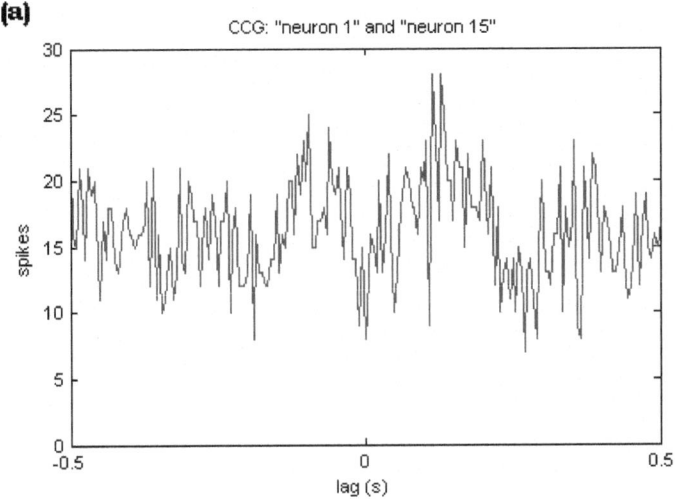

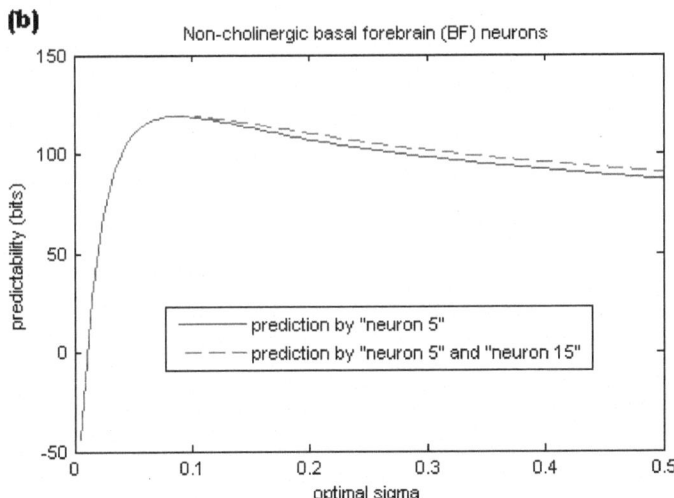

Figure 1.6: (a): Cross-correlogram of neuron 1 and neuron 15. There is no significant relation between the target neuron and predicting neuron. (b): Predictability from only one peer (neuron 5) and from two peers (neuron 5 and neuron 15).

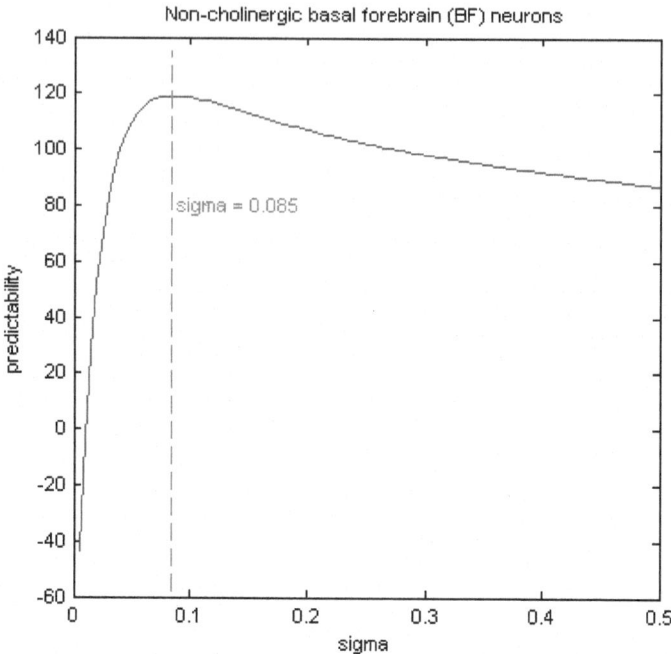

Figure 1.7: Predictability versus peer prediction timescale for peer neuron 5. The optimal bandwidth for prediction is 0.085 seconds.

1.5 M-FILES FOR SIMULATION

```
% gauss1D.m
function [s]=gauss1D(T_train,  spk2_train ,sigma)
  persistent size1 size2  a x y temp temp3
  size1 = length(T_train);
  size2 = length(spk2_train);
  s = zeros(1,size1);
  a = zeros(1,size1);

  for i=1:size1,
      x=T_train(i);
      y=spk2_train';
      temp=repmat(x,size2,1);
      temp3=exp(-(temp-y).^2/(2*sigma^2));
      a(i)=sum(temp3);
      s(i)=a(i)/(sqrt(2*pi)*sigma);
  end
```

```
end


%gauss2D.m
function [S]=gauss2D(T_train, spk2_train, spk3_train, sigma,n)
persistent size1 size2 SPK2_train s a x y temp temp3
  S = cell(1,n);
  size1 = length(T_train);
  size2 = [length(spk2_train) length(spk3_train)];
  SPK2_train = cell(1,n);
  SPK2_train{1} = spk2_train;
  SPK2_train{2} = spk3_train;

  for j=1:n,
    s = zeros(1,size1);
    a = zeros(1,size1);

    for i=1:size1,
      x=T_train(i);
      y=SPK2_train{j}';
      temp=repmat(x,size2(j),1);
      temp3=exp(-(temp-y).^2/(2*sigma^2));
      a(i)=sum(temp3);
      s(i)=a(i)/(sqrt(2*pi)*sigma);
    end
    S{j}=s;
  end

end


% predict1D.m
function [T_test,g]=predict1D(wmin,zoom,spk2_test,sigma,T_test)
  persistent s2 a sum1 size2 sizeT temp temp3
  size2=length(spk2_test);
  sizeT=length(T_test);
  s2 = zeros(1,sizeT);
  a = zeros(1,sizeT);
  for i=1:sizeT,
      x=T_test(:,i);
      y=spk2_test';
        temp=repmat(x,size2,1);
        temp3=exp(-(temp-y).^2/(2*sigma^2));
        a(i)=sum(temp3);
        s2(i)=a(i)/(sqrt(2*pi)*sigma);
```

```
  end


g=zeros(1,sizeT);
for t=1:sizeT,
  sum1 = s2(t)*wmin;
  if sum1>=0,
     g(t) = sum1+1;
  else g(t) = exp(sum1);
  end
end

end


% predict2D.m

function [T_test,g]=predict2D(wmin,zoom,T_test,spk2_test, ...
                    spk3_test, sigma, n)
  persistent S2 SPK2_test sizeT size2 s a x y temp temp3 sum1
  S2 = cell(1,n);
  sizeT=length(T_test);
  SPK2_test = cell(1,n);
  SPK2_test{1} = spk2_test;
  SPK2_test{2} = spk3_test;
  size2=[length(SPK2_test{1}) length(SPK2_test{2})];
  for j=1:n,
    s = zeros(1,sizeT);
    a = zeros(1,sizeT);
    for i=1:sizeT,
        x=T_test(i);
        y=SPK2_test{j}';
        temp=repmat(x,size2(j),1);
        temp3=exp(-(temp-y).^2/(2*sigma^2));
        a(i)=sum(temp3);
        s(i)=a(i)/(sqrt(2*pi)*sigma);
    end
    S2{j}=s;
  end

g=zeros(1,sizeT);
for t=1:sizeT,
  sum1 = S2{1}(t)*wmin(1)+S2{2}(t)*wmin(2);
  if sum1>=0,
     g(t) = sum1+1;
```

```
   else g(t) = exp(sum1);
   end
 end

end


% weights1D.m

function [wmin,fmin]=weights1D(s,T_train, spk1_train, delta)
global Cnt
persistent T_lo T_hi

  T_lo = T_train-(delta/2);
  T_hi = T_train+(delta/2);
  for i=1:length(T_train),
    Cnt(i) = sum(T_lo(i)<spk1_train & spk1_train<T_hi(i));
  end
    w0 = 0;
   options=optimset('GradObj','on');
   [wmin,fmin]= fminunc(@myfunc1,w0,options);

end



%myfunc1.m

function [f,g]=myfunc1(w)
global T_train s Cnt
persistent l size1 gg m
l=0;
delta=0.005;
size1 = length(T_train);
m=zeros(1,size1);
for t=1:size1,
    if s(t)*w>=0,
        m(t)=s(t)*w+1;
    else m(t)=exp(s(t)*w);
    end
  l=l-m(t)*delta+Cnt(t)*log(m(t));
end
    f=-l+(1/4)*(w^2);
if nargout > 1
    gg=zeros(1,size1);
    for t=1:size1,
      if s(t)*w>=0,
```

```
        gg(t)=-s(t)*delta+(Cnt(t)*s(t)/m(t));
    else gg(t)=-(m(t)*s(t)*delta)+(Cnt(t)*s(t));
    end
    g=-sum(gg)+(w/2);
  end
end
end


% myfunc2Dg.m

function [f,g]=myfunc2Dg(w)
global T_train S Cnt
%persistent l size1 m
persistent l size1 gg1 gg2 m
l=0;
delta=0.005;
size1 = length(T_train);
m=zeros(1,size1);

for t=1:size1,
    if S{1}(t)*w(1)+S{2}(t)*w(2)>=0,
        m(t)=S{1}(t)*w(1)+S{2}(t)*w(2)+1;
    else m(t)=exp(S{1}(t)*w(1)+S{2}(t)*w(2));
    end
    l=l-m(t)*delta+Cnt(t)*log(m(t));
end
    f=-l+(1/4)*(w(1)^2+w(2)^2);
if nargout > 1
    gg1=zeros(1,size1); gg2=zeros(1,size1);
    for t=1:size1,
    if S{1}(t)*w(1)+S{2}(t)*w(2)>=0,
        gg1(t)=-S{1}(t)*delta+(Cnt(t)*S{1}(t)/m(t));
        gg2(t)=-S{2}(t)*delta+(Cnt(t)*S{2}(t)/m(t));
    else gg1(t)=-(m(t)*S{1}(t)*delta)+(Cnt(t)*S{1}(t));
         gg2(t)=-(m(t)*S{2}(t)*delta)+(Cnt(t)*S{2}(t));
    end
    g(1)=-sum(gg1)+(w(1)/2);
    g(2)=-sum(gg2)+(w(2)/2);
    end
end
end

% weights2DQg.m
```

```
function [wmin,fmin]=weights2DQg(S, T_train, spk1_train, delta)
global Cnt
persistent T_lo T_hi

    T_lo = T_train-(delta/2);
    T_hi = T_train+(delta/2);
    for i=1:length(T_train),
        Cnt(i) = sum(T_lo(i)<spk1_train & spk1_train<T_hi(i));
    end
    w0 = [1 1];
    options=optimset('GradObj','on','MaxFunEvals',5000);
    [wmin,fmin]=fminunc(@myfunc2Dg,w0,options);

end

% inho_simulation.m

clear all;

n=1;
w = 3;
T = 50;
J = 1000;
delta = T/J;
%SPK1=spk_out('TONE31','20070403');
%hist(diff(SPK1.TS{n}(1:500)),100)

%%%%%%%%%%%%%%%%%%%%%%%%%%%%%%%%%%%%%%%%%%%%%%%%%%%%%%%%%%%%%%%%
% Generate inhomogeneous rate spike train by time-rescaling %
%%%%%%%%%%%%%%%%%%%%%%%%%%%%%%%%%%%%%%%%%%%%%%%%%%%%%%%%%%%%%%%%
    spike1_train = cell(1,100);
    spike2_train = cell(1,100);
    spike1_test = cell(1,100);
    spike2_test = cell(1,100);

for j=1:100


  r = zeros(1,J);
  k1 = zeros(1,J);
  k2 = zeros(1,J);
  k3 = zeros(1,J);
  k4 = zeros(1,J);

  for i=1:J
```

```
    r(i)=(T*(i*delta)-(i*delta)^2)/200;

    if (binornd(1,(r(i)*delta))==1)
        k2(i) = i;
    end;
    if (binornd(1,(r(i)*delta))==1)
        k4(i) = i;
    end;
    if (binornd(1,((r(i)*w+1)*delta))==1)
        k1(i) = i;
    end;
    if (binornd(1,((r(i)*w+1)*delta))==1)
        k3(i) = i;
    end;
  end

  t1=k1(find(k1>0))*delta;
  t2=k2(find(k2>0))*delta;
  t3=k3(find(k3>0))*delta;
  t4=k4(find(k4>0))*delta;

  spike1_train{j} = t1;
  spike2_train{j} = t2;
  spike1_test{j} = t3;
  spike2_test{j} = t4;
end


for j=1:100
  tn_train(j) = max(max(spike1_train{j}),max(spike2_train{j}));
  tn_test(j) = max(max(spike1_test{j}),max(spike2_test{j}));
end;

tmax_train = max(tn_train);
tmax_test = max(tn_test);
T_train = delta:delta:T;
T_test = delta:delta:T;

save('./spikes','spike1_train','spike2_train','spike1_test',...
    'spike2_test')
save('./bigw','bigw')
save('./predictability','sigma2','predL1')
save('./prediction','T_test','bigs','bigg','r')
%%%%%%%%%%%%%%%%%%%%%%%%%%%%%%%%%%%%%%%%%%%%%%%%%%%%%%%%%%%%%%%%%%%%%%%
```

```
% Plot the point process(raster plot) and ISI            %
%%%%%%%%%%%%%%%%%%%%%%%%%%%%%%%%%%%%%%%%%%%%%%%%%%%%%%%%%%%%%
for i=1:length(spike1_train{1})
    plot([spike1_train{1}(i) spike1_train{1}(i)], [-50 0]);
    hold on; plot(spike1_train{1}(i),i);hold on;
end;
xlabel('time (s)');
ylabel('cumulative number of events');
title('Counting process of neuron','FontSize',18)

hist(diff(spike1_train{1}),50);xlabel('time');title('ISI',...
    'FontSize',18)


%%%%%%%%%%%%%%%%%%%%%%%%%%%%%%%%%%%%%%%%%%%%%%%%%%%%%%%%%%%%%
% Start Prediction                                        %
%%%%%%%%%%%%%%%%%%%%%%%%%%%%%%%%%%%%%%%%%%%%%%%%%%%%%%%%%%%%%

sigma2 = 0.1 :0.1:6;
bigw=zeros(60,100);

for k = 1:length(sigma2)
sigma = sigma2(k);
for j=1:100

  spk1_train = spike1_train{j};
  spk2_train = spike2_train{j};
  spk1_test = spike1_test{j};
  spk2_test = spike2_test{j};


%%%%%%%%%%%%%%%%%%%%%%%%%%%%%%%%%%%%%%%%%%%%%%%%%%%%%%%%%%%%%
% Calculate weight using gauss1P and weights1P            %
%%%%%%%%%%%%%%%%%%%%%%%%%%%%%%%%%%%%%%%%%%%%%%%%%%%%%%%%%%%%%

    global T_train s Cnt
    s = gauss1D(T_train,  spk2_train ,sigma);
    [wmin,fmin] = weights1D(s,T_train, spk1_train, delta);

%%%%%%%%%%%%%%%%%%%%%%%%%%%%%%%%%%%%%%%%%%%%%%%%%%%%%%%%%%%%%
% Predict the intensity function g via predict1P
%%%%%%%%%%%%%%%%%%%%%%%%%%%%%%%%%%%%%%%%%%%%%%%%%%%%%%%%%%%%%


    [T_test,pred]=predict1D(wmin,20,spk2_test, sigma, T_test);
```

```
%%%%%%%%%%%%%%%%%%%%%%%%%%%%%%%%%
% Intensity function of spk1_test %
%%%%%%%%%%%%%%%%%%%%%%%%%%%%%%%%%%%%

  bigw(k,j) = wmin;

  Cnt2 = zeros(1,length(T_test));
  T_lo = T_test-(delta/2);
  T_hi = T_test+(delta/2);

    for i=1:length(T_test),
        Cnt2(i) = sum(T_lo(i)<spk1_test & spk1_test<T_hi(i));
    end

  PredRate = pred;
  fRate = 2;
  L_error=-(PredRate-fRate)*delta + ...
    (log(PredRate)-log(fRate)).*Cnt2;
  predL1(k,j) = sum(L_error)/log(2); % make it to base 2.

  clear global s Cnt
end

end

for j=1:100
    max1=max(predL1(:,j));
    sigmax(j)=sigma2(find(predL1(:,j)==max1));
end
[fs1,xis1] = ksdensity(sigmax);
figure(1);clf
plot(xis1,fs1);hold on;plot([5.35 5.35],[0 0.5],'--',...
    'Color','r');
m=median(sigmax);
title('Kernel density of the optimal sigma by maximizing ' ...
    'Predictability', 'FontSize',18);
text(3,0.3,['Median: sigma = 5.35'],'FontSize',15,'Color','r');
xlabel('optimal sigma')
for i=1:length(sigma2)
  [f1(i,:),xi1(i,:)] = ksdensity(bigw(i,:));
end
```

```
figure(2);clf
plot(xi1(53,:),f1(53,:),'Color','r')
 hold on;
plot(xi1(11,:),f1(11,:),'--','Color','b')
 hold on;
plot(xi1(60,:),f1(60,:),':','Color','b')
legend('sigma=5.3','sigma=1.1','sigma=6')
 hold on;
plot([3,3],[2,0],'--','Color','r')
 xlabel('weight')
 text(3,1.7,[' sigma = 5.3'],'FontSize',15,'Color','r');
 text(2.5,1.6,[' sigma = 1'],'FontSize',15,'Color','b');
 text(3.3,1.3,[' sigma = 6'],'FontSize',15,'Color','b');
title('Kernel density estimation of w','FontSize',18);


sigma=m;
for j=1:100

  spk1_train = spike1_train{j};
  spk2_train = spike2_train{j};
  spk1_test = spike1_test{j};
  spk2_test = spike2_test{j};


%%%%%%%%%%%%%%%%%%%%%%%%%%%%%%%%%%%%%%%%%%%%%%%%%%%%%%%%%%%%%%%%%%%
% Calculate weight using gauss1P and weights1P
%%%%%%%%%%%%%%%%%%%%%%%%%%%%%%%%%%%%%%%%%%%%%%%%%%%%%%%%%%%%%%%%%%%

    global T_train s Cnt
    s = gauss1D(T_train,  spk2_train ,sigma);
    [wmin,fmin] = weights1D(s,T_train, spk1_train, delta);

%%%%%%%%%%%%%%%%%%%%%%%%%%%%%%%%%%%%%%%%%%%%%%%%%%%%%%%%%%%%%%%%%%%
% Predict the intensity function g via predict1P
%%%%%%%%%%%%%%%%%%%%%%%%%%%%%%%%%%%%%%%%%%%%%%%%%%%%%%%%%%%%%%%%%%%


    [T_test,pred]=predict1D(wmin,20,spk2_test, sigma, T_test);

%%%%%%%%%%%%%%%%%%%%%%%%%%%%%%%%%%%%%%%%%%%%%%%%%%
% Intensity function of spk1_test %
%%%%%%%%%%%%%%%%%%%%%%%%%%%%%%%%%%%%%%%%%%%%%%%%%%
```

```
size1 = length(spk1_test);
sizeT = length(T_test);
s2 = zeros(1,sizeT);
a = zeros(1,sizeT);
for i=1:sizeT,
    x=T_test(:,i);
    y=spk1_test';
        temp=repmat(x,size1,1);
        temp3=exp(-(temp-y).^2/(2*sigma^2));
        a(i)=sum(temp3);
        s2(i)=a(i)/(sqrt(2*pi)*sigma);
end

bigg(j,:) = pred;
bigs(j,:) = s2;

clear s Cnt
end


gmean=mean(bigg,1);
smean=mean(bigs,1);

gmax=prctile(bigg,95);
gmin=prctile(bigg,5);
smax=prctile(bigs,95);
smin=prctile(bigs,5);

figure(3);clf
  plot(T_test,smean,':',T_test,smin,'-.','Color','k')
  hold on;
  plot(T_test,gmean,'--',T_test,gmin,'-.', 'Color','r')
  hold on;
  plot(delta:delta:T,r*w+1,'Color','b')
h = legend('mean intensity function', ...
'5% & 95% boundary','mean predicted intensity',...
'5% & 95% predicted boundary', ...
'true rate function','Location','South');
title('Prediction by sigma=5.35','FontSize',18)
xlabel('time (s)')
ylabel('intensity')
  hold on;
  plot(T_test,smax,'-.','Color','k')
  hold on;
```

```
  plot(T_test,gmax,'-.', 'Color','r')

figure(1);clf
plot(T_test, bigs(1,:))
hold on;
plot(T_test,bigg(1,:), 'Color','r')
hold on;
plot(delta:delta:T,r, 'Color','k')
ylim([-1,6])


% inho_simulation2D_Q.m

clear all;
n=2;
w1 = 1.5; w2 = 1.5; w = w1 +w2;
T = 50;
J = 1000;
delta = T/J;
%SPK1=spk_out('TONE31','20070403');
%hist(diff(SPK1.TS{n}(1:500)),100)

%%%%%%%%%%%%%%%%%%%%%%%%%%%%%%%%%%%%%%%%%%%%%%%%%%%%%%%%%%%%%%%%%
% Generate inhomogeneous rate spike train by time-rescaling   %
%%%%%%%%%%%%%%%%%%%%%%%%%%%%%%%%%%%%%%%%%%%%%%%%%%%%%%%%%%%%%%%%%
  spike1_train = cell(1,100);
  spike2_train = cell(1,100);
  spike3_train = cell(1,100);
  spike1_test = cell(1,100);
  spike2_test = cell(1,100);
  spike3_test = cell(1,100);


for j=1:100

  r = zeros(1,J);
  k1 = zeros(1,J);
  k2 = zeros(1,J);
  k3 = zeros(1,J);
  k4 = zeros(1,J);
  k5 = zeros(1,J);
  k6 = zeros(1,J);

  for i=1:J
    r(i)=(T*(i*delta)-(i*delta)^2)/200;
```

```
    if (binornd(1,(r(i)*delta))==1)
        k2(i) = i;
    end;
    if (binornd(1,(r(i)*delta))==1)
        k4(i) = i;
    end;
    if (binornd(1,((r(i)*w+1)*delta))==1)
        k1(i) = i;
    end;
    if (binornd(1,((r(i)*w+1)*delta))==1)
        k3(i) = i;
    end;
    if (binornd(1,(r(i)*delta))==1)
        k5(i) = i;
    end;
    if (binornd(1,(r(i)*delta))==1)
        k6(i) = i;
    end;
  end

  t1=k1(find(k1>0))*delta;
  t2=k2(find(k2>0))*delta;
  t3=k3(find(k3>0))*delta;
  t4=k4(find(k4>0))*delta;
  t5=k5(find(k5>0))*delta;
  t6=k6(find(k6>0))*delta;

  spike1_train{j} = t1;
  spike2_train{j} = t2;
  spike3_train{j} = t5;
  spike1_test{j} = t3;
  spike2_test{j} = t4;
  spike3_test{j} = t6;
end


for j=1:100
  tn_train(j) = max([max(spike1_train{j}) ...
                    max(spike2_train{j}) ...
                    max(spike3_train{j})]);
  tn_test(j) = max([max(spike1_test{j}) max(spike2_test{j}) ...
                  max(spike3_test{j})]);
end;
```

```
tmax_train = max(tn_train);
tmax_test = max(tn_test);
T_train = delta:delta:T;
T_test = delta:delta:T;

save('C:\2D\spikes','spike1_train','spike2_train',...
        'spike3_train', 'spike1_test','spike2_test',...
        'spike3_test')
save('C:\2D\bigw','bigw1','bigw2')
save('C:\2D\predictibility','sigma1','predL')
save('C:\2D\prediction','T_test','bigs','bigg','r')
%%%%%%%%%%%%%%%%%%%%%%%%%%%%%%%%%%%%%%%%%%%%%%%%%%%%%%%%%%%%%%%%
% Start select optimal sigma value                             %
%%%%%%%%%%%%%%%%%%%%%%%%%%%%%%%%%%%%%%%%%%%%%%%%%%%%%%%%%%%%%%%%

%%warning off all
bigw1=zeros(50,100);
bigw2=zeros(50,100);
predL = zeros(50,100);

sigma1=1:0.1:6;
%sigma=1;
%bigs=zeros(100,length(T_test));
%bigg=zeros(100,length(T_test));
for k=1:length(sigma1)
sigma=sigma1(k);
for j=1:100
  spk1_train = spike1_train{j};
  spk2_train = spike2_train{j};
  spk3_train = spike3_train{j};
  spk1_test = spike1_test{j};
  spk2_test = spike2_test{j};
  spk3_test = spike3_test{j};
%%%%%%%%%%%%%%%%%%%%%%%%%%%%%%%%%%%%%%%%%%%%%%%%%%%%%%%%%%%%%%%%
% Calculate weight using gauss2D and weights2D
%%%%%%%%%%%%%%%%%%%%%%%%%%%%%%%%%%%%%%%%%%%%%%%%%%%%%%%%%%%%%%%%
  global T_train S
  S=gauss2D(T_train, spk2_train, spk3_train, sigma,n);
  [wmin,fmin]=weights2DQg(S, T_train, spk1_train, delta);

%%%%%%%%%%%%%%%%%%%%%%%%%%%%%%%%%%%%%%%%%%%%%%%%%%%%%%%%%%%%%%%%
% Predict the intensity function g via predict2D
%%%%%%%%%%%%%%%%%%%%%%%%%%%%%%%%%%%%%%%%%%%%%%%%%%%%%%%%%%%%%%%%
```

```
[T_test,pred]=predict2D(wmin,zoom,T_test,spk2_test, ...
                        spk3_test, sigma, n);


%%%%%%%%%%%%%%%%%%%%%%%%%%%%%%%%%%%%%%%%
% Intensity function of spk1_test %
%%%%%%%%%%%%%%%%%%%%%%%%%%%%%%%%%%%%%%%%


  bigw1(k,j) = wmin(1);
  bigw2(k,j) = wmin(2);

  Cnt2 = zeros(1,length(T_test));
  T_lo = T_test-(delta/2);
  T_hi = T_test+(delta/2);

    for i=1:length(T_test),
        Cnt2(i) = sum(T_lo(i)<spk1_test & spk1_test<T_hi(i));
    end

  PredRate = pred;
  fRate = 2;
  L_error=-(PredRate-fRate)*delta + ...
              (log(PredRate)-log(fRate)).*Cnt2;
  predL(k,j) = sum(L_error)/log(2);

clear global S Cnt
end


end
%%%%%%%%%%%%%%%%%%%%%%%%%%%%%%%%%%%%%%%%%%%%%%%%%%%%%%%%%%%%%%%%
% Plot the kernel density of 100 optimal sigma         %
%%%%%%%%%%%%%%%%%%%%%%%%%%%%%%%%%%%%%%%%%%%%%%%%%%%%%%%%%%%%%%%%
for j=1:100
    max1=max(predL(:,j));
    sigmax(j)=sigma1(find(predL(:,j)==max1));
end
[fs1,xis1] = ksdensity(sigmax);
figure(4);clf;
m=median(sigmax);
plot(xis1,fs1);hold on;
plot([4.5 4.5],[0 0.25],'--','Color','r')
title('Density of the optimal sigma by max Predictability',
```

```
        'FontSize',18);
text(4.5,0.2,['Median: sigma = 4.5'],'FontSize',15,...
        'Color','r');
xlabel('optimal sigma')
%ylabel('density')
%%%%%%%%%%%%%%%%%%%%%%%%%%%%%%%%%%%%%%%%%%%%%%%%%%%%%%%%%%%%%
% Plot the predictibilty of 100 simulation        %
%%%%%%%%%%%%%%%%%%%%%%%%%%%%%%%%%%%%%%%%%%%%%%%%%%%%%%%%%%%%%
figure(5);clf;
plot(sigma1,predL(:,4),sigma1,predL(:,6),sigma1,predL(:,8));
hold on;
plot([2.6 2.6],[352 370], '--',[3.1 3.1] ,[352 370],'--',...
        [4.2 4.2],[352 370],'--')
title('Predictibility with optimal sigma','FontSize',18);
text(2.6,369,[' sigma = 2.6'],'FontSize',15,'Color','b');
text(3.1,367,[' sigma = 3.1'],'FontSize',15,'Color','g');
text(4.2,364,[' sigma = 4.2'],'FontSize',15,'Color','r');
xlabel('sigma')
ylabel('predictability')


%%%%%%%%%%%%%%%%%%%%%%%%%%%%%%%%%%%%%%%%%%%%%%%%%%%%%%%%%%%%%
% Start Prediction                                         %
%%%%%%%%%%%%%%%%%%%%%%%%%%%%%%%%%%%%%%%%%%%%%%%%%%%%%%%%%%%%%

sigma=4.5;
bigs=zeros(1,length(T_test));
bigg=zeros(1,length(T_test));
%for k=1:length(sigma1)
%sigma=sigma1(k);
for j=1:100
  spk1_train = spike1_train{j};
  spk2_train = spike2_train{j};
  spk3_train = spike3_train{j};
  spk1_test = spike1_test{j};
  spk2_test = spike2_test{j};
  spk3_test = spike3_test{j};
%%%%%%%%%%%%%%%%%%%%%%%%%%%%%%%%%%%%%%%%%%%%%%%%%%%%%%%%%%%%%
% Calculate weight, using gauss2D and weights2D
%%%%%%%%%%%%%%%%%%%%%%%%%%%%%%%%%%%%%%%%%%%%%%%%%%%%%%%%%%%%%
  global T_train S
  S=gauss2D(T_train, spk2_train, spk3_train, sigma,n);
  [wmin,fmin]=weights2DQg(S, T_train, spk1_train, delta);
```

```
%%%%%%%%%%%%%%%%%%%%%%%%%%%%%%%%%%%%%%%%%%%%%%%%%%%%%%%%%%%%%%
% Predict the intensity function g via predict2D
%%%%%%%%%%%%%%%%%%%%%%%%%%%%%%%%%%%%%%%%%%%%%%%%%%%%%%%%%%%%%%
%      wmin=[bigw1(45,j),bigw2(45,j)];
   [T_test,pred]=predict2D(wmin,zoom,T_test,spk2_test, ...
                           spk3_test, sigma, n);


%%%%%%%%%%%%%%%%%%%%%%%%%%%%%%%%%%%%%%%
% Intensity function of spk1_test %
%%%%%%%%%%%%%%%%%%%%%%%%%%%%%%%%%%%%%%%

   size1 = length(spk1_test);
   sizeT = length(T_test);
   s2 = zeros(1,sizeT);
   a = zeros(1,sizeT);
   for i=1:sizeT,
       x=T_test(i);
       y=spk1_test';
       temp=repmat(x,size1,1);
       temp3=exp(-(temp-y).^2/(2*sigma^2));
       a(i)=sum(temp3);
       s2(i)=a(i)/(sqrt(2*pi)*sigma);
   end

   bigg(j,:) = pred;
   bigs(j,:) = s2;

clear global S Cnt
end

%end

gmean=mean(bigg,1);
smean=mean(bigs,1);

gmax=prctile(bigg,95);
gmin=prctile(bigg,5);
smax=prctile(bigs,95);
smin=prctile(bigs,5);

figure(6);clf
plot(T_test,smean,':',T_test,smin,'-.','Color','k')
   hold on;
```

```
  plot(T_test,gmean,'--',T_test,gmin,'-.', 'Color','r')
  hold on;
  plot(delta:delta:T,r*w+1,'Color','b')
h = legend('mean intensity function','5% & 95% boundary',...
            'mean predicted intensity',...
            '5% & 95% predicted boundary',...
            'true rate function', 'Location','South');
title('Prediction by sigma=4.5','FontSize',18)
xlabel('time (s)')
ylabel('intensity')
  hold on;
  plot(T_test,smax,'-.','Color','k')
  hold on;
  plot(T_test,gmax,'-.', 'Color','r')




%Likelihood2D(S,wmin,fmin)
%%%%%%%%%%%%%%%%%%%%%%%%%%%%%%%%%%%%%%%%%%%
% Calculate prediction log likelihood %
%%%%%%%%%%%%%%%%%%%%%%%%%%%%%%%%%%%%%%%%%%%

    PredRate = g;
    fRate = r*w+1;
    L_error=-(PredRate-fRate) + (log(PredRate)-log(fRate));
    LogL_error = sum(L_error)/log(2); % make it to base 2.


%%%%%%%%%%%%%%%%%%%%%%%%%%%%%%%%%%%%%%%%%%%
% Calculate the optimal sigma          %
%%%%%%%%%%%%%%%%%%%%%%%%%%%%%%%%%%%%%%%%%%%

  sigma1 = 0.1:0.1:2;

  [LogL_error] = optsigma2D(sigma1, r, w, T_train, ...
                            T_test, spk1_train,
     spk1_test, spk2_train, spk2_test, spk3_train, spk3_test,n);
  clf;
plot(sigma1, LogL_error)



%%%%%%%%%%%%%%%%%%%%%%%%%%%%%%%%%%%%%%%%%%%%%%%%%%%%%%%%%%%%%
```

```
function [S]=gauss2D(T_train, spk2_train, spk3_train, sigma,n)
persistent size1 size2 SPK2_train s a
  S = cell(1,length(n));
  size1 = length(T_train);
  size2 = [length(spk2_train) length(spk3_train)];
  SPK2_train = cell(1,length(n));
  SPK2_train{1} = spk2_train;
  SPK2_train{2} = spk3_train;

  for j=1:n,
    s = zeros(1,size1);
    a = zeros(1,size1);

    for i=1:size1,
      x=T_train(i);
      y=SPK2_train{j}';
      temp=repmat(x,size2(j),1);
      temp3=exp(-(temp-y).^2/(2*sigma^2));
      a(i)=sum(temp3);
      s(i)=a(i)/(sqrt(2*pi)*sigma);
    end
    S{j}=s;
  end

end
%%%%%%%%%%%%%%%%%%%%%%%%%%%%%%%%%%%%%%%%%%%%%%%%%%%%%%%%%%%%%%%%%%
%p=struct('s',s,'tt1',tt1,'nt',nt);
function [wmin,fmin]=weights2D(S, T_train, spk1_train, ...
                              sigma, n)
  global T_train S Cnt
  persistent T_lo T_hi
    Cnt = zeros(1,length(T_train));
    T_lo = T_train-(sigma/2);
    T_hi = T_train+(sigma/2);
    for i=1:length(T_train),
       Cnt(i) = sum(T_lo(i)<spk1_train & spk1_train<T_hi(i));
    end

    w0 = [0 0];
    options=optimset('GradObj','off');
    options = optimset('Hessian','on');
    [wmin,fmin]= fminunc(@myfunc2D,w0);

end
```

```
%%%%%%%%%%%%%%%%%%%%%%%%%%%%%%%%%%%%%%%%%%%%%%%%%%%%%%%%%%%
function f=myfunc2D(w)
global T_train S Cnt
persistent l size1
l=0;
size1 = length(T_train);
for t=1:size1,
    if S{1}(t)*w(1)+S{2}(t)*w(2)>=0,
        m=S{1}(t)*w(1)+S{2}(t)*w(2)+1;
    else m=exp(S{1}(t)*w(1)+S{2}(t)*w(2));
    end
    l=l-m+Cnt(t)*log(m);
end
    f=-l+(1/4)*(w(1)^2+w(2)^2);

end


%%%%%%%%%%%%%%%%%%%%%%%%%%%%%%%%%%%%%%%%%%%%%%%%%%%%%%%%%%%%%
function [T_test,g]=predict2D(wmin,zoom,T_test,spk2_test, ...
                       spk3_test, sigma, n)
  persistent S2 SPK2_test sizeT size2 s a x y temp temp3 sum1
  S2 = cell(1,length(n));
  sizeT=length(T_test);
  SPK2_test = cell(1,length(n));
  SPK2_test{1} = spk2_test;
  SPK2_test{2} = spk3_test;
  size2=[length(SPK2_test{1}) length(SPK2_test{2})];
  for j=1:n,
    s = zeros(1,sizeT);
    a = zeros(1,sizeT);
    for i=1:sizeT,
        x=T_test(i);
        y=SPK2_test{j}';
        temp=repmat(x,size2(j),1);
        temp3=exp(-(temp-y).^2/(2*sigma^2));
        a(i)=sum(temp3);
        s(i)=a(i)/(sqrt(2*pi)*sigma);
    end
    S2{j}=s;
  end

  g=zeros(1,sizeT);
  for t=1:sizeT,
    sum1 = S2{1}(t)*wmin(1)+S2{2}(t)*wmin(2);
```

```
    if sum1>=0,
       g(t) = sum1+1;
    else g(t) = exp(sum1);
    end
  end

  %subplot(2,1,1);plot(T,g)

  %index=find(T<max(T)/zoom);
  %pT=T(index);
  %pg=g(index);

  %subplot(2,1,2);plot(pT,pg)

end




%%%%%%%%%%%%%%%%%%%%%%%%%%%%%%%%%%%%%%%%%%%%%%%%%%%%%%%%%%%%%%%%%%%%%%%%

function [LogL_error]=optsigma2D(sigma1, r, w, T_train, ...
                                 T_test, spk1_train, ...
                                 spk1_test, spk2_train, ...
                                 spk2_test, spk3_train, ...
                                 spk3_test, n)
  persistent sigma  L_error Cnt2 T_lo T_hi S2 wmin2 fmin2 g2
  LogL_error=zeros(1,length(sigma1));
  for i=1:length(sigma1)
    sigma=sigma1(i);

    S2=gauss2D(T_train, spk2_train, spk3_train, sigma,n);
    [wmin2,fmin2]=weights2D(S2, T_train, spk1_train, sigma, n);

    [T_test,g2]=predict2D(wmin2,zoom,T_test,spk2_test, ...
                          spk3_test, sigma, n);

    Cnt2 = zeros(1,length(T_test));
    T_lo = T_test-(sigma/2);
    T_hi = T_test+(sigma/2);

      for j=1:length(T_test),
          Cnt2(j) = sum(T_lo(j)<spk1_test & spk1_test<T_hi(j));
      end
```

```
      PredRate = g2;
      fRate = r*w+1;
      L_error=-(PredRate-fRate) + ...
                 (log(PredRate)-log(fRate)).*Cnt2;
      LogL_error(i) = sum(L_error)/log(2); % make it to base 2.

   end
end


%%%%%%%%%%%%%%%%%%%%%%%%%%%%%%%%%%%%%%%%%%%%%%%%%%%%%%%%%%%%%%%%%%
function [wdif]=optsigma1Dw(sigma1, w, T_train, spk1_train, ...
                               spk2_train )
   wdif=zeros(1,length(sigma1));
   for i=1:length(sigma1)
     sigma=sigma1(i)


     s2 = gauss1D(T_train,  spk2_train ,sigma);
     [wmin2,fmin2] = weights1D(s2,T_train, spk1_train, sigma);


     wdif(i) = wmin2-w;
   end
end
```

1.6   M-FILES FOR REAL DATA

```
% realdata2D.m

   clear all;
   RAT='TONE31'; EXPDATE='20070403';
   SPK=spk_out(RAT,EXPDATE);
   T1=SPK.EVENT{find(strcmp(SPK.EVENTNAME,'T1ON'))};
   T2=SPK.EVENT{find(strcmp(SPK.EVENTNAME,'T2ON'))};
   HL=SPK.EVENT{find(strcmp(SPK.EVENTNAME,'HLOFF'))};
   lick=SPK.EVENT{find(strcmp(SPK.EVENTNAME,'IROpen'))};
   Suc=SPK.EVENT{find(strcmp(SPK.EVENTNAME,'Sucrose'))};
   Qui=SPK.EVENT{find(strcmp(SPK.EVENTNAME,'Quin'))};

save('predictability','sigma2','LogL_error1','LogL_error2')
save('predictability2','sigma2','LogL_error1','LogL_error3')


%%%%%%%%%%%%%%%%%%%%%%%%%%%%%%%%%%%%%%%%%%%%%%%%%%%%%%%%%%%%%%%%%
% Plot the poin process(raster plot) and ISI                    %
```

```
%%%%%%%%%%%%%%%%%%%%%%%%%%%%%%%%%%%%%%%%%%%%%%%%%%%%%%%%%%%%%%%%%%%%%%%%%
clf;
EVENT={T1,T2,HL,lick, Suc, Qui};
for j=1:6
 event=EVENT{j};
  for i=1:length(event)
      plot([event(i) event(i)], [j (j+1)]);hold on;
  end
end


%%%%%%%%%%%%%%%%%%%%%%%%%%%%%%%%%%%%%%%%%%%%%%%%%%%%%%%%%%%%%%%%%%%%%%%%%
% Correlation plot                                                     %
%%%%%%%%%%%%%%%%%%%%%%%%%%%%%%%%%%%%%%%%%%%%%%%%%%%%%%%%%%%%%%%%%%%%%%%%%
n=[1 5 4 15];
spike1 = SPK.TS{n(1)};
spike2 = SPK.TS{n(2)};
spike3 = SPK.TS{n(3)};
spike4 = SPK.TS{n(4)};

[N1,lags1]=sparsexcorr2(spike1,spike3,'maxlag',0.5, ...
                    'minlag', -0.5,'b',delta);
plot(lags1,N1);
xlim([-.5 .5])

for k=1:10

c1=(k-1)*50;c2=k*50;c3=400;c4=450;
spike1_train = spike1(find(c1<spike1 & spike1<=c2))-c1;
spike1_test = spike1(find(c3<spike1 & spike1<=c4))-c3;
spike2_train = spike2(find(c1<spike2 & spike2<=c2))-c1;
spike2_test = spike2(find(c3<spike2 & spike2<=c4))-c3;

%hist(diff(spike1_train))

tn_train = max([max(spike1_train) max(spike2_train)]);
tn_test = max([max(spike1_test) max(spike2_test)]);

delta = 0.01;

T_train = delta:delta:tn_train;
T_test = delta:delta:tn_test;


  s1_train = zeros(1,length(T_train));
```

```
a1_train = zeros(1,length(T_train));
for i=1:length(T_train),
    x=T_train(:,i);
    y=spike1_train';
      temp=repmat(x,length(spike1_train),1);
      temp3=exp(-(temp-y).^2/(2*sigma^2));
      a1_train(i)=sum(temp3);
      s1_train(i)=a1_train(i)/(sqrt(2*pi)*sigma);
end

s2_train = zeros(1,length(T_train));
a2_train = zeros(1,length(T_train));
for i=1:length(T_train),
    x=T_train(:,i);
    y=spike2_train';
      temp=repmat(x,length(spike2_train),1);
      temp3=exp(-(temp-y).^2/(2*sigma^2));
      a2_train(i)=sum(temp3);
      s2_train(i)=a2_train(i)/(sqrt(2*pi)*sigma);
end

s3_train = zeros(1,length(T_train));
a3_train = zeros(1,length(T_train));
for i=1:length(T_train),
    x=T_train(:,i);
    y=spike3_train';
      temp=repmat(x,length(spike3_train),1);
      temp3=exp(-(temp-y).^2/(2*sigma^2));
      a3_train(i)=sum(temp3);
      s3_train(i)=a3_train(i)/(sqrt(2*pi)*sigma);
end

figure(k);clf;
plot(T_train,s1_train,T_train,s3_train)
end

%%%%%%%%%%%%%%%%%%%%%%%%%%%%%%%%%%%%%%%%%%%%%%%%%%%%%%%%%%%%%%%%%
% Plot the poin process(raster plot) and ISI                   %
%%%%%%%%%%%%%%%%%%%%%%%%%%%%%%%%%%%%%%%%%%%%%%%%%%%%%%%%%%%%%%%%%
figure(1);clf;
for i=1:length(spike1_train)
    plot([spike1_train(i) spike1_train(i)], [-50 0]);hold on;
    plot(spike1_train(i),i);hold on;
end;
```

```
xlabel('time (s)');ylabel('cumulative number of events');
title('Counting process of neuron','FontSize',18)

figure(2);clf;
hist(diff(spike1_train),50);
xlabel('time');title('ISI','FontSize',18)

figure(3);clf;
for i=1:length(spike1_train)
    plot([spike1_train(i) spike1_train(i)], [-50 -5]);hold on;
end;
for i=1:length(spike2_train)
    plot([spike2_train(i) spike2_train(i)], [5 50]);hold on;
end;

k=6;
c1=(k-1)*50;c2=k*50;c3=400;c4=450;
spike1_train = spike1(find(c1<spike1 & spike1<=c2))-c1;
spike1_test = spike1(find(c3<spike1 & spike1<=c4))-c3;
spike2_train = spike2(find(c1<spike2 & spike2<=c2))-c1;
spike2_test = spike2(find(c3<spike2 & spike2<=c4))-c3;
spike3_train = spike3(find(c1<spike3 & spike3<=c2))-c1;
spike3_test = spike3(find(c3<spike3 & spike3<=c4))-c3;
spike4_train = spike4(find(c1<spike4 & spike4<=c2))-c1;
spike4_test = spike4(find(c3<spike4 & spike4<=c4))-c3;

tn_train = max([max(spike1_train) max(spike2_train) ...
            max(spike4_train)]);
tn_test = max([max(spike1_test) max(spike2_test) ...
            max(spike4_test)]);

delta = 0.005;

T_train = delta:delta:tn_train;
T_test = delta:delta:tn_test;

figure(1);clf;

[N1,lags1]=sparsexcorr2(spike1_train,spike4_train, ...
            'maxlag',0.5, 'minlag',-0.5,'b',delta);
subplot(2,1,1), plot(lags1,N1);
xlim([-.5 .5])
title('CCG: "neuron 1" and "neuron 15"','FontSize',18)
xlabel('lag (s)')
```

```
ylabel('spikes')
subplot(2,1,2),plot(sigma2,LogL_error1,sigma2,LogL_error3)
xlabel('sigma');ylabel('predictability (bits)');
%text(0.085,80,[' sigma = 0.085'],'FontSize',18,'Color','r');
title('Non-cholinergic basal forebrain (BF) neurons',...
        'FontSize',18);
legend('prediction by one peer: "neuron 5"',...
        'prediction by two peers: "neuron 5" and "neuron 15"');
figure(2);clf;
[N2,lags2]=sparsexcorr2(spike1_train,spike4_train,...
                'maxlag',0.5, 'minlag',-0.5,'b',delta);
plot(lags2,N2);

hist(diff(spike1_train),50);
xlabel('time');title('ISI','FontSize',18)

sigma2=0.005:0.005:0.1;
%bigw=zeros(length(sigma2),2);
%LogL_error3=zeros(1,length(sigma2));
for j = 1:length(sigma2)
  sigma = sigma2(j);
%sigma=.3
  spk1_train = spike1_train;
  spk2_train = spike2_train;
  spk1_test = spike1_test;
  spk2_test = spike2_test;
  spk3_train = spike4_train;
  spk3_test = spike4_test;
%%%%%%%%%%%%%%%%%%%%%%%%%%%%%%%%%%%%%%%%%%%%%%%%%%%%%%%%%%%%%%%%%
% Calculate weight with gauss2D and weights2D
%%%%%%%%%%%%%%%%%%%%%%%%%%%%%%%%%%%%%%%%%%%%%%%%%%%%%%%%%%%%%%%%%
  global T_train S
  S=gauss2D(T_train, spk2_train, spk3_train, sigma,2);
  [wmin,fmin]=weights2DQg(S, T_train, spk1_train, delta);

%%%%%%%%%%%%%%%%%%%%%%%%%%%%%%%%%%%%%%%%%%%%%%%%%%%%%%%%%%%%%%%%%
% Predict the intensity function g via predict2D
%%%%%%%%%%%%%%%%%%%%%%%%%%%%%%%%%%%%%%%%%%%%%%%%%%%%%%%%%%%%%%%%%

  [T_test,pred]=predict2D(wmin,zoom,T_test,spk2_test, ...
                                spk3_test, sigma, 2);

%%%%%%%%%%%%%%%%%%%%%%%%%%%%%%%%%%%%%%%%%%
% Intensity function of spk1_test %
```

```
%%%%%%%%%%%%%%%%%%%%%%%%%%%%%%%%%%%%%%

  bigw(j,:) = wmin;

  Cnt2 = zeros(1,length(T_test));
  T_lo = T_test-(delta/2);
  T_hi = T_test+(delta/2);

    for i=1:length(T_test),
        Cnt2(i) = sum(T_lo(i)<spk1_test & spk1_test<T_hi(i));
    end

  PredRate = pred;
    fRate = 2;
  L_error=-(PredRate-fRate)*delta + ...
            (log(PredRate)-log(fRate)).*Cnt2;
  LogL_error3(j) = sum(L_error)/log(2); % make it to base 2.

clear S Cnt
end

figure(6);clf;
plot(sigma2,LogL_error1);hold on;
plot([0.085  0.085],[ -60 135], '--','Color','r')
xlabel('optimal sigma');ylabel('predictability');
text(0.085,80,[' sigma = 0.085'],'FontSize',18,'Color','r');
title('Non-cholinergic basal forebrain (BF) neurons',...
        'FontSize',18);

%%%%%%%%%%%%%%%
figure(7);clf;
subplot(2,1,1);
[N1,lags1]=sparsexcorr2(spike1_train,spike4_train,...
                'maxlag',0.5, 'minlag',-0.5,'b',delta);
plot(lags1,N1);
xlim([-.5 .5]);
xlabel('lag (s)');
ylabel('spikes');
title('CCG: "neuron 1" and "neuron 15"','FontSize',10);
subplot(2,1,2);
plot(sigma2,LogL_error1,sigma2,LogL_error3,'--');
ylim([-50 150])
xlabel('optimal sigma');ylabel('predictability (bits)');
legend('prediction by "neuron 5"', ...
```

```
          'prediction by "neuron 5" and "neuron 15"');
title('Non-cholinergic basal forebrain (BF) neurons',...
          'FontSize',10);

plot(sigma2,-LogL_error1/3200,[.15 .15],[ 0 3.5])
text(.16,2,[' sigma = 0.15'],'FontSize',18,'Color','r');
title('Non-cholinergic basal forebrain (BF) neurons',...
          'FontSize',18);
xlabel('sigma');ylabel('Predictability');

clear global T_train T_test

tn_train = max([max(spike1_train) max(spike2_train) ...
                 max(spike3_train)]);
tn_test = max([max(spike1_test) max(spike2_test) ...
                 max(spike3_test)]);

delta = 0.005;

T_train = delta:delta:tn_train;
T_test = delta:delta:tn_test;

sigma=0.06;

  spk1_train = spike1_train;
  spk2_train = spike2_train;
  spk1_test = spike1_test;
  spk2_test = spike2_test;
  spk3_train = spike3_train;
  spk3_test = spike3_test;

%%%%%%%%%%%%%%%%%%%%%%%%%%%%%%%%%%%%%%%%%%%%%%%%%%%%%%%%%%%%%%%%%%%%
% Calculate weight with gauss2D and weights2D
%%%%%%%%%%%%%%%%%%%%%%%%%%%%%%%%%%%%%%%%%%%%%%%%%%%%%%%%%%%%%%%%%%%%
  global T_train S
  S=gauss2D(T_train, spk2_train, spk3_train, sigma,2);
  [wmin,fmin]=weights2DQg(S, T_train, spk1_train, delta);

%%%%%%%%%%%%%%%%%%%%%%%%%%%%%%%%%%%%%%%%%%%%%%%%%%%%%%%%%%%%%%%%%%%%
% Predict the intensity function g via predict2D
%%%%%%%%%%%%%%%%%%%%%%%%%%%%%%%%%%%%%%%%%%%%%%%%%%%%%%%%%%%%%%%%%%%%

  [T_test,pred]=predict2D(wmin,zoom,T_test,spk2_test, ...
                             spk3_test, sigma, 2);
```

```
%%%%%%%%%%%%%%%%%%%%%%%%%%%%%%%%%%%%%%
% Intensity function of spk1_test %
%%%%%%%%%%%%%%%%%%%%%%%%%%%%%%%%%%%%%%

    size1 = length(spk1_test);
    sizeT = length(T_test);
    s2 = zeros(1,sizeT);
    a = zeros(1,sizeT);
    for i=1:sizeT,
        x=T_test(:,i);
        y=spk1_test';
            temp=repmat(x,size1,1);
            temp3=exp(-(temp-y).^2/(2*sigma^2));
            a(i)=sum(temp3);
            s2(i)=a(i)/(sqrt(2*pi)*sigma);
    end

clear S Cnt

clf
plot( T_test,s2);hold on;plot(T_test,pred,'Color','r');
legend('intensity function of testing set', ...
                'predicted intensity function')
xlabel('time (s)')
ylabel('firing rate')
title('Prediction by two peer, "neuron 4" and "neuron 5"',...
        'FontSize',18)

gmean=mean(bigg,1);
smean=mean(bigs,1);

gmax=prctile(bigg,95);
gmin=prctile(bigg,5);
smax=prctile(bigs,95);
smin=prctile(bigs,5);

figure(1);clf
  plot(T_test,smean,T_test,smin,'--',T_test,smax,'--',...
        'Color','k')
  hold on;
  plot(T_test,gmean,T_test,gmin,'--',T_test,gmax,'--', ...
        'Color','r')
  hold on;
```

```
  plot(delta:delta:T,r*w+1,'Color','b')
  ylim([-1,16])
h = legend('mean intensity','5% intensity','95% intensity',
'mean predicted intensity','5% predicted intensity',
'95% predicted intensity','true rate function','Location',
'EastOutside');
```

1.7   R FILES FOR REAL DATA

```
library(gdata)

beta1_hat = read.table('beta1_hat',header = FALSE)
beta2_hat = read.table('beta2_hat',header = FALSE)
b_hat = read.table('b_hat',header = FALSE)
theta_hat = read.table('theta_hat',header = FALSE)
a_hat = read.table('a_hat',header = FALSE)

beta1_hat = as.numeric(unmatrix(beta1_hat))
beta2_hat = as.numeric(unmatrix(beta2_hat))
b_hat = as.numeric(unmatrix(b_hat))
a_hat = as.numeric(unmatrix(a_hat))

par( mfcol= c(2, 2))

plot(density(theta_hat[,1]),xlim = c(-5,0),
                 xlab='theta[1]',main='')
lines(c(-3,-3),c(0,10),col=2,lty="dashed",lwd=2)
title('Kernel estimation of theta[1]')
legend("topright",inset=0.05,cex=.8,c("theta[1] = -3"),
        lty="dashed",lwd=2, col = 2)

plot(density(theta_hat[,2]),xlim = c(0,9),
                 xlab='theta[2]',main='')
lines(c(4,4),c(0,10),col=2,lty="dashed",lwd=2)
title('Kernel estimation of theta[2]')
legend("topright",inset=0.05,cex=.8,c("theta[2] = 4"),
        lty="dashed",lwd=2, col = 2)

plot(density(theta_hat[,3]),xlim = c(-5,2),
                 xlab='theta[3]',main='')
lines(c(-2,-2),c(0,8),col=2,lty="dashed",lwd=2)
title('Kernel estimation of theta[3]')
```

```
legend("topright",inset=0.05,cex=.8,c("theta[3] = -2"),
        lty="dashed",lwd=2, col = 2)

plot(density(theta_hat[,4]),xlim = c(-11,20),
                xlab='theta[4]',main='')
lines(c(3,3),c(0,10),col=2,lty="dashed",lwd=2)
title('Kernel estimation of theta[4]')
legend("topright",inset=0.05,cex=.8,c("theta[1] = 3"),
            lty="dashed",lwd=2, col = 2)

par( mfcol= c(1, 2))

plot(density(a_hat),xlim = c(-8,5),xlab='a',main='')
lines(c(a,a),c(0,3),col=2,lty="dashed",lwd=2)
title('Kernel estimation of a')
legend("topright",inset=0.05,cex=.8,c("a = -4"),
            lty="dashed",lwd=2, col = 2)

plot(density(b_hat),xlim = c(-3,9),xlab='b',main='')
lines(c(b,b),c(0,3),col=2,lty="dashed",lwd=2)
title('Kernel estimation of b')
legend("topright",inset=0.05,cex=.8,c("b = 2"),
            lty="dashed",lwd=2, col = 2)

plot(density(beta1_hat),xlim = c(0,6),xlab='beta1',main='')
lines(c(beta1,beta1),c(0,10),col=2,lty="dashed",lwd=2)
title('Kernel estimation of beta1')
legend("topright",inset=0.05,cex=.8,c("beta1 = 2"),
            lty="dashed",lwd=2, col = 2)

plot(density(beta2_hat),xlim = c(-1,4),xlab='beta2',main='')
lines(c(beta2,beta2),c(0,10),col=2,lty="dashed",lwd=2)
title('Kernel estimation of beta2')
legend("topright",inset=0.05,cex=.8,c("beta2 = 1"),
            lty="dashed",lwd=2, col = 2)


rm(list = ls(all = TRUE))
library(polspline)
library(splines)
library(gdata)

# Generate peer1 spike train : deterministic gap = 1
spk1 = c(0,cumsum(rep(1,600)))
spk2 = c(0.5,cumsum(rep(1,600))+0.5)
```

```
# Generate the target spike train:
# log(lam) = a + b*t + theta*basis + beta1*[K-(t-tao_star)]
   # Truncated cubic power function
   bas3 <- function(t,knot){
     ((abs(t - knot) + t - knot)/2)^3
   }


#Set up parameters
a = -4; b=2; theta = c(-3,4,-2,3); beta1 = 2; beta2 = 1
knots = c(0.2,0.4,0.6,0.8)
nknots = length(knots)
K=2
Delta = 0.001


# Initialize
t =  c(0,seq(Delta,600, by=Delta))
niter = 1   # 100
spk0 = matrix(rep(0,niter*length(spk1)*2),nrow = niter)
taostar1 = matrix(rep(0,niter*length(spk1)*2),nrow = niter)
taostar2 = matrix(rep(0,niter*length(spk1)*2),nrow = niter)


# Start iteration
ptm <- proc.time()
   for (p in 1:niter){
       tao0 =t[1]
       log_lam = rep(0,length(t))
       t_taostar1 = rep(0,length(t))
       t_taostar2 = rep(0,length(t))
       K_cov1 = rep(0,length(t))
       K_cov2 = rep(0,length(t))
       j = 1
       for (i in 1:length(t)){
          taostar11 = 0
          taostar22 = 0.5
          gamma_t = t[i] - tao0
           if (any((t[i]-spk1)>0)){
               taostar11 = spk1[max(which((t[i]-spk1)>0))]
               if (taostar11 >= tao0){
                  t_taostar1[i] = t[i]-taostar11
                  K_cov1[i] = K-t_taostar1[i]
               }
           }
           if (any((t[i]-spk2)>0)){
```

```
                    taostar22 = spk2[max(which((t[i]-spk2)>0))]
                    if (taostar22 >= tao0){
                        t_taostar2[i] = t[i]-taostar22
                        K_cov2[i] = K-t_taostar2[i]
                    }
                }
                sp1 = 0
                for (k in 1:nknots){
                    sp1 = sp1 + theta[k]*bas3(gamma_t,knots[k])
                }
                log_lam[i] = a+b*gamma_t+ sp1 +beta1*(K_cov1[i])
                              + beta2*(K_cov2[i])
                p_t = exp(log_lam[i])*Delta
                if (rbinom(1,1,p_t)==1){
                    spk0[p,j] = t[i]
                    taostar1[p,j] = t[i]-t_taostar1[i]
                    taostar2[p,j] = t[i]-t_taostar2[i]
                    j = j+1
                    tao0 = t[i]
                }
            }
        }
        p
    }
proc.time() - ptm

test = spk0[,1:650]
test2 = taostar1[,1:650]
test3 = taostar2[,1:650]
write.table(test,file="spk0.txt",row.names=F)
write.table(test2,file="taostar1.txt",row.names=F)
write.table(test3,file="taostar2.txt",row.names=F)



rm(list = ls(all = TRUE))
library(polspline)
library(splines)
library(gdata)

# Generate peer1 spike train : deterministic gap = 1
spk1 = c(0,cumsum(rep(1,600)))
spk2 = c(0.5,cumsum(rep(1,600))+0.5)

# Generate the target spike train:
#    log(lam) = a + b*t + theta*basis + beta1*[K-(t-tao_star)]
```

```
# Truncated cubic power function
bas3 <- function(t,knot){
  ((abs(t - knot) + t - knot)/2)^3
}

#Set up parameters
a = -4; b=2; theta = c(-3,4,-2,3); beta1 = 2; beta2 = 1
knots = c(0.2,0.4,0.6,0.8)
nknots = length(knots)
K=2
Delta = 0.001

# Initialize
t =  c(0,seq(Delta,600, by=Delta))
niter = 100
spk0 = matrix(rep(0,niter*length(spk1)*2),nrow = niter)
taostar1 = matrix(rep(0,niter*length(spk1)*2),nrow = niter)
taostar2 = matrix(rep(0,niter*length(spk1)*2),nrow = niter)

   # Start iteration
   for (p in 1:niter){
       tao0 =t[1]
       log_lam = rep(0,length(t))
       t_taostar1 = rep(0,length(t))
       t_taostar2 = rep(0,length(t))
       K_cov1 = rep(0,length(t))
       K_cov2 = rep(0,length(t))
       j = 1
       for (i in 1:length(t)){
          taostar11 = 0
          taostar22 = 0.5
          gamma_t = t[i] - tao0
           if (any((t[i]-spk1)>0)){
              taostar11 = spk1[max(which((t[i]-spk1)>0))]
              if (taostar11 >= tao0){
                 t_taostar1[i] = t[i]-taostar11
                 K_cov1[i] = K-t_taostar1[i]
              }
           }
           if (any((t[i]-spk2)>0)){
              taostar22 = spk2[max(which((t[i]-spk2)>0))]
              if (taostar22 >= tao0){
                 t_taostar2[i] = t[i]-taostar22
                 K_cov2[i] = K-t_taostar2[i]
```

```
                }
            }
            sp1 = 0
            for (k in 1:nknots){
                sp1 = sp1 + theta[k]*bas3(gamma_t,knots[k])
            }
            log_lam[i] = a+b*gamma_t+ sp1 +beta1*(K_cov1[i])+
                            beta2*(K_cov2[i])
            p_t = exp(log_lam[i])*Delta
            if (rbinom(1,1,p_t)==1){
                spk0[p,j] = t[i]
                taostar1[p,j] = t[i]-t_taostar1[i]
                taostar2[p,j] = t[i]-t_taostar2[i]
                j = j+1
                tao0 = t[i]
            }
        }
        p
    }

test = spk0[,1:650]
test2 = taostar1[,1:650]
test3 = taostar2[,1:650]
write.table(test,file="spline_b1(K-t+tao)_spk0.txt",
                row.names=F)

write.table(test2,file="spline_b1(K-t+tao)_taostar1.txt",
                row.names=F)

write.table(test3,file="spline_b1(K-t+tao)_taostar2.txt",
                row.names=F)
```

Bibliography

1. E. N. Brown, L. M. Frank, D. Tang, M. C. Quirk, and M. A. Wilson. A statistical paradigm for neural spike train decoding applied to position prediction from ensemble firing patterns of rat hippocampal place cells. *Journal of Neuroscience*, 18:7411–7425, 1998.

2. E. N. Brown, R. E. Kass, and P. P. Mitra. Multiple neural spike train data analysis: State-of-the-art and future challenges. *Nature Neuroscience*, 7(5): 456–461, 2004.

3. G. Casella and R. L. Berger. *Statistical Inference*. Belmont, CA: Duxbury, 1990.

4. David R. Cox. *Renewal Theory*, volume 4. Methuen London, 1962.

5. D. Daley and Vere-Jones. *An Introduction to the Theory of Point Process (2nd ed.)*. Springer-Verlag, New York, 2003.

6. K. J. Friston. Functional and effective connectivity in neuroimaging: A synthesis. *Human Brain Mapping*, 2:56–78, 1994.

7. W. Gerstner and W. M. Kistler. *Spiking Neuron Models: Single Neurons, Populations, Plasticity*. Cambridge University Press, 2002.

8. C. Gu. Smoothing noisy data via regularization: Statistical perspectives. *Inverse Problems*, 24:034002, 2008.

9. K. D. Harris, J. Csicsvari, H. Hirase, G. Dragoi, and G. Buzsaki. Organization of cell assemblies in the hippocampus. *Nature*, 424:552–556, 2003.

10. N. G. Hatsopoulos, C. L. Ojakangas, L. Paninski, and J. P. Donoghue. Information about movement direction obtained from synchronous activity of motor cortical neurons. *Proceedings of the National Academy of Sciences of the United States of America*, 95:15706–15711, 1998.

11. S-C Lin and M. Nicolelis. Neuronal ensemble bursting in the basal forebrain encodes salience irrespective of valence. *Neuron*, 59(1):138–149, 2008.

12. A. Luczak, P. Bartho, S. Marguet, G. Buzsaki, and K. Harris. Sequential structure of neocortical spontaneous activity in vivo. *Proceedings of the National Academy of Sciences of the United States of America*, 104:347–352, 2007.

13. M. Nawrot, A. Aertsen, and S. Rotter. Single-trial estimation of neuronal firing rates: From single-neuron spike trains to population activity. *Journal of Neuroscience Methods*, 94:81–92, 1999.

14. M. Nicolelis, D. Dimitrov, J. M. Carmena, R. Crist, G. Lehew, J. D. Kralik, and S. P. Wise. Chronic, multisite, multielectrode recordings in macaque monkeys. *Proceedings of the National Academy of Sciences of the United States of America*, 100(19):11041–11046, 2003.

15. J. Pillow, J. Shlens, L. Paninski, A. Sher, A. Litke, E. Chichilnisky, and E. Simoncelli. Spatiotemporal correlations and visual signaling in a complete neuronal population. *Nature*, 454:995–999, 2008.

16. B. Richmond, L. Optican, and H. Spitzer. Temporal encoding of two-dimensional patterns by single units in primate primary visual cortex. *Journal of Neurophysiology*, 64:351–369, 1999.

17. G. Santhanam, S. I. Ryu, B. M. Yu, A. Afshar, and K. V.Shenoy. A high-performance brain-computer interface. *Nature*, 442:195–198, 2006.

18. H. Shimazaki and S. Shinomoto. A method for selecting the bin size of a time histogram. *Neural Computation*, 19(6):1503–1527, 2007.

19. H. Shimazaki and S. Shinomoto. Kernel bandwidth optimization in the spike rate estimation. *Journal of Computational Neuroscience*, 29(1–2):171–182, 2009.

20. I. H. Stevenson, J. M. Rebesco, N. G. Hatsopoulos, Z. Haga, L. E. Miller, and K. P. Kording. Bayesian inference of functional connectivity and network structure from spikes. *Neural Systems and Rehabilitation Engineering*, 17(3):203 – 213, 2009.

21. W. Truccolo, U. T. Eden, M. R. Fellows, J. P. Donoghue, and E. N. Brown. A point process framework for relating neural spiking activity to spiking history, neural ensemble and extrinsic covariate effects. *Journal of Neurophysiology*, 93:1074–1089, 2005.

22. F. Varela, J.P. Lachaux, E. Rodriguez, and J. Martinerie. The brainweb: Phase synchronization and large-scale integration. *Nature Reviews Neuroscience*, 2 (4):229–239, 2001.

23. Y. Yoshimura and E. M. Callaway. Fine-scale specificity of cortical networks depends on inhibitory cell type and connectivity. *Nature Neuroscience*, 8: 1552–1555, 2005.

24. R. Zhang. Likelihood analysis of neuro spike train. PhD thesis, Department of Statistics & Operations Research, University of North Carolina at Chapel Hill, 2011.

2 Regression Spline Model for Neural Spike Train Data

Ruiwen Zhang
SAS Institute

Shih-Chieh Lin
NIH

Haipeng Shen
University of Hong Kong, China

Young K. Truong
University of North Carolina at Chapel Hill

CONTENTS

Neuroscience experiments and neural spike train data have special features that present novel and exciting challenges for statistical researches. Several standard statistical procedures, widely used in other fields of science, have found their way into mainstream application in neuroscience data analysis. Given the firing times of an ensemble of neurons and their stimulating and inhibitory inputs from several regions, an integrated model is introduced based on the conditional intensity function approach. This is different from the existing methods where the intensity function is approximated by discretization with the sampling intervals chosen arbitrarily. In this chapter, we model the log conditional intensity function directly by employing a polynomial spline function for the target or response spike train and a tensor product of splines for the peer or predictor spike trains. The parameters are defined by those used in constructing the polynomial splines, and they will be estimated by the maximum likelihood method. The statistical properties of this procedure will be evaluated using both a simulated experiment and a real data set involving 15 peers of neural spike trains. Our model captures the underlying spontaneous firing of the target as well as the stimulus inputs from its peers, both in continuous time.

2.1   INTRODUCTION

Brillinger [2] formulated an approach for analyzing interacting nerve cells, which extends [4] to the case of several spike train inputs. The approach articulated the spontaneous firing rate and interacting effects from arbitrary numbers of neurons, and offered high flexibility and efficiency for estimating interacting neuron effects in a big network. The method begins with a discretization process by sampling the time over small intervals. Within each interval, the firing probability or the intensity function is modeled as a probit function of the target neuron's lapse time plus the firing times of the associated neurons. Flexibility was enhanced by introducing a polynomial function for the lapse time. This can be described more systematically using generalized linear models (GLM) in the framework of point processes [8].

Recall that the conditional intensity function is a history-dependent generalization of the rate function for a neural spike train. The parametric estimation of the conditional intensity function is a good start, although the strong assumption of the underlying probability model may be inappropriate in most cases. The field of non-parametric estimation has advanced with the use of computing technology and we focus on the numerical approximation of spontaneous firing of neurons using spline functions.

In this chapter, the discretization and parametric component will be relaxed by modeling the log conditional intensity function directly using polynomial spline functions. For the target neuron, a cubic spline is used to model its effect in the spirit of HEFT [6], whereas linear splines and their tensor products are used to model the associated neurons or peers' effects. This method is similar to the HARE methodology for hazard regression [6] with the exception that it is a time-dependent covariate approach. The advantage here is that the flexibility of the model will allow us to examine directly the impacts of other neurons within the neural network, either stimulating or inhibiting. The model specification will be described in detail in Section 2.2.

To estimate our approximated model of the log conditional intensity function, we turn to the problem of estimating the parameters introduced for the construction of the spline functions. This will be carried out using the maximum likelihood method, which will be discussed more thoroughly in Section 2.3. We also employ an adaptive methodology for knot selection in a manner similar to HEFT and HARE as described in [6].

Our implementation of the model is written in C and now is readily available with an R interface. The method has been numerically evaluated using a simulated experiment. The results showed that our estimate is consistent and asymptotically normal, hence the inference part of our approach can be implemented easily. We further applied the method to analyze a real data set from noncholinergic basal forebrain (BF) neurons [7], and obtained some interesting novel findings. Specifically, our model captured the underlying spontaneous firing of the target as well as the stimulus inputs from its peers, both in continuous time.

2.2 LINEAR MODELS FOR THE CONDITIONAL LOG-INTENSITY FUNCTION

For one particular neuron, the spontaneous firing rate depends on its own natural characteristics. Also, the firings of other neurons or peers within the neural network have an impact on the target neuron. Suppose there are M such peers and their firing times before time t are given by $\mathbf{x}(t) = (x_1(t),..,x_M(t))$. Let T be a positive random variable whose distribution may depend on the peers $\mathbf{x}(t)$.

Let $\lambda(\cdot|\mathbf{x}(s), s > 0)$ denote the conditional intensity function of T given $\mathbf{x}(s)$ so that

$$\lambda(t|\mathbf{x}(t))\,dt = \text{prob}(T \in (t, t+dt)|\mathbf{x}(t)).$$

Let $\alpha(\cdot|\mathbf{x}(t))$ denote the log conditional intensity function. To simplify the discussion, we assume the conditional intensity function at time t depends only on the value of the covariates up to that time; that is, we assume that $\lambda(t|\mathbf{x}(s), s > 0) = \lambda(t|\mathbf{x}(s), 0 \leq s \leq t)$ and hence that $\alpha(t|\mathbf{x}(s), s > 0) = \alpha(t|\mathbf{x}(s), 0 \leq s \leq t)$.

In this chapter, we model the log conditional intensity function via

$$\alpha(t|\mathbf{x}(t)) = \mu(t) + \beta \cdot \mathbf{h}(\mathbf{x}(t)), \quad t \geq 0, \tag{2.1}$$

where $\mu(t)$ is a cubic spline function, $\beta = (\beta_1,...,\beta_M)'$ are the coefficients for M covariates, and $\mathbf{h}(\cdot)$ is a vector of M functions corresponding to the covariates for the effects of the interacting neurons. For examples, step functions have been used to approximate the peer effects on the target neuron [3]. In our method, the M components of $\mathbf{h}$ are all continuous or smooth functions whose explicit forms will be discussed in detail in Section 2.3.

The idea of (2.1) is straightforward in the sense that we take into account two basic facts. First, a nerve cell has a spontaneous firing rate across time without any intrinsic or extrinsic stimuli that is considered as the baseline firing rate. Second, under the proportionality assumption, any type of input from the peer activities would have a cumulative effect on the conditional log-intensity function.

Let $1 \leq p < \infty$ and let G be a p-dimensional linear space of functions on $[0, \infty)$ spanned by the B-spline functions: $B_1,...,B_p$. Model (2.1) is now written as

$$\alpha(t|\mathbf{x}(t)) = \sum_{i=1}^{p} \theta_i B_i(t) + \beta \cdot \mathbf{h}(\mathbf{x}(t)), \quad t \geq 0, \tag{2.2}$$

where $\theta = (\theta_1, ..., \theta_p)'$ are the coefficients for the p basis functions.

Our approach is certainly a step forward compared to the point process–GLM framework [8], which has to use the discrete time representation to approximate the joint probability density of a continuous time point process. For example, a simple version of (2.2) with only one covariate can be written explicitly as

$$\alpha(t|\mathbf{x}(t)) = \sum_{i=1}^{p} \theta_i B_i(t) + \beta_1 h(x_1(t)), \quad t \geq 0. \tag{2.3}$$

We will add multiple covariates as a more realistic case later. For simplicity of illustration, (2.3) will be used in the next two sections.

2.3 MAXIMUM LIKELIHOOD ESTIMATION

Neural spike train data are collected as a time series of electrical impulses generated by individual neurons, and the typical form of a neural spike train is a temporal point process that shows precisely the times of firing. The time between two consecutive firings as shown in Figure 2.1 are referred to as the *interspike interval* (ISI), and here we denote the interspike intervals for a spike train with N spikes as $\{u_k\}_{k=1}^{N}$.

Given the ISI data with the vector of covariates $\mathbf{x}(t)$ and a set $B_1, ..., B_p$ of basis functions, we will estimate the coefficients in (2.3) using the maximum likelihood estimation (MLE).

The partial log-likelihood for a neuron can be written as

$$\phi(t|\mathbf{x}(t)) = \log \lambda(t|\mathbf{x}(t)) - \int_0^t \lambda(s|\mathbf{x}(s))ds = \alpha(t|\mathbf{x}(t)) - \Lambda(t|\mathbf{x}(t)). \tag{2.4}$$

We take the partial derivatives of $\phi(\cdot)$ to examine its concavity and maximize the log likelihood. Set

$$D_j^\theta = \frac{\partial \Lambda(t|\mathbf{x}(t))}{\partial \theta_j} = \int_0^t B_j(s) \exp \left(\sum_{i=1}^{p} \theta_i B_i(s) + \beta_1 h(x_1(s)) \right) ds,$$

$$D_1^\beta = \frac{\partial \Lambda(t|\mathbf{x}(t))}{\partial \beta_1} = \int_0^t h(x_1(s)) \exp \left(\sum_{i=1}^{p} \theta_i B_i(s) + \beta_1 h(x_1(s)) \right) ds, \tag{2.5}$$

and

$$E_{j,k}^\theta = \frac{\partial^2 \Lambda(t|\mathbf{x}(t))}{\partial \theta_j \partial \theta_k} = \int_0^t B_j(s) B_k(s) \exp \left(\sum_{i=1}^{p} \theta_i B_i(s) + \beta_1 h(x_1(s)) \right) ds,$$

$$E_{j,1}^{\theta,\beta} = \frac{\partial^2 \Lambda(t|\mathbf{x}(t))}{\partial \theta_j \partial \beta_1} = \int_0^t B_j(s) h(x_1(s)) \exp \left(\sum_{i=1}^{p} \theta_i B_i(s) + \beta_1 h(x_1(s)) \right) ds, \tag{2.6}$$

$$E_{1,1}^\beta = \frac{\partial^2 \Lambda(t|\mathbf{x}(t))}{\partial \theta_k^2} = \int_0^t h^2(x_1(s)) \exp \left(\sum_{i=1}^{p} \theta_i B_i(s) + \beta_1 h(x_1(s)) \right) ds.$$

Then

$$\frac{\partial \phi}{\partial \theta_j} = B_j(t) - D_j^\theta, \quad \frac{\partial \phi}{\partial \beta_1} = h(x_1(t)) - D_1^\beta,$$

$$\frac{\partial^2 \phi}{\partial \theta_j \partial \theta_k} = -E_{j,k}^\theta, \quad \frac{\partial^2 \phi}{\partial \theta_j \partial \beta_1} = -E_{j,1}^{\theta,\beta}, \quad \frac{\partial^2 \phi}{\partial \beta_1^2} = -E_{1,1}^\beta. \tag{2.7}$$

It follows (2.7) that $\phi(t|\mathbf{x}(t))$ is a concave function.

The log-likelihood function is a sum for all N consecutive ISIs $\{u_k\}_{k=1}^N$ given by

$$l(\theta, \beta_1) = \sum_{k=1}^N \alpha(u_k|\mathbf{x}(u_k)) - \sum_{k=1}^N \int_0^{u_k} \lambda(s|\mathbf{x}(s))ds. \tag{2.8}$$

The maximum likelihood estimate $\widehat{\theta}$ and $\widehat{\beta_1}$ will be obtained by maximizing the log-likelihood, $l(\theta, \beta_1)$.

2.4   SIMULATION STUDIES

2.4.1   CASE 1: FIXED NUMBER OF BASIS FUNCTIONS

We follow the principle ideas from the "integrate-and-fire" model by Lapicque [1]. Stimuli from peers in the form of current inputs are cumulative until the very next firing of the target cell. After triggering that firing, these prior stimuli would have no impact on any coming events, similar to resetting a timer and counting again from zero. In the meantime, the voltage of the membrane naturally decays over time, referred to as *leaky integrate-and-fire*, so that the impact of peer firing diminishes gradually. Based on these known neurobiological phenomena, we proposed a model for interspike interval (ISI) data that follows (2.3) when h is a decreasing function of time.

For two interacting neural spike trains, the target spike train has N spikes or N ISIs, $\{u_k\}_{k=1}^N$, Each spike of the peer then belongs to a specific ISI of the target spike train and only has its effect within that interval. Thus, for any u_k, we denote the distance from the beginning of the interval to a particular spike of a peer within the interval as v_k, which then is the covariate for the k-th observation (Figure 2.1).

For two interacting neural spike trains, the target spike train has N spikes or N ISIs, $\{u_k\}_{k=1}^N$, then each spike of the peer belongs to a specific ISI of the target spike train and only has its affect within that interval. So, for any u_k, we denote the distance from the beginning of the interval to a particular spike of the peer within the interval as v_k, which is then the covariate for the kth observation (see Figure 2.1). For the case where the peer fires more than once within an interspike interval of the target, we only consider the last one as the covariate.

The conditional log-intensity function for the kth observation is

$$\alpha(t|v_k) = \sum_{i=1}^p \theta_i B_i(t) + \beta_1 [K - (t - v_k)] 1_{t \geq v_k}, \quad t \geq 0. \tag{2.9}$$

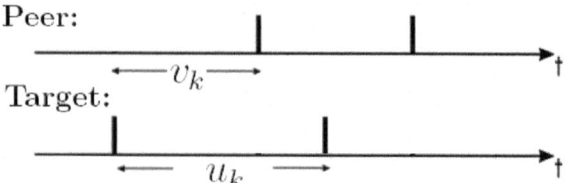

Figure 2.1: Target and peer spike trains are recorded simultaneously. Between two consecutive spikes of the target, the distance is our observation and u_k is for the interval ending at the kth spike. The peer may fire within the interval, and then we record the distance from the beginning of u_k to the peer firing time as our covariate v_k.

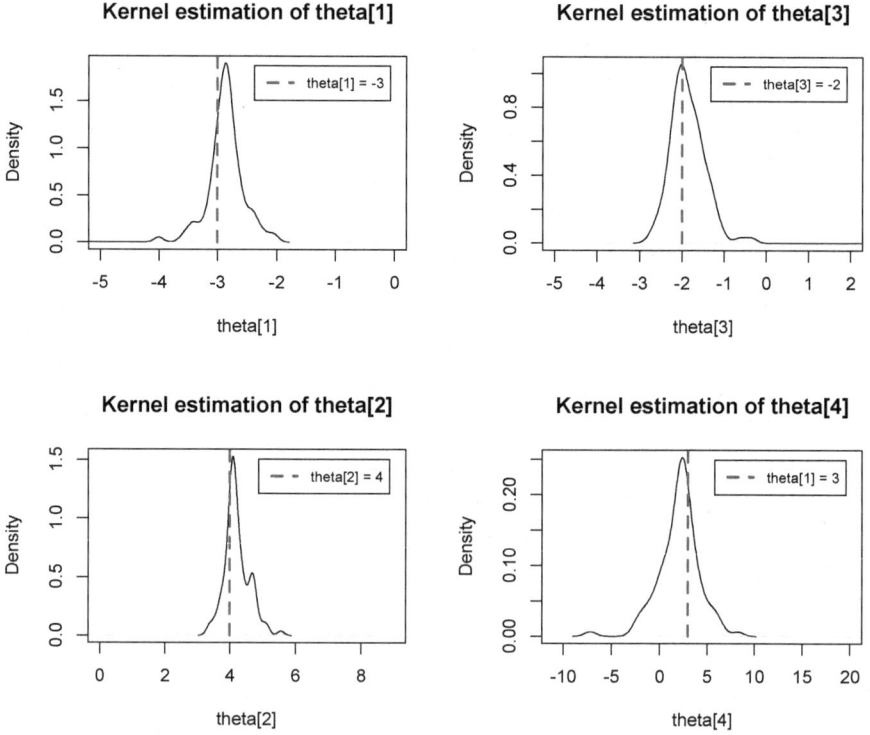

Figure 2.2: Kernel density estimations of θ's from 100 simulations. In the setup, the spline knot locations are fixed as $\{0.2, 0.4, 0.6, 0.8\}$, and two covariates are considered in the model. The estimates have unbiased means and fewer variations.

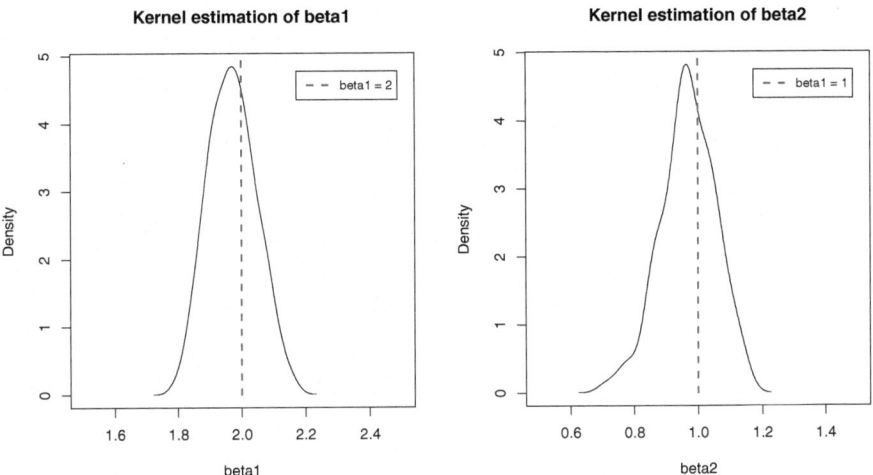

Figure 2.3: Kernel density estimations of β_1 and β_2 from 100 simulations. In the setup, the spline knot locations are fixed as $\{0.2, 0.4, 0.6, 0.8\}$, and two covariates are considered in the model. The estimates have unbiased means and fewer variations.

where $1_{t \geq v_k}$ is an indicator function and K is a constant that is deterministic for each target spike train. In the model, the target spike train keeps its spontaneous firing rate before it is stimulated by the peer firing and therefore the covariate term is zero until $t \geq v_k$.

By setting up the conditional log-intensity function as shown in (2.9), we obtained consistent estimates of the coefficients. Nevertheless, our method is very flexible in handling the influence of multiple peers simultaneously or multiple spiking stimulations from a single peer. In other words, in the model we proposed, the function h could be not only a simple linear term of one covariate as shown in (2.9) but also a summation of functions for various covariates.

To take this one step further, we add a second linear term for another peer spike, and the conditional log-intensity function is then

$$\alpha(t|v_k) = \sum_{i=1}^{p} \theta_i B_i(t) + \beta_1 [K - (t - v_k^{(1)})] 1_{t \geq v_k^{(1)}} + \beta_2 [K - (t - v_k^{(2)})] 1_{t \geq v_k^{(2)}}, \quad t \geq 0.$$
(2.10)

We certainly can expand the model by adding more covariate terms.

Following the model formulated in (2.10), we simulated 100 spike trains, each having 650 spike times given a deterministic peer spike train. In Figure 2.2 and Figure 2.3, the estimators of θ_1 to θ_4 corresponding to knot locations $\{0.2, 0.4, 0.6, 0.8\}$ and the estimators of β_1 and β_2 are shown, respectively, after 100 simulations.

2.4.2 CASE 2: ADAPTIVE KNOTS SELECTION

When modeling the log-intensity function with a linear model (see Equation (2.2)), the remaining issue to be resolved is the choice of G.

Initially, we use minimal allowable space to model $\alpha(t|\mathbf{x}(t))$, so that it does not depend on t or the vector $\mathbf{x}$ of covariates. Then we proceed with stepwise addition, successively replacing a $(p-1)$-dimensional space G_0 by a p-dimensional space G containing G_0 as a subspace. The candidates for the new basis function (a function that together with a basis of G_0 spans G) depends on which functions are already in G. When multiple regression is possible to evaluate candidates for a new basis function, we choose among the various candidates using a heuristic search that maximizes the Rao statistics.

The addition will be stopped when one of the following conditions is satisfied [6]:

1. The number P of basis functions equals P_{max}, where the default value for P_{max} is $\min(4n^{1/5}, n/4, 30)$.
2. The search algorithm yields no possible new basis function.
3. $\hat{l}_P - \hat{l}_p < \frac{1}{2}(P-p) - 0.5$ for some p with $3 \leq p \leq P-3$, where $\hat{l}_p$ is the log-likelihood for the model with p basis functions.

Upon stopping the stepwise addition stage, we proceed to stepwise deletion by successively replacing the p-dimensional allowable space G by a $(p-1)$-dimensional allowable subspace G_0 until we arrive at the minimal allowable space. For each step, the basis function that would be removed in going from G to G_0 has the smallest Wald statistic in space G.

During the combination of stepwise addition and deletion, we get a sequence of models. Eventually, we select the model that minimizes Bayesian information criterion (BIC).

As was the case in Case 1, we do not fix the knots locations for the following regression so that the baseline function would be fitted by an adaptive sequence of basis. The same datasets as in Case 1 are used here and yield the results through model selection. The mean curve of the baseline function from 100 simulations is plotted in Figure 2.4 along with its 95% boundary and the true function, as well as the estimators of β_1 and β_2.

2.5 DATA ANALYSIS

We conduct our studies on some real data, the 16 neural spike trains that were recorded simultaneously from noncholinergic basal forebrain (BF) neurons [7]. Tables 2.1 and 2.2 contain the estimation results from the single peer model and the multiple peers model, respectively.

When we only consider an individual spike train as a peer for one particular target, then most of the 16 neurons have excitatory impacts on others. In Table 2.1, a cell entry in black color implies a significantly positive coefficient β_1. The rows index the neural spike trains that are treated as targets in each round, whereas the columns index the peers in the single peer model, one peer at a time. For example, cell $(1,2)$,

Table 2.1

The rows index the neural spike trains that are treated as targets in each round, while the columns index the peers in the single peer model, one peer at a time. For example, cell $(1, 2)$ is for the regression with $spk1$ as the target and $spk2$ as the peer. Estimates of β_1 printed in black are statistically significantly positive ($p < 0.05$). Numbers in blue are significantly negative estimates, and values printed in red color are not statistically significant.

	$spk1$	$spk2$	$spk3$	$spk4$	$spk5$	$spk6$
$spk1$		0.4405	1.5218	0.4426	0.6858	0.5609
$spk2$	0.7724		1.7642	0.4584	0.8318	0.8411
$spk4$	0.6949	0.5599	1.4620		0.713	0.6766
$spk5$	0.4603	0.3722	1.61305	0.4074		0.4707
$spk6$	0.2876	0.2727	1.4639	0.2352	0.3802	
$spk7$	0.5312	0.4479	1.6335	0.3893	0.6256	0.7252
$spk8$	0.7347	0.4688	1.7880	0.4529	0.7677	0.6891
$spk9$	0.4152	0.2839	1.4446	0.2916	0.4966	0.6176
$spk10$	0.4090	0.3107	1.5790	0.3742	0.5206	0.5174
$spk11$	0.8383	0.6765	1.6248	0.5460	0.8397	0.7982
$spk12$	0.4072	0.2924	1.2444	0.3428	0.4287	0.4279
$spk13$	0.4726	0.3996	1.4336	0.3414	0.5145	0.5073
$spk14$	1.1003	0.6702	1.8320	0.6069	1.0699	0.8568
$spk16$	0.9234	0.7045	1.7724	0.5893	0.9077	0.9070

	$spk7$	$spk8$	$spk9$	$spk10$	$spk11$	$spk12$
$spk1$	0.0005	0.4712	0.5176	0.6579	0.5184	0.7541
$spk2$	-0.0001	0.6173	0.8114	0.9995	0.6451	0.9584
$spk4$	-0.0009	0.5978	0.63944	0.7528	0.5967	0.8150
$spk5$	-0.0006	0.3421	0.4096	0.4819	0.4125	0.5461
$spk6$	0.0005	0.3409	0.3610	0.3408	0.1917	0.4338
$spk7$		0.4900	0.6801	0.6842	0.4441	0.7566
$spk8$	0.0015		0.5856	0.7185	0.4895	0.8842
$spk9$	0.0008	0.3755		0.4746	0.3112	0.5595
$spk10$	-0.0006	0.3445	0.4695		0.424	0.6411
$spk11$	0.0002	0.6644	0.7263	1.0991		1.0071
$spk12$	0.0033	0.3252	0.4007	0.4740	0.3433	
$spk13$	0.0012	0.3790	0.5464	0.5204	0.4069	0.6227
$spk14$	9.6E-05	0.8345	0.7720	0.8912	0.6933	1.0000
$spk16$	0.0033	0.7816	0.7962	1.1255	0.7521	1.2817

	$spk13$	$spk14$	$spk15$	$spk16$
$spk1$	0.6440	0.3396	0.6496	0.40782
$spk2$	0.8476	0.3563	0.8414	0.52440
$spk4$	0.6802	0.4389	0.7933	0.51952
$spk5$	0.5062	0.3214	0.5948	0.34374
$spk6$	0.3596	0.1452	0.5456	0.14997
$spk7$	0.6694	0.2953	0.7232	0.39546
$spk8$	0.7024	0.3934	0.7524	0.42563
$spk9$	0.4805	0.2212	0.6245	0.26352
$spk10$	0.4854	0.2658	0.6661	0.27316
$spk11$	0.7932	0.3924	0.8429	0.59255
$spk12$	0.4518	0.2200	0.5298	0.29791
$spk13$		0.2572	0.58706	0.32570
$spk14$	0.8727		0.89328	0.63851
$spk16$	0.8397	0.4361	0.8989	

Table 2.2

The rows index the neural spike trains that are treated as targets in each round, while the columns index the peers in the multiple peers model, 15 peers each round. So, for example, the first row is for the regression with $spk1$ as the target and $spk2$ through $spk16$ as the peers. Estimates of β_1 printed in black are statistically significantly positive ($p < 0.05$). Numbers in blue are significantly negative estimates, and values printed in red color are not significant.

	$spk1$	$spk2$	$spk3$	$spk4$	$spk5$	$spk6$
$spk1$		0.0344	1.1295	0.0902	0.1906	0.1152
$spk2$	0.1451		1.1892	0.0541	0.1433	0.1452
$spk4$	0.1150	0.0284	1.0511		0.1393	0.1119
$spk5$	0.1205	0.0033	1.3866	0.0504		0.0612
$spk6$	0.0538	-0.0244	1.3263	0.0386	0.0767	
$spk7$	0.0378	0.0451	1.3486	0.0400	0.0117	0.0810
$spk8$	0.1732	0.0119	1.3445	0.0491	0.1276	0.0962
$spk9$	0.0457	0.0236	1.2441	0.0279	0.0735	0.1974
$spk10$	0.0650	0.0463	1.3018	0.0862	0.1155	0.1304
$spk11$	0.1973	0.0931	1.0516	0.1217	0.1453	0.1636
$spk12$	0.1087	0.0130	1.2274	0.0597	0.0859	0.1145
$spk13$	0.1061	0.0664	1.1725	0.0499	0.0865	0.0724
$spk14$	0.3291	-0.0484	1.3661	0.0507	0.2425	-0.0064
$spk16$	0.1771	0.0845	1.1085	0.0772	0.1472	0.1662

	$spk7$	$spk8$	$spk9$	$spk10$	$spk11$	$spk12$
$spk1$	-1.5E-05	0.1162	0.0771	0.1309	0.0882	0.2604
$spk2$	-0.0001	0.0653	0.1646	0.2668	0.1040	0.2523
$spk4$	-0.0006	0.0709	0.0974	0.1507	0.0676	0.2481
$spk5$	-0.0007	0.0369	0.0739	0.0782	0.0312	0.1313
$spk6$	0.0001	0.0166	0.1395	0.0827	-0.0233	0.1505
$spk7$		-0.0225	0.1083	0.1031	0.0881	0.1395
$spk8$	0.0008		0.0767	0.1086	0.0325	0.2504
$spk9$	0.0001	0.0221		0.0740	0.0043	0.1527
$spk10$	-0.0001	0.0348	0.0736		0.1215	0.2230
$spk11$	-6.0E-05	0.0658	0.1060	0.3841		0.3357
$spk12$	0.0014	0.0490	0.0673	0.1157	0.0485	
$spk13$	0.0007	0.0267	0.1392	0.0841	0.0564	0.1478
$spk14$	-0.0003	0.0596	0.0224	0.0377	-0.0103	0.1585
$spk16$	0.0009	0.1215	0.1283	0.2434	0.1371	0.4807

	$spk13$	$spk14$	$spk15$	$spk16$
$spk1$	0.1760	0.1039	0.2287	0.0564
$spk2$	0.2317	-0.0037	0.2518	0.0707
$spk4$	0.0994	0.0173	0.2846	0.0301
$spk5$	0.1229	0.0401	0.2191	0.0014
$spk6$	0.0814	-0.0122	0.2531	-0.0063
$spk7$	0.1203	-0.0561	0.1123	-0.0239
$spk8$	0.1255	0.0230	0.2203	0.0374
$spk9$	0.1329	-0.0135	0.2432	-1.9E-05
$spk10$	0.1218	-0.0171	0.2619	0.0421
$spk11$	0.1732	0.0036	0.2641	0.1061
$spk12$	0.1111	-0.0110	0.1994	0.0336
$spk13$		-0.0089	0.1635	0.0207
$spk14$	0.0962		0.1354	-0.0410
$spk16$	0.1433	-0.0069	0.2689	

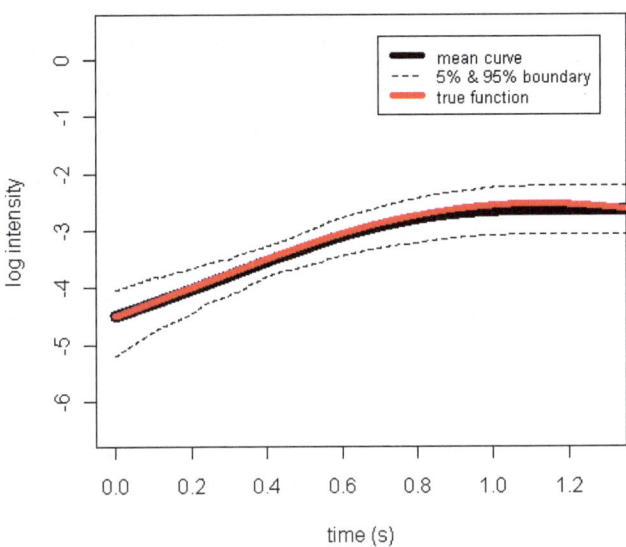

Figure 2.4: The black solid line is the mean curve of the baseline intensity function from 100 simulations, while the black dashed line is its 95% boundary. The red solid line is the true function. The knot locations for the estimates are adaptively selected as described, and two covariates are considered in the model. The estimates have unbiased means and fewer variations.

is for the regression with $spk1$ as the target and $spk2$ as its peer. The numbers in blue color imply a significantly negative estimator and those in red indicate no significant effect.

For example, neuron 7 displays some interesting behavior when it acts as a peer. It excites neurons 1, 6, 8, 9, 12, 13 and 16, but inhibits neurons 4, 5 and 10; and as for neurons 2, 11 and 14, neuron 7 barely has any effect on them.

What if we consider 15 peers at a time and look at the covariates out of the multiple peers model. Table 2.2 contains all the estimation results for each individual target spike train corresponding to its 15 peers. The rows still index the neural spike trains that are treated as targets in each round, whereas the columns index the peers in the multiple peers model with 15 peers each round. Thus the first row is for the regression with $spk1$ as the target and $spk2$ through $spk16$ as the peers. The numbers in bold show significantly positive estimators, underlining denotes significantly negative estimator, and no bolding or underlining indicate no significant effect.

Besides neuron 7, there are more negative coefficients estimated under the multiple peers model. Cell (2,14), for example, is positive using the single peer model, but here turns negative. One possible explanation is that given all of the positive stimuli from other peers, neuron 14 prevents the peer stimulation from getting too large, which is a natural mechanism in biology.

Tables 2.1 and 2.2 are informative in the sense that we can distinguish the positively or negatively correlated neurons. However, peers with the same sign for the coefficients cannot be ordered based on the absolute value of the coefficients. For completeness, we also perform the log-likelihood ratio test to see how the peers benefit the log likelihood and thus infer which peer may have stronger impact on the target.

For any target, the spontaneous firing is modeled according to its own history, and so treated as the baseline without any peer effects. The baseline intensity function in our model can be explicitly written as $\omega(t) = \sum_{i=1}^{p} \theta_i B_i(t)$. The likelihood of the baseline is L_0, and the likelihood of the complete model (2.9) is L_1. So the log-likelihood ratio is $-2 * log(\frac{L_0}{L_1})$. For the single-peer model, the log ratios are computed for all the target and peer combinations (see Figure 2.5); large values of the log ratio correspond to small p-values*. The cells are also colored to better visualize the peer effects; warm colors represent large ratios. $spk3$ benefits the likelihood the most among all the 15 peers; in contrast, $spk7$ raises the likelihood the least and sometimes even reduces it.

2.6 CONCLUSION

2.6.1 A PARAMETRIC MODEL FOR INTERACTING NEURONS

In the framework of the integrate-and-fire model, each presynaptic spike generates a postsynaptic current pulse, and the total input current to a integrate-and-fire neuron is the sum over all current pulses [5]. That is, the presynaptic currents are additive until an action potential is triggered when crossing its particular threshold.

Taking the fact that the current for each firing is relatively stable, the first model considers a neuron as a non-leaky integrator with excitatory probability depends on the number of the presynaptic spikes. In this approach, spike trains are replaced by a 0-1 time series that essentially discretizes the continuous time domain into finite numbers of small intervals for computation purposes. The computations are realized by the generalized linear models (GLIM). For example, in a simple network that includes only two neurons A and B, the ISI density function of neuron A is composed of the history behavior of its own and the influence from neuron B. The influence, of course, can be both excitatory and inhibitory. The firing rate or probability of the neuron A is given by

$$p_t = \Phi(\sum_{u=0}^{\gamma_t-1} b_u \Gamma_{t-u} + \theta_1 \gamma_t + \theta_2 \gamma_t^2 + \theta_3 \gamma_t^3 - \theta), \tag{2.11}$$

where $\Phi(\cdot)$ is for the normal cumulative function, γ_t is the time elapsed since the last firing of neuron A, and Γ_t is a 0-1 time series that represents the spike times of neuron B. The unknown parameters are estimated by maximizing the likelihood function [3]. In general, the target firing rate is modeled using its interacting nerve cells. The approach is flexible for arbitrary numbers of the input cells, and it allows biological interpretation of the results. Computer programs are widely available for fitting this model.

Some drawbacks of Brillinger's approach [2] were discussed briefly. Here we will systematically compare the performance of Brillnger's GLIM method and our proposed regression spline model.

First of all, the GLIM method has to discretize the experimental time into small sub-intervals with size Δ, so it limits the program capacity. Take one of our real spike train data as an example for the problem. For $spk1$, the first 100 spikes occur in about 123 seconds. When $\Delta = 0.005s$, the time then spans to 24,639 sub-intervals, which is a huge expansion and causes computers to run out of memory easily.

Secondly, the choice of the size Δ is subject to the data, and so it is data specific. As we explained above, the discrete likelihood function of a point process assumes the partition of the time to be sufficiently small so that there is at most one spike in any sub-interval. In other words, Δ must be no larger than the minimal ISI, which is the optimal choice in most cases. The smaller Δ value not only leads to computational crisis but also significantly affects the likelihood function (we will provide more details in the simulation study).

Thirdly, for peer effects, Brillinger's method traces back 13 bins and aggregates the counts of peer spikes within each bin. The bin size is not adjustable over the entire spike train. That is not a flexible way to handle the peer effects since the ISIs vary in length. The coverage of the 13 bins could be too much when the ISI is smaller, or too few when the ISI is larger.

Lastly, we conduct a simulation study to compare the results from the GLIM method and our proposed regression spline model in terms of predictability. Here, only the spontaneous firing based on the history is considered to make a fair comparison between the two methods. Brillinger's probability model then can be expressed as

$$p_t = \Phi(\mu(\gamma_t)) = \Phi(\theta_1 \gamma_t + \theta_2 \gamma_t^2 + \theta_3 \gamma_t^3 - \theta), \qquad (2.12)$$

where $\Phi(\cdot)$ is the normal cumulative function and γ_t is the time elapsed since the last firing of the neuron. Through a probit link function, the coefficients are estimated by maximizing the likelihood function.

Each sub-interval with size Δ is a Bernoulli trial with probability p_t, and ultimately a spike train is generated as a Bernoulli process. For the simulation study, the parameters are set: $\theta = -3$, $\theta_1 = 3$, $\theta_2 = 3$, and $\theta_3 = -3$. Both Brillinger's method and our model are applied to the generated spike train data in 100 simulations. To verify the unbiasedness of the estimation, the procedure was repeated 100 times and the estimated function $\widehat{p}_t$ was plotted along with the true function highlighted in black (Figure 2.9).

---

*χ^2 statistic (df = 1): 3.841 for p-value<0.05

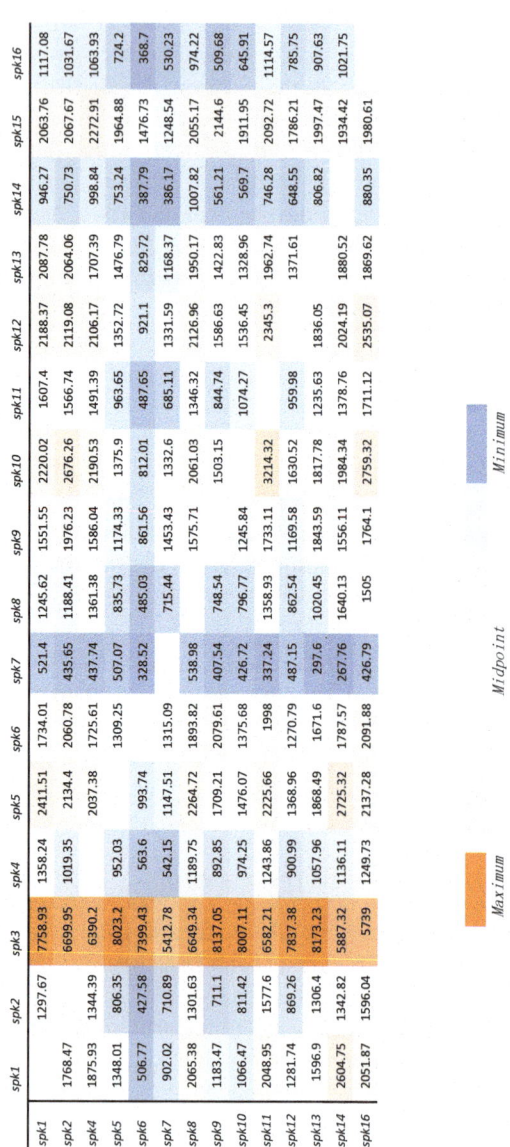

Figure 2.5: The log-likelihood ratios for all the target and peer combinations. Warm colors represent large ratios. *spk*3 benefits the likelihood the most among all the 15 peers; in contrast, *spk*7 raises the likelihood the least.

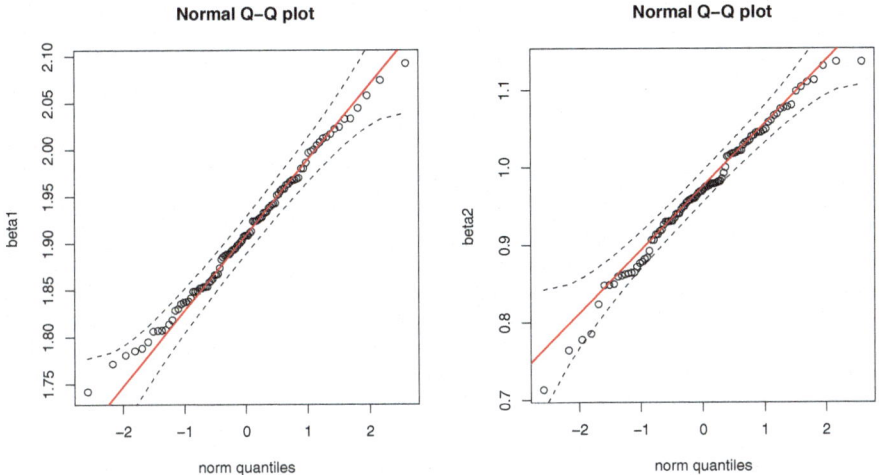

Figure 2.6: Q-Q plot for the β_1 and β_2 estimates. The envelope is the 95% boundary and the estimates follow approximately a normal distribution.

When Δ is small enough, the conditional intensity function, which the regression spline model estimates, multiplied by Δ gives the probability of a spike event in a small time interval Δ.

Brillinger's method has less bias in this simulation study due to the setup; the 95% envelope indicates fewer variation of Brillinger's method, especially when γ_t is less than 0.5 second.

According to the definition of predictability in Section 1.4, the mean predictability of Brillinger's method is 836.9 while the mean predictability of the regression spline model is 781.33. However, the predictability of Brillinger's method is sensitive to the sub-interval size Δ, which can be understood intuitively. The response variable will have more 0's when the Δ tends to be smaller. In other words, the process is more likely to fail, and so the change would be reflected by the smaller likelihood, which is obviously another disadvantage of Brillinger's method.

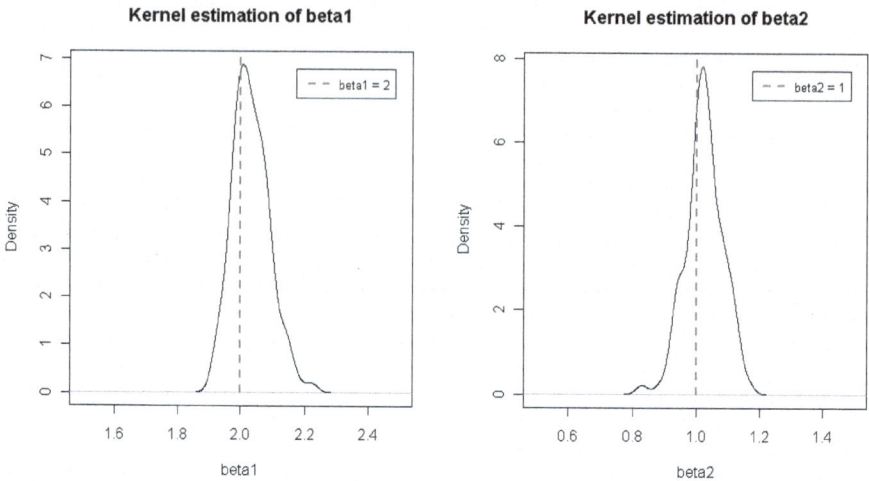

Figure 2.7: Kernel density estimations of β_1 and β_2 from 100 simulations. The knot locations for the estimates are adaptively selected as described, and two covariates are considered in the model. The estimates have unbiased means and fewer variations.

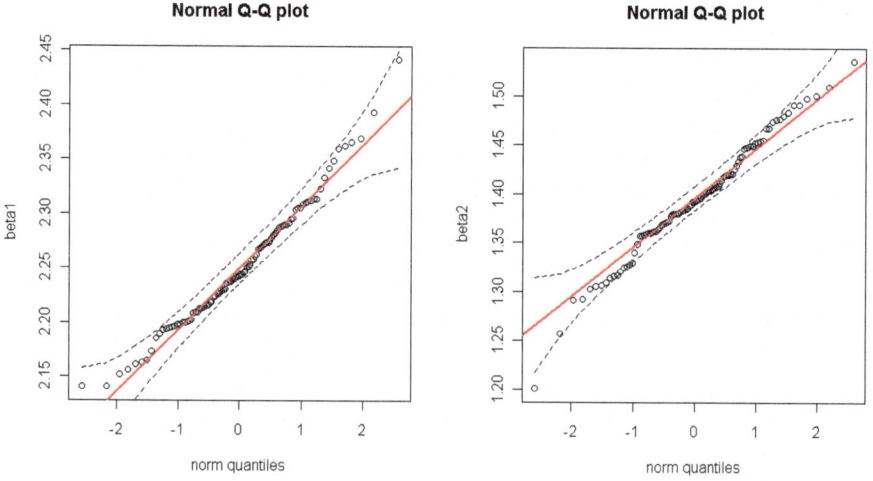

Figure 2.8: Q-Q plot for the β_1 and β_2 estimates. The envelope is the 95% boundary and the estimates follow approximately a normal distribution.

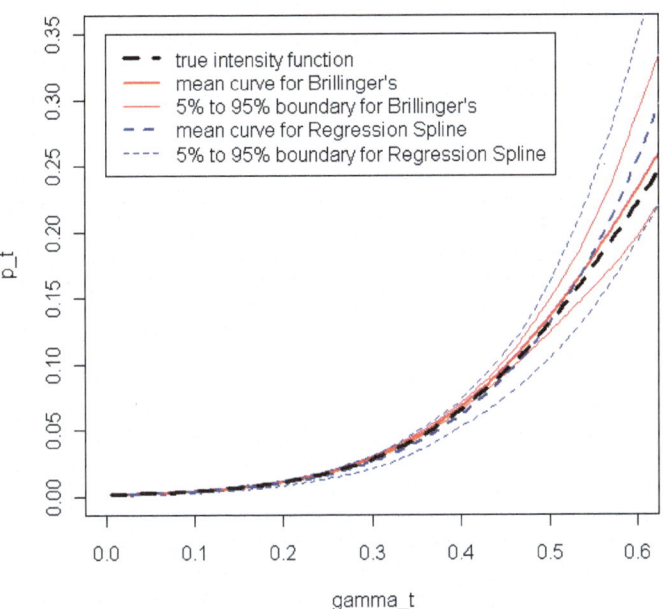

Estimation from Brillinger's and Regression Spline Model

Figure 2.9: The red solid lines are the mean and 95% boundary of the estimation by GLM with probit link function; the blue dashed lines are the mean and 95% boundary of the estimation by regression spline model. The black bold dashed line is the true probability function when the parameters are set, $\theta = -3$, $\theta_1 = 3$, $\theta_2 = 3$, and $\theta_3 = -3$.

2.7   R CODE FOR REAL DATA ANALYSIS

```
# realdata.R

rm(list = ls(all = TRUE))
library(polspline)
library(splines)
library(gdata)

#### Truncated cubic power function
   bas3 <- function(t,knot){
    ((abs(t - knot) + t - knot)/2)^3
   }

### target: spk0
```

```
spk0 = read.table('header = FALSE)
spk1 = read.table('header = FALSE)
spk2 = read.table('header = FALSE)
spk3 = read.table(',header = FALSE)
#spk4 = read.table('D:/Research/20100406/spike4',header = FALSE)
#spk5 = read.table('D:/Research/20100406/spike5',header = FALSE)
spk0 = as.numeric(unmatrix(spk0))
spk1 = as.numeric(unmatrix(spk1))
spk2 = as.numeric(unmatrix(spk2))
spk3 = as.numeric(unmatrix(spk3))
spk4 = as.numeric(unmatrix(spk4))
spk5 = as.numeric(unmatrix(spk5))


npeers = 13
rang = c(1,2,4:14,16)
### sort out peer1 peer2 peer3
for (p in rang){
  spk0 = spk[p,]
  taostar_tao0_all = matrix(rep(0,npeers*length(spk0)),
  ncol=npeers)
  pp=1

  for (j in rang){
      if (j!=p){
          spk1 = spk[j,]
          tao0 = 0;
          taostar1 = rep(0,length(spk0))
          for (i in 1:length(spk0)){

            if (any((spk0[i]-spk1)>0)){
               taostar11 = spk1[max(which((spk0[i]-spk1)>0))]
               if (taostar11 >= tao0){taostar1[i] = taostar11}
               if (taostar11 < tao0){taostar1[i] = spk0[i]}
             }

          tao0 = spk0[i]
          }

          dif_spk0 = diff(c(0,spk0))
          delta = rep(1,length(dif_spk0))

          taostar_tao0 = dif_spk0- (spk0-taostar1)
```

```
        source("heft4_m.R")
        dyn.load("allpack.dll")
        dyn.load("heftall_c3_m.dll")

        h1 = heft4(dif_spk0,delta,taostar_tao0,leftlin=TRUE,...
                      leftlog=0,rightlog=0,silent=TRUE)

        dyn.unload("allpack.dll")
        dyn.unload("heftall_c3_m.dll")
        taostar_tao0_all[,pp] = taostar_tao0
        pp = pp+1
    }
    source("heft4_m.R")
        dyn.load("allpack.dll")
        dyn.load("heftall_c3_m.dll")

        h2 = heft4(dif_spk0,delta,taostar_tao0_all,leftlin=TRUE,
                      leftlog=0,rightlog=0,silent=TRUE)

        dyn.unload("allpack.dll")
        dyn.unload("heftall_c3_m.dll")


    }
}


##################################################
# METHOD 1: Calculate the MLE beta1 beta0
##################################################

  b1 = seq(-10,-0.1,by=.1)

    spk_0 = c(0,spk0)
    inter11 = rep(0, length(spk_0))
    for (i in 2:length(spk_0)){
        if (any((spk_0[i]-spk1)>0)){
            taostar = spk1[max(which((spk_0[i]-spk1)>0))]
            if (taostar >= spk_0[(i-1)]){
                inter11[i] = spk_0[i]-taostar
            }
        }
    }
    ind2 = which(inter11!=0)
    temp2 = inter11[ind2]
```

```
  N = length(temp2)
 f = rep(0,length(b1))
 for (i in 1: length(b1)){
     part1 = sum(exp(b1[i]*temp2)*temp2)
     part2 = sum(exp(b1[i]*temp2)-1)
     f[i] = (part1-part2/b1[i])*N/part2-sum(temp2)
 }

 b1_hat = b1[which(abs(f)==min(abs(f)))]
 b0_hat = -log(sum(exp(b1_hat*temp2)-1)/b1_hat/N)

###################################################
# Results: n3 - n15
# > b1_hat
# [1] 2.1
# > b0_hat
# [1] 3.886787


# Results: n3 - n6
# > b1_hat
# [1] 3.4
# > b0_hat
# [1] 3.516183

# Results: n3 - n1
# > b1_hat
# [1] -3.4
# > b0_hat
# [1] 3.687158


#####################################################
# METHOD 2: loglam = beta0 + beta1*(taostar-tao0)
#####################################################

 b1 = seq(-100,-1,by=0.1)
 N = length(spk0)
   spk_0 = c(0,spk0)
   inter11 = rep(0, length(spk_0))
   inter12 = rep(0, length(spk_0))
   for (i in 2:length(spk_0)){
       if (any((spk_0[i]-spk1)>0)){
           taostar = spk1[max(which((spk_0[i]-spk1)>0))]
           if (taostar >= spk_0[(i-1)]){
```

```
                inter11[i] = taostar - spk_0[(i-1)]
                inter12[i] = spk_0[i] - taostar
            }
        }
    }
        temp3 = inter12[-1]
        temp2 = inter11[-1]
    f = rep(0,length(b1))
    for (i in 1: length(b1)){
        part1 = sum(temp3*temp2*exp(b1[i]*temp2))
        part2 = sum(temp3*exp(b1[i]*temp2))
        f[i] = part1*N/part2-sum(temp2)

    }

    b1_hat = b1[which(abs(f)==min(abs(f)))]
    b0_hat[p] = -log(sum(exp(b1_hat[p]*temp2)*temp3)/N)




plot(tt[-1],(b0_hat[1]+b1_hat[1]*t_tao0[1,-1]),type='l')
lines(tt[-1],true_loglam,col=3,lty="solid",lwd=2)
sum2 = b0_hat[1]+b1_hat[1]*t_tao0[1,-1]
for (p in 2:84){
#lines(tt[-1],(b0_hat[p]+b1_hat[p]*t_tao0[p,-1]))
sum2 = sum2+b0_hat[p]+b1_hat[p]*t_tao0[p,-1]
}

##########################################
# Estimate loglam by HEFT+HARE, Covariates: taostar_tao0
##########################################


    spk_0 = c(0,spk0)
    dat1 = diff(spk_0)
    #t = seq(Delta,max(dat1),by = Delta)
    #t_tao0 = rep(0,length(t))
    #t_taostar1 = rep(0,length(t))
    #inter11 = rep(0, length(spk_0))
    taostar_p1 = rep(0, length(spk_0))
    for (i in 2:length(spk_0)){
        if (any((spk_0[i]-spk1)>0)){
            taostar1 = spk1[max(which((spk_0[i]-spk1)>0))]
```

```
        if (taostar1 >= spk_0[(i-1)]){
          # inter11[i] = spk_0[i]-taostar1
            taostar_p1[i] = taostar1 - spk_0[i-1]
          }
      }
  }
taostar_p2 = rep(0, length(spk_0))
for (i in 2:length(spk_0)){
    if (any((spk_0[i]-spk2)>0)){
        taostar1 = spk2[max(which((spk_0[i]-spk2)>0))]
        if (taostar1 >= spk_0[(i-1)]){
          # inter11[i] = spk_0[i]-taostar1
            taostar_p2[i] = taostar1 - spk_0[i-1]
          }
      }
  }


heft_p1 = heft(dat1,rep(1,length(dat1)))
q_0_p1 = -log(1-pheft(dat1,heft_p1))
cov1 = taostar_p1[-1]
hare_p1 = hare(q_0_p1,rep(1,length(q_0_p1)),
                cov1,linear = c(1))

cov2 = taostar_p2[-1]
hare_p2 = hare(q_0_p1,rep(1,length(q_0_p1)),
                cov2,linear = c(1))


#############################################
# Calculate the estimation of loglam(t)
#############################################

tt = seq(Delta,max(dat1),by = Delta)
taostar1 = 0.07
tt_taostar1 = apply(matrix(c((tt-taostar1),
                rep(0,length(tt))), ncol=2),1,max)

lam0_p1 = hheft(tt,heft_p1)
q0_0_p1 = -log(1-pheft(tt,heft_p1))
lam1_p1 = hhare(q0_0_p1,tt_taostar1,hare_p1)
#lam2[[p]] = hhare(t_tao0[p,-1],t_taostar1[p,-1],hare2)
#lam2_2[[p]] = hhare(t_tao0[p,-1],t_taostar1[p,-1],hare2_2)
 dat2 = diff(c(0,spk1))
 h2 = heft(dat2,rep(1,length(dat2)))
```

```
lam2 = hheft(tt,h2)
dat3 = diff(c(0,spk2))
h3 = heft(dat3,rep(1,length(dat3)))
lam3 = hheft(tt,h3)




plot(tt,(log(lam0_p1)+log(lam1_p1)),type='l',
        xlab="time (s)", ylab="log intensity",main="",lwd=3)
lines(tt,log(lam0_p1),col=2,lwd=3)
lines(tt,log(lam2),col=1,lty = "dashed",lwd=2)
lines(tt,log(lam3),col=1,lty = "dotted",lwd=2)

legend(0.1,-2,c( "Target (Neuron 3) with peer (Neuron 15)
            spike at 0.07s","Target with no peer spike",
            "Peer (Neuron 15)", "Peer (Neuron 1)"),
            col=c('black','blue'),lty=c("solid","solid"),
            merge = TRUE,lwd=c(1,2))
legend(2.7,9,c("target spike train","peer1 spike train"),
            col=c('red','green'), pch = c(3, 1),lwd=c(2,2))
title('Target: Neuron 3, Peer: Neuron 1')




#########################################
# Estimate loglam by HEFT+HARE, Covariates: tao1-taostar
#########################################


spk_0 = c(0,spk0)
dat1 = diff(spk_0)
#t = seq(Delta,max(dat1),by = Delta)
#t_tao0 = rep(0,length(t))
#t_taostar1 = rep(0,length(t))
tao0_taostar_p1 = rep(0, length(spk_0))
taostar_p1 = rep(0, length(spk_0))
for (i in 2:length(spk_0)){
    if (any((spk_0[i]-spk1)>0)){
        taostar1 = spk1[max(which((spk_0[i]-spk1)>0)))]
        if (taostar1 >= spk_0[(i-1)]){
            tao0_taostar_p1[i] = spk_0[i]-taostar1
            taostar_p1[i] = taostar1 - spk_0[i-1]
        }
```

```
      }
    }
tao0_taostar_p2 = rep(0, length(spk_0))
taostar_p2 = rep(0, length(spk_0))
for (i in 2:length(spk_0)){
    if (any((spk_0[i]-spk2)>0)){
        taostar1 = spk2[max(which((spk_0[i]-spk2)>0))]
        if (taostar1 >= spk_0[(i-1)]){
            tao0_taostar_p2[i] = spk_0[i]-taostar1
            taostar_p2[i] = taostar1 - spk_0[i-1]
        }
    }
}

heft_p1 = heft(dat1,rep(1,length(dat1)))
q_0_p1 = -log(1-pheft(dat1,heft_p1))
cov1 = taostar_p1[-1]
hare_p1 = hare(q_0_p1,rep(1,length(q_0_p1)),
               cov1,linear = c(1))

cov2 = taostar_p2[-1]
hare_p2 = hare(q_0_p1,rep(1,length(q_0_p1)),
               cov2,linear = c(1))

#hare2=hare(dat1,rep(1,length(dat1)),cov[[p]],
#                         linear=c(0,1))
#hare2_2=hare(dat1,rep(1,length(dat1)),cov[[p]])
#################################################
# Calculate the estimation of loglam(t)
#################################################

tt = seq(Delta,max(dat1),by = Delta)
taostar1 = 0.07
tt_taostar1=apply(matrix(c((tt-taostar1),
                rep(0,length(tt))), ncol=2),1,max)

lam0_p1 = hheft(tt,heft_p1)
q0_0_p1 = -log(1-pheft(tt,heft_p1))
lam1_p1 = hhare(q0_0_p1,tt_taostar1,hare_p1)
#lam2[[p]] = hhare(t_tao0[p,-1],t_taostar1[p,-1],hare2)
#lam2_2[[p]] = hhare(t_tao0[p,-1],t_taostar1[p,-1],hare2_2)
 dat2 = diff(c(0,spk1))
 h2 = heft(dat2,rep(1,length(dat2)))
 lam2 = hheft(tt,h2)
```

```
dat3 = diff(c(0,spk2))
h3 = heft(dat3,rep(1,length(dat3)))
lam3 = hheft(tt,h3)
```

```
plot(tt,(log(lam0_p1)+log(lam1_p1)),type='l',xlab="time (s)",
            ylab="log intensity",main="",lwd=3)
lines(tt,log(lam0_p1),col=2,lwd=3)
lines(tt,log(lam2),col=1,lty = "dashed",lwd=2)
lines(tt,log(lam3),col=1,lty = "dotted",lwd=2)

legend(0.1,-2,c( "Target (Neuron 3) with peer (Neuron 15)
                spike at 0.07s","Target with no peer spike",
                "Peer (Neuron 15)", "Peer (Neuron 1)"),
                col=c('black','blue'),lty=c("solid","solid"),
                merge = TRUE,lwd=c(1,2))
legend(2.7,9,c("target spike train","peer1 spike train"),
        col=c('red','green'), pch = c(3, 1),lwd=c(2,2))
title('Target: Neuron 3, Peer: Neuron 1')
```

```
sum = log(lam0[[1]])+log(lam1[[1]])
for(p in 2:57){
sum = sum + log(lam0[[p]])+log(lam1[[p]])
}
legend(0,-25,c("true log intensity : -5 + 10*(t-tao_star)"),
        col=3,lty="dashed",lwd=3)
title("HEFT+HARE, with option = linear")

  true_loglam = -5+10*(t_taostar1[1,-1])
 for (p in 2:57){
   true_loglam = true_loglam-5+10*(t_taostar1[p,-1])
   #lines(t[-1],true_loglam,col=3,lty="solid",lwd=1)
 }

plot(t[-1], (sum/57),col='red',lty="solid",lwd=2,type='l',
            xlab="time (s)", ylab="log intensity",main="",
                ylim=c(-10,5),xlim=c(0,50))
plot(t[-1],(sum2/57),col = 4,lty="solid",lwd=2,type='l',
```

```
                xlab="time (s)", ylab="log intensity",main="",
                ylim=c(-10,5),xlim=c(0,50))

        lines(t[-1],(true_loglam/57),col=3,lty="solid",lwd=1)
        legend(0,-8,c("mean curve of MLE","mean curve of HEFT+HARE",
                "true log intensity : -5 + 10*(t-tao_star)"),
                col=c('red',4,3),lty=c("solid","solid","solid"),
                lwd=c(2,2,2))
        title("HEFT+HARE, with option = linear")


        plot(tt[-1],(log(lam0[[1]])+log(lam1_1[[1]])),type='l',
            xlab="time (s)", ylab="log intensity",main="")
        sum1 = log(lam0[[1]])+log(lam1_1[[1]])
        for(p in 1:84){
        lines(tt[-1],(log(lam0[[p]])+log(lam1_1[[p]])),type='l')
        sum1 = sum1 + log(lam0[[p]])+log(lam1_1[[p]])
        }
        legend(0,-25,c("true log intensity : -5 + 10*(t-tao_star)"),
            col=3,lty="dashed",lwd=3)
        title("HEFT+HARE, without option = linear")


        plot(tt[-1], (sum1/84),col='red',lty="solid",lwd=2,
                type='l', xlab="time (s)", ylab="log intensity",
                main="",ylim=c(-10,5))
        true_loglam = -5+10*(t_taostar1[p,-1])
        lines(tt[-1],true_loglam,col=3,lty="solid",lwd=2)
        legend(0,-8,c("mean curve",
            "true log intensity : -5 + 10*(t-tao_star)"),
            col=c('red',3),lty=c("solid","solid"),lwd=c(2,2))
        title("HEFT+HARE, without option = linear")


plot(tt,lam0_0[1,], ylim = c(-2,4),
    xlab='gap time (s)',ylab='log intensity',type='l')
for(p in 1:100){
lines(tt,lam0_0[p,],col=1)
}
m_lam0=colSums(lam0_0)/100
lines(tt, m_lam0,col='red',lty="dashed",lwd=3) # mean function
lines(c(0,1.2),c(log(2),log(2)+min1),lty="dashed",lwd=3, col=3)
legend(0.5, 4,c("mean intensity function",
    "true function"),lty="dashed",lwd=3, col=c('red','green'))
legend(0.5,3,c("beta0 = log(2), beta1 = 1 "))
```

```
kern = density(-beta[which(beta!=0)])
plot(kern,xlim = c(0,5),xlab='beta1',main='')
lines(c(beta1,beta1),c(0,3),col=2,lty="dashed",lwd=2)
title('Kernel estimation of beta1')
legend(3, 1.5,c("beta1 = 2"),lty="dashed",lwd=2, col = 2)


plot(cumsum(xx[[1]]),(log(lam0[[1]])+log(lam1[[1]])+
        tao_star[[1]]), type = 'l',xlab='time (s)',
        ylab = 'log_intensity',main='')
for(p in list1){
lines(cumsum(xx[[p]]),
    (log(lam0[[p]])+log(lam1[[p]])+tao_star[[p]]),type='l')
}
lines(c(0,200),c(log(2),log(2)+400),col=2,lty="dashed",lwd=3)
title('beta0 + beta1 * t')
legend(0, 380,c("true intensity = log(2) + 2* t"),
    lty="dashed",lwd=3, col = 2)

save(lam0,beta,file = "simulation_2_lam.Rdata")
```

2.8 R CODE FOR SIMULATION

```
library(gdata)

beta1_hat = read.table('beta1_hat',header = FALSE)
beta2_hat = read.table('beta2_hat',header = FALSE)
b_hat = read.table('b_hat',header = FALSE)
theta_hat = read.table('theta_hat',header = FALSE)
a_hat = read.table('a_hat',header = FALSE)

beta1_hat = as.numeric(unmatrix(beta1_hat))
beta2_hat = as.numeric(unmatrix(beta2_hat))
b_hat = as.numeric(unmatrix(b_hat))
a_hat = as.numeric(unmatrix(a_hat))

par( mfcol= c(2, 2))

plot(density(theta_hat[,1]),xlim = c(-5,0),xlab='theta[1]',
```

```
        main='')
lines(c(-3,-3),c(0,10),col=2,lty="dashed",lwd=2)
title('Kernel estimation of theta[1]')
legend("topright",inset=0.05,cex=.8,
c("theta[1] = -3"),lty="dashed",lwd=2, col = 2)

plot(density(theta_hat[,2]),xlim = c(0,9),xlab='theta[2]',
        main='')
lines(c(4,4),c(0,10),col=2,lty="dashed",lwd=2)
title('Kernel estimation of theta[2]')
legend("topright",inset=0.05,cex=.8,
c("theta[2] = 4"),lty="dashed",lwd=2, col = 2)


plot(density(theta_hat[,3]),xlim = c(-5,2),xlab='theta[3]',
        main='')
lines(c(-2,-2),c(0,8),col=2,lty="dashed",lwd=2)
title('Kernel estimation of theta[3]')
legend("topright",inset=0.05,cex=.8,
c("theta[3] = -2"),lty="dashed",lwd=2, col = 2)


plot(density(theta_hat[,4]),xlim = c(-11,20),xlab='theta[4]',
        main='')
lines(c(3,3),c(0,10),col=2,lty="dashed",lwd=2)
title('Kernel estimation of theta[4]')
legend("topright",inset=0.05,cex=.8,
c("theta[1] = 3"),lty="dashed",lwd=2, col = 2)


par( mfcol= c(1, 2))

plot(density(a_hat),xlim = c(-8,5),xlab='a',main='')
lines(c(a,a),c(0,3),col=2,lty="dashed",lwd=2)
title('Kernel estimation of a')
legend("topright",inset=0.05,cex=.8,
c("a = -4"),lty="dashed",lwd=2, col = 2)

plot(density(b_hat),xlim = c(-3,9),xlab='b',main='')
lines(c(b,b),c(0,3),col=2,lty="dashed",lwd=2)
title('Kernel estimation of b')
legend("topright",inset=0.05,cex=.8,
c("b = 2"),lty="dashed",lwd=2, col = 2)
```

```
plot(density(beta1_hat),xlim = c(0,6),xlab='beta1',main='')
lines(c(beta1,beta1),c(0,10),col=2,lty="dashed",lwd=2)
title('Kernel estimation of beta1')
legend("topright",inset=0.05,cex=.8,
c("beta1 = 2"),lty="dashed",lwd=2, col = 2)

plot(density(beta2_hat),xlim = c(-1,4),xlab='beta2',main='')
lines(c(beta2,beta2),c(0,10),col=2,lty="dashed",lwd=2)
title('Kernel estimation of beta2')
legend("topright",inset=0.05,cex=.8,
c("beta2 = 1"),lty="dashed",lwd=2, col = 2)


rm(list = ls(all = TRUE))
library(polspline)
library(splines)
library(gdata)

# Generate peer1 spike train : deterministic gap = 1
spk1 = c(0,cumsum(rep(1,600)))
spk2 = c(0.5,cumsum(rep(1,600))+0.5)

# Generate the target spike train:
# log(lam) = a + b*t + theta*basis + beta1*[K-(t-tao_star)]
    # Truncated cubic power function
    bas3 <- function(t,knot){
     ((abs(t - knot) + t - knot)/2)^3
     }

#Set up parameters
a = -4; b=2; theta = c(-3,4,-2,3); beta1 = 2; beta2 = 1
knots = c(0.2,0.4,0.6,0.8)
nknots = length(knots)
K=2
Delta = 0.001

# Initialize
t =  c(0,seq(Delta,600, by=Delta))
niter = 1   # 100
spk0 = matrix(rep(0,niter*length(spk1)*2),nrow = niter)
taostar1 = matrix(rep(0,niter*length(spk1)*2),nrow = niter)
taostar2 = matrix(rep(0,niter*length(spk1)*2),nrow = niter)
```

```
# Start interation
ptm <- proc.time()
   for (p in 1:niter){
       tao0 =t[1]
       log_lam = rep(0,length(t))
       t_taostar1 = rep(0,length(t))
       t_taostar2 = rep(0,length(t))
       K_cov1 = rep(0,length(t))
       K_cov2 = rep(0,length(t))
       j = 1
       for (i in 1:length(t)){
          taostar11 = 0
          taostar22 = 0.5
          gamma_t = t[i] - tao0
           if (any((t[i]-spk1)>0)){
              taostar11 = spk1[max(which((t[i]-spk1)>0))]
              if (taostar11 >= tao0){
                 t_taostar1[i] = t[i]-taostar11
                 K_cov1[i] = K-t_taostar1[i]
              }
          }
          if (any((t[i]-spk2)>0)){
              taostar22 = spk2[max(which((t[i]-spk2)>0))]
              if (taostar22 >= tao0){
                 t_taostar2[i] = t[i]-taostar22
                 K_cov2[i] = K-t_taostar2[i]
              }
          }
          sp1 = 0
          for (k in 1:nknots){
              sp1 = sp1 + theta[k]*bas3(gamma_t,knots[k])
          }
          log_lam[i] = a+b*gamma_t+ sp1 +beta1*(K_cov1[i])
                          +beta2*(K_cov2[i])
          p_t = exp(log_lam[i])*Delta
          if (rbinom(1,1,p_t)==1){
              spk0[p,j] = t[i]
              taostar1[p,j] = t[i]-t_taostar1[i]
              taostar2[p,j] = t[i]-t_taostar2[i]
              j = j+1
              tao0 = t[i]
          }
       }
   }
```

```
        p
    }
proc.time() - ptm




test = spk0[,1:650]
test2 = taostar1[,1:650]
test3 = taostar2[,1:650]
write.table(test,file="spk0.txt",row.names=F)
write.table(test2,file="taostar1.txt",row.names=F)
write.table(test3,file="taostar2.txt",row.names=F)




rm(list = ls(all = TRUE))
library(polspline)
library(splines)
library(gdata)

# Generate peer1 spike train : deterministic gap = 1
spk1 = c(0,cumsum(rep(1,600)))
spk2 = c(0.5,cumsum(rep(1,600))+0.5)

# Generate the target spike train:
# log(lam) = a + b*t + theta*basis + beta1*[K-(t-tao_star)]
    # Truncated cubic power function
    bas3 <- function(t,knot){
      ((abs(t - knot) + t - knot)/2)^3
    }

#Set up parameters
a = -4; b=2; theta = c(-3,4,-2,3); beta1 = 2; beta2 = 1
knots = c(0.2,0.4,0.6,0.8)
nknots = length(knots)
K=2
Delta = 0.001

# Initialize
t =  c(0,seq(Delta,600, by=Delta))
niter = 100
spk0 = matrix(rep(0,niter*length(spk1)*2),nrow = niter)
taostar1 = matrix(rep(0,niter*length(spk1)*2),nrow = niter)
taostar2 = matrix(rep(0,niter*length(spk1)*2),nrow = niter)
```

```
# Start interation
for (p in 1:niter){
    tao0 =t[1]
    log_lam = rep(0,length(t))
    t_taostar1 = rep(0,length(t))
    t_taostar2 = rep(0,length(t))
    K_cov1 = rep(0,length(t))
    K_cov2 = rep(0,length(t))
    j = 1
    for (i in 1:length(t)){
        taostar11 = 0
        taostar22 = 0.5
        gamma_t = t[i] - tao0
         if (any((t[i]-spk1)>0)){
            taostar11 = spk1[max(which((t[i]-spk1)>0))]
            if (taostar11 >= tao0){
                t_taostar1[i] = t[i]-taostar11
                K_cov1[i] = K-t_taostar1[i]
            }
        }
        if (any((t[i]-spk2)>0)){
            taostar22 = spk2[max(which((t[i]-spk2)>0))]
            if (taostar22 >= tao0){
                t_taostar2[i] = t[i]-taostar22
                K_cov2[i] = K-t_taostar2[i]
            }
        }
        sp1 = 0
        for (k in 1:nknots){
            sp1 = sp1 + theta[k]*bas3(gamma_t,knots[k])
        }
        log_lam[i] = a+b*gamma_t+ sp1 +beta1*(K_cov1[i])
                        +beta2*(K_cov2[i])
        p_t = exp(log_lam[i])*Delta
        if (rbinom(1,1,p_t)==1){
            spk0[p,j] = t[i]
            taostar1[p,j] = t[i]-t_taostar1[i]
            taostar2[p,j] = t[i]-t_taostar2[i]
            j = j+1
            tao0 = t[i]
        }
    }
    p
```

```
    }

test = spk0[,1:650]
test2 = taostar1[,1:650]
test3 = taostar2[,1:650]
write.table(test,file="spline_b1(K-t+tao)_spk0.txt",row.names=F)

write.table(test2,file="spline_b1(K-t+tao)_taostar1.txt",
                row.names=F)

write.table(test3,file="spline_b1(K-t+tao)_taostar2.txt",
                row.names=F)



% Truncated cubic power function (save as m file)
  function [f]=bas3 (t,knot)
      f=((abs(t - knot) + t - knot)./2).^3;
  end

% Copy of est_spike_model_2peers_pos.m

  theta_hat = zeros(niter,nknots);
  a_hat = zeros();
  b_hat = zeros();
  beta1_hat = zeros();
  beta2_hat = zeros();
  f_hat = zeros();

  for i=1:niter
      global dif_spk0 tao1_taostar1 tao1_taostar2 u u_star1
      global u_star2  N knots K
      knots = [0.2,0.4,0.6,0.8];
      K = 2;

      spk_0 = [0,spk0(i,1:nspikes)];
      tao1_taostar1 =  spk0(i,1:nspikes)-taostar1(i,1:nspikes);
      tao1_taostar2 =  spk0(i,1:nspikes)-taostar2(i,1:nspikes);

      dif_spk0 = diff(spk_0);
      taostar1_tao0 = dif_spk0-tao1_taostar1;
      taostar2_tao0 = dif_spk0-tao1_taostar2;

      N = length(dif_spk0);
      precis = 0.005;
```

```
u = zeros();
u_star1 = zeros();
u_star2 = zeros();

seq1 = 0:precis:dif_spk0(1);
u(1:length(seq1)) = seq1;
u_star1(1:length(seq1)) = seq1-taostar1_tao0(1);
u_star2(1:length(seq1)) = seq1-taostar2_tao0(1);

for k=2:N,

    seq1 = 0:precis:dif_spk0(k);
    u((length(u)+1):(length(u)+length(seq1))) = seq1;
    u_star1((length(u_star1)+1):(length(u_star1)
            +length(seq1))) = seq1-taostar1_tao0(k);
    u_star2((length(u_star2)+1):(length(u_star2)
            +length(seq1))) = seq1-taostar2_tao0(k);

end


options = optimset('GradObj','on','Hessian','on',
            'MaxFunEvals', 5000,'MaxIter',5000,'TolFun',
            0.00001,'TolX',0.00001);
[bmin,fmin]=fminunc(@loglam_spline_beta6_2,para,options);
theta_hat(i,:) = bmin(1:nknots);
b_hat(i) = bmin(nknots+1);
beta1_hat(i) = bmin(nknots+2);
beta2_hat(i) = bmin(nknots+4);
a_hat(i) = bmin(nknots+3);
f_hat(i) = fmin;
i
clear global dif_spk0 tao1_taostar1 tao1_taostar2 u
clear global u_star1 u_star2  N knots K
end

save('beta1_hat_a1.txt','beta1_hat','-ascii', '-append')
save('beta2_hat_a1.txt','beta2_hat','-ascii', '-append')
save('b_hat_a1.txt','b_hat','-ascii', '-append')
save('theta_hat_a1.txt','theta_hat','-ascii', '-append')
save('a_hat_a1.txt','a_hat','-ascii', '-append')
```

```
theta_hat = zeros(niter,nknots);
a_hat = zeros();
b_hat = zeros();
beta1_hat = zeros();
beta2_hat = zeros();
f_hat = zeros();

for i=1:niter
    global dif_spk0 tao1_taostar1 tao1_taostar2 u u_star1
    global u_star2  N knots K
    knots = [0.2,0.4,0.6,0.8];
    K = 2;

    spk_0 = [0,spk0(i,1:nspikes)];
    tao1_taostar1 =  spk0(i,1:nspikes)-taostar1(i,1:nspikes);
    tao1_taostar2 =  spk0(i,1:nspikes)-taostar2(i,1:nspikes);

    dif_spk0 = diff(spk_0);
    taostar1_tao0 = dif_spk0-tao1_taostar1;
    taostar2_tao0 = dif_spk0-tao1_taostar2;

    N = length(dif_spk0);
    precis = 0.005;

    u = zeros();
    u_star1 = zeros();
    u_star2 = zeros();

    seq1 = 0:precis:dif_spk0(1);
    u(1:length(seq1)) = seq1;
    u_star1(1:length(seq1)) = seq1-taostar1_tao0(1);
    u_star2(1:length(seq1)) = seq1-taostar2_tao0(1);

    for k=2:N,

        seq1 = 0:precis:dif_spk0(k);
        u((length(u)+1):(length(u)+length(seq1))) = seq1;
        u_star1((length(u_star1)+1):(length(u_star1)
            +length(seq1))) = seq1-taostar1_tao0(k);
        u_star2((length(u_star2)+1):(length(u_star2)
            +length(seq1))) = seq1-taostar2_tao0(k);

    end
```

```
    options = optimset('GradObj','on','Hessian','on',...
       'MaxFunEvals',5000,'MaxIter',5000,'TolFun',0.00001,...
       'TolX',0.00001);
    [bmin,fmin]=fminunc(@loglam_spline_beta6_2,para,options);
    %[bmin,fmin]=fminsearch(@loglam_1,beta0,options);
    theta_hat(i,:) = bmin(1:nknots);
    b_hat(i) = bmin(nknots+1);
    beta1_hat(i) = bmin(nknots+2);
    beta2_hat(i) = bmin(nknots+4);
    a_hat(i) = bmin(nknots+3);
    f_hat(i) = fmin;
    i
    clear global  dif_spk0 tao1_taostar1 tao1_taostar2 u u_star1
    clear global  u_star2  N knots K
 end

save('beta1_hat_a1','beta1_hat','-ascii')
save('beta2_hat_a1','beta2_hat','-ascii')
save('b_hat_a1','b_hat','-ascii')
save('theta_hat_a1','theta_hat','-ascii')
save('a_hat_a1','a_hat','-ascii')

% Copy of gen_spike_model_2peers_pos.m

% Truncated cubic power function (save as m file)
%   function [f]=bas3 (t,knot)
%     f=((abs(t - knot) + t - knot)./2).^3;
%   end

clear;
% Generate peer 1 and peer 2 spike train : deterministic gap = 1
nspikes = 600;    % 600
spk1 = [0,cumsum(ones(1,nspikes))];
spk2 = [0.5,cumsum(ones(1,nspikes))+0.5];

% Generate the target spike train: log(lam) = a + b*t +
%   theta*basis+beta1*[K-(t-tao_star1)]+beta2*[K-(t-tao_star2)]


% Set up parameters
    %a=-2 ;b=2; theta = c(-5,8,-3,3); beta1 = 2;
    %theta = c(-5,8,2,-5)
para = [-3 4 -2 3 2 2 -4 1];
```

```
a = para(7); b=para(5); theta = para(1:4);
beta1 = para(6); beta2 = para(8);
knots = [0.2,0.4,0.6,0.8];
nknots = length(knots);
K=2;
Delta = 0.001;


%tt = [0,Delta:Delta:5];
%sp = matrix(rep(0,nknots*length(tt)),nrow = nknots)
%for( i in 1:nknots){
%   sp[i,] = bas3(tt,knots[i])
%}
%f = a +b*tt+apply(sp*theta,2,sum)
%plot(tt,f,xlim=c(0,2),ylim=c(a-10,a+10))

t =  0:Delta:nspikes;
niter = 1;  % 100
spk0 = zeros();
taostar1 = zeros();
taostar2 = zeros();

tStart=tic;      % start the stopwatch

for p = 1:niter
    tao0 =t(1);
    log_lam = zeros();
    t_taostar1 = zeros(1,length(t));
    t_taostar2 = zeros(1,length(t));
    K_cov1 = zeros(1,length(t));
    K_cov2 = zeros(1,length(t));
    j = 1;
    for i =1:length(t)
        taostar11 = 0;
        taostar22 = 0.5;
        gamma_t = t(i) - tao0;
        if any((t(i)-spk1)>0)
            taostar11 = spk1(find((t(i)-spk1)>0, 1, 'last' ));
            if taostar11 >= tao0
                t_taostar1(i) = t(i)-taostar11;
                K_cov1(i) = K-t_taostar1(i);
            end
        end
        if any((t(i)-spk2)>0)
```

```
            taostar22 = spk2(find((t(i)-spk2)>0, 1, 'last' ));
            if taostar22 >= tao0
                t_taostar2(i) = t(i)-taostar22;
                K_cov2(i) = K-t_taostar2(i);
            end
        end
        spl = 0;
        for k = 1:nknots
            spl = spl + theta(k)*bas3(gamma_t,knots(k));
        end
        log_lam(i) = a+b*gamma_t+ spl +beta1*K_cov1(i)
                        +beta2*K_cov2(i);
        p_t = exp(log_lam(i))*Delta;
        if random('bino',1,p_t,1,1)
            spk0(p,j) = t(i);
            taostar1(p,j) = t(i)-t_taostar1(i);
            taostar2(p,j) = t(i)-t_taostar2(i);
            j = j+1 ;
            tao0 = t(i);
        end
    end
    p
end


save('all.txt', 'spk0','spk1','spk2','t_taostar1','t_taostar2',...
        'taostar1','taostar2','K_cov1','K_cov2','-ascii');

toc(tStart)




% Truncated cubic power function (save as m file)
%   function [f]=bas3 (t,knot)
%     f=((abs(t - knot) + t - knot)./2).^3;
%   end

clear;
% Generate peer 1 and peer 2 spike train : deterministic gap = 1
nspikes = 600;
spk1 = [0,cumsum(ones(1,nspikes))];
spk2 = [0.5,cumsum(ones(1,nspikes))+0.5];
```

```
% Generate the target spike train:
% log(lam) = a + b*t + theta*basis +
%    beta1*[K-(t-tao_star1)]+ beta2*[K-(t-tao_star2)]


% Set up parameters
    %a=-2 ;b=2; theta = c(-5,8,-3,3); beta1 = 2;
    %theta = c(-5,8,2,-5)
para = [-3 4 -2 3 2 2 -4 1];
a = para(7); b=para(5); theta = para(1:4);
beta1 = para(6); beta2 = para(8);
knots = [0.2,0.4,0.6,0.8];
nknots = length(knots);
K=2;
Delta = 0.001;

t =  0:Delta:nspikes;
niter = 100;
spk0 = zeros();
taostar1 = zeros();
taostar2 = zeros();

for p = 1:niter
    tao0 =t(1);
    log_lam = zeros();
    t_taostar1 = zeros(1,length(t));
    t_taostar2 = zeros(1,length(t));
    K_cov1 = zeros(1,length(t));
    K_cov2 = zeros(1,length(t));
    j = 1;
    for i =1:length(t)
        taostar11 = 0;
        taostar22 = 0.5;
        gamma_t = t(i) - tao0;
        if any((t(i)-spk1)>0)
            taostar11 = spk1(find((t(i)-spk1)>0, 1, 'last' ));
            if taostar11 >= tao0
                t_taostar1(i) = t(i)-taostar11;
                K_cov1(i) = K-t_taostar1(i);
            end
        end
        if any((t(i)-spk2)>0)
            taostar22 = spk2(find((t(i)-spk2)>0, 1, 'last' ));
```

```
              if taostar22 >= tao0
                  t_taostar2(i) = t(i)-taostar22;
                  K_cov2(i) = K-t_taostar2(i);
              end
          end
          spl = 0;
          for k = 1:nknots
              spl = spl + theta(k)*bas3(gamma_t,knots(k));
          end
          log_lam(i) = a+b*gamma_t+ spl +beta1*K_cov1(i)
                          +beta2*K_cov2(i);
          p_t = exp(log_lam(i))*Delta;
          if random('bino',1,p_t,1,1)
              spk0(p,j) = t(i);
              taostar1(p,j) = t(i)-t_taostar1(i);
              taostar2(p,j) = t(i)-t_taostar2(i);
              j = j+1 ;
              tao0 = t(i);
          end
      end
      p
end

save('all','spk0','spk1','spk2','t_taostar1','t_taostar2',...
      'taostar1','taostar2','K_cov1','K_cov2',-ascii);



function [f,g,H]=loglam_spline_beta6_2(beta)
global dif_spk0 tao1_taostar1 tao1_taostar2 u
global u_star1 u_star2  N knots  K

persistent  nknots temp sumbas precis

precis = 0.005;
nknots = length(knots);
temp = 0;
for i=1:nknots,
    temp=temp+beta(i)*(((abs(u - knots(i))
            + u - knots(i))/2).^3);
end

sumbas = 0;
for i=1:nknots,
    sumbas = sumbas+beta(i)*sum(((abs(dif_spk0 - knots(i)) ...
```

```
                + dif_spk0 - knots(i))/2).^3);
end
f=-(N*beta(nknots+3) + beta(nknots+1)*sum(dif_spk0) +sumbas
    +beta(nknots+2)*sum((K-tao1_taostar1).*(tao1_taostar1>0))
    +beta(nknots+4)*sum((K-tao1_taostar2).*(tao1_taostar2>0))
    -(sum(exp(beta(nknots+3)
    +temp+beta(nknots+1)*u+(K-u_star1).*(u_star1>0)
    *beta(nknots+2)+(K-u_star2).*(u_star2>0)*beta(nknots+4)))
    *precis));

if nargout > 1,
  for i=1:nknots,
  g(i) = sum(exp(beta(nknots+3)+temp
            +beta(nknots+1)*u+(K-u_star1).
    *(u_star1>0)*beta(nknots+2)+(K-u_star2).*(u_star2>0)
    *beta(nknots+4)).*(((abs(u-knots(i))+u-knots(i))/2).^3))
    *precis-sum(((abs(dif_spk0 - knots(i)) + dif_spk0
    - knots(i))/2).^3);
  end
  g(nknots+1) = sum(exp(beta(nknots+3)+temp+beta(nknots+1)*u
    +(K-u_star1).*(u_star1>0)*beta(nknots+2)+(K-u_star2).
    *(u_star2>0)*beta(nknots+4)).*u)*precis-sum(dif_spk0);
  g(nknots+2) = sum(exp(beta(nknots+3)+temp+beta(nknots+1)*u
    +(K-u_star1).*(u_star1>0)*beta(nknots+2)+(K-u_star2).
    *(u_star2>0)*beta(nknots+4)).*((K-u_star1).*(u_star1>0)))
    *precis-sum((K-tao1_taostar1).*(tao1_taostar1>0));
  g(nknots+3) = sum(exp(beta(nknots+3)+temp+beta(nknots+1)*u
    +(K-u_star1).*(u_star1>0)*beta(nknots+2)+(K-u_star2).
    *(u_star2>0)*beta(nknots+4)))*precis-N;
  g(nknots+4) = sum(exp(beta(nknots+3)+temp+beta(nknots+1)*u
    +(K-u_star1).*(u_star1>0)*beta(nknots+2)+(K-u_star2).
    *(u_star2>0)*beta(nknots+4)).*((K-u_star2).*(u_star2>0)))
    *precis-sum((K-tao1_taostar2).*(tao1_taostar2>0));

end

if nargout>2,
    for i=1:nknots,
        H(i,i) = sum(exp(beta(nknots+3)+temp+beta(nknots+1)*u
            +(K-u_star1).*(u_star1>0)*beta(nknots+2)+(K-u_star2).
            *(u_star2>0)*beta(nknots+4)).*(((abs(u - knots(i))
            + u - knots(i))/2).^3).*(((abs(u - knots(i)) + u
            - knots(i))/2).^3))*precis;
        for j=(i+1): nknots,
```

```
        H(j,i) = sum(exp(beta(nknots+3)+temp+beta(nknots+1)
        *u+(K-u_star1).*(u_star1>0)*beta(nknots+2)
        +(K-u_star2).*(u_star2>0)*beta(nknots+4)).
        *(((abs(u - knots(i)) + u - knots(i))/2).^3).
        *(((abs(u - knots(j)) + u
        - knots(j))/2).^3))*precis;
    end;
    H((nknots+1),i) = sum(exp(beta(nknots+3)
      +temp+beta(nknots+1)*u+(K-u_star1).*(u_star1>0)
      *beta(nknots+2)+(K-u_star2).*(u_star2>0)
      *beta(nknots+4)).*u.*(((abs(u-knots(i))+u
      -knots(i))/2).^3))*precis;
    H((nknots+2),i) = sum(exp(beta(nknots+3)+temp
      +beta(nknots+1)*u+(K-u_star1).
      *(u_star1>0)*beta(nknots+2)+(K-u_star2).
      *(u_star2>0)*beta(nknots+4)).*((K-u_star1).
      *(u_star1>0)).*(((abs(u - knots(i))
      + u - knots(i))/2).^3))*precis;
    H((nknots+3),i) = sum(exp(beta(nknots+3)+temp
      +beta(nknots+1)*u+(K-u_star1).*(u_star1>0)
      *beta(nknots+2)+(K-u_star2).*(u_star2>0)
      *beta(nknots+4)).*(((abs(u - knots(i))
      + u - knots(i))/2).^3))*precis;
    H((nknots+4),i) = sum(exp(beta(nknots+3)+temp
      +beta(nknots+1)*u+(K-u_star1).*(u_star1>0)
      *beta(nknots+2)+(K-u_star2).*(u_star2>0)
      *beta(nknots+4)).*((K-u_star2).
      *(u_star2>0)).*(((abs(u - knots(i))
      + u - knots(i))/2).^3)) *precis;
end;
H((nknots+1),(nknots+1)) = sum(exp(beta(nknots+3)+temp
+beta(nknots+1)*u+(K-u_star1).*(u_star1>0)*beta(nknots+2)
+(K-u_star2).*(u_star2>0)*beta(nknots+4)).*u.*u)*precis;
H((nknots+2),(nknots+1)) = sum(exp(beta(nknots+3)+temp
+beta(nknots+1)*u+(K-u_star1).*(u_star1>0)*beta(nknots+2)
+(K-u_star2).*(u_star2>0)*beta(nknots+4)).*((K-u_star1).
*(u_star1>0)).*u)*precis;
H((nknots+2),(nknots+2)) = sum(exp(beta(nknots+3)+temp
+beta(nknots+1)*u+(K-u_star1).*(u_star1>0)*beta(nknots+2)
+(K-u_star2).*(u_star2>0)*beta(nknots+4)).*((K-u_star1).
*(u_star1>0)).*((K-u_star1).*(u_star1>0)))*precis;
H((nknots+3),(nknots+1)) = sum(exp(beta(nknots+3)+temp
+beta(nknots+1)*u+(K-u_star1).*(u_star1>0)*beta(nknots+2)
+(K-u_star2).*(u_star2>0)*beta(nknots+4)).*u)*precis;
```

```
        H((nknots+3),(nknots+2)) = sum(exp(beta(nknots+3)+temp
        +beta(nknots+1)*u+(K-u_star1).*(u_star1>0)*beta(nknots+2)
        +(K-u_star2).*(u_star2>0)*beta(nknots+4)).*((K-u_star1).
        *(u_star1>0)))*precis;
        H((nknots+3),(nknots+3)) = sum(exp(beta(nknots+3)+temp
        +beta(nknots+1)*u+(K-u_star1).*(u_star1>0)*beta(nknots+2)
        +(K-u_star2).*(u_star2>0)*beta(nknots+4)))*precis;
        H((nknots+4),(nknots+1)) = sum(exp(beta(nknots+3)+temp
        +beta(nknots+1)*u+(K-u_star1).*(u_star1>0)*beta(nknots+2)
        +(K-u_star2).*(u_star2>0)*beta(nknots+4)).*((K-u_star2).
        *(u_star2>0)).*u)*precis;
        H((nknots+4),(nknots+2)) = sum(exp(beta(nknots+3)+temp
        +beta(nknots+1)*u+(K-u_star1).*(u_star1>0)*beta(nknots+2)
        +(K-u_star2).*(u_star2>0)*beta(nknots+4)).*((K-u_star2).
        *(u_star2>0)).*((K-u_star1).*(u_star1>0)))*precis;
        H((nknots+4),(nknots+3)) = sum(exp(beta(nknots+3)+temp
        +beta(nknots+1)*u+(K-u_star1).*(u_star1>0)*beta(nknots+2)
        +(K-u_star2).*(u_star2>0)*beta(nknots+4)).*((K-u_star2).
        *(u_star2>0)))*precis;
        H((nknots+4),(nknots+4)) = sum(exp(beta(nknots+3)+temp
        +beta(nknots+1)*u+(K-u_star1).*(u_star1>0)*beta(nknots+2)
        +(K-u_star2).*(u_star2>0)*beta(nknots+4)).*((K-u_star2).
        *(u_star2>0)).*((K-u_star2).*(u_star2>0)))*precis;
end


end
```

Bibliography

1. L. F. Abbott. Lapique's introduction of the integrate-and-fire model neuron. *Brain Research Bulletin*, 50(5–6):303–304, 1999.
2. D. R. Brillinger. Some statistical methods for random process data from seismology and neurophysiology. *Annals of Statistics*, 16(1):1–54, 1988.
3. D. R. Brillinger. Maximum likelihood analysis of spike trains of interacting nerve cells. *Biological Cybernetics*, 59:189–200, 1988.
4. D. R. Brillinger and J. P. Segundo. Empirical examination of the: threshold model of neuron firing. *Biological Cybernetics*, 35:213–220, 1979.
5. W. Gerstner and W. M. Kistler. *Spiking Neuron Models: Single Neurons, Populations, Plasticity*. Cambridge University Press, 2002.
6. C. Kooperberg, C. J. Stone, and Y. K. Truong. Hazard regression. *Journal of the American Statistical Association*, 90:78–94, 1995.

7. S-C Lin and M. Nicolelis. Neuronal ensemble bursting in the basal forebrain encodes salience irrespective of valence. *Neuron*, 59(1):138–149, 2008.
8. W. Truccolo, U. T. Eden, M. R. Fellows, J. P. Donoghue, and E. N. Brown. A point process framework for relating neural spiking activity to spiking history, neural ensemble and extrinsic covariate effects. *Journal of Neurophysiology*, 93: 1074–1089, 2005.

Part II

Statistical Analysis of fMRI Data

3 A Hypothesis Testing Approach for Brain Activation

Wenjie Chen
AIG, New York

Haipeng Shen
University of Hong Kong, China

Young K. Truong
University of North Carolina at Chapel Hill

CONTENTS

Functional magnetic resonance imaging (fMRI) is based on the blood-oxygen-level-dependent (BOLD) principle. When neurons are active, they consume oxygen, which then leads to increased blood flow to the activated area. As a result, neural activity can be inferred from the BOLD signal. The relationship between the experimental stimulus and the BOLD signal involves the hemodynamic response function (HRF). Standard fMRI analysis packages such as Statistical Parametric Mapping (SPM) [11] and FMRIB Software Library (FSL) [18] usually assume the same predefined parametric HRF for every subject throughout the whole brain. Nonparametric estimation of HRF is an important approach in fMRI that allows distinct HRFs for different subjects and brain regions. This problem has been studied under event-related designs because the design paradigm allows the HRF to return to baseline or recover after every trial. In this chapter, we introduce the transfer function estimate (TFE), a spectrum domain nonparametric method, to extract the HRF from any kind of experiment design. Furthermore, the hypothesis testing under TFE is able to detect BOLD-related activation and validate the linearity assumption. Based on the simulation and real data analysis, TFE enables a more accurate reflection of the brain's BOLD mechanics.

3.1 INTRODUCTION

The BOLD signal from fMRI data has been shown to be linked closely to neural activity [25]. Through a process called the *hemodynamic response*, blood releases oxygen to active neurons at a greater rate than to inactive ones. The difference in magnetic susceptibility between oxyhemoglobin and deoxyhemoglobin, and thus oxygenated or deoxygenated blood, leads to magnetic signal variation that can be detected using an MRI scanner. The relationship between the study stimulus and the BOLD signal involves the HRF. Estimating or determining the HRF is important for the correct interpretation of neurological studies.

The variability of the hemodynamic response exists across time, subjects, and regions of the brain. As such, assuming homogeneity of the HRF for each time point, subject, and brain region would seem counterintuitive. Current data analysis methods for fMRI data, however, generally use cross-correlation and t tests using the general linear model [8, 30] and apply the same HRF across time and for each subject and brain region. There are two common fMRI data analysis packages, SPM and FSL, for data preprocessing and statistical modeling. In SPM, the canonical HRF is a typical BOLD impulse response characterized by two gamma functions, one modeling the peak and one modeling the undershoot or post-dip after the peak [13, see Figure 3.1]. By default, FSL uses a single gamma function. The use of the canonical function decreases the variation across subjects, but does not change the variation across the regions of the brain. To allow the variation of HRFs at different voxels or volume elements of the brain, [12] and [14] propose to model the HRF and BOLD response voxel-wise using a Bayesian framework by defining a certain number of parameters and assuming prior distributions on parameters. The introduction of the prior distribution leads to an estimation bias and thus some nonparametric studies on fMRI have been proposed [32, 26]. In Bai's method, the authors accounted for the variation across regions of the brain by using a nonparametric estimation of the frequency domain, a method that captures the different HRF shapes [1]. In this chapter, we adjusted this method in order to make it applicable in a block design experiment results in an excellent estimation of the hemodynamic response.

An early and influential study by [2] provided evidence supporting the idea that the fMRI BOLD signal results from a linear transformation of neuronal activity.

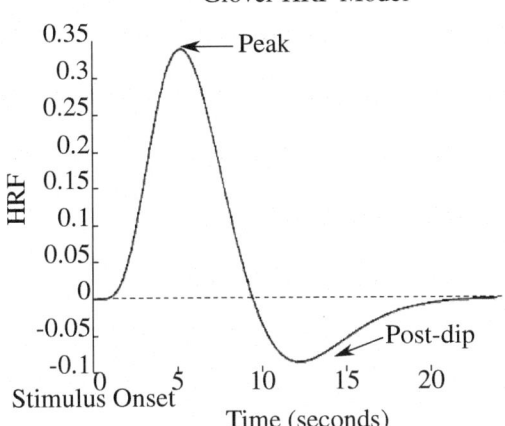

Figure 3.1: Typical hemodynamic response function. This is a typical HRF as double gamma functions, called Glover's HRF [13]. It is composed of the peak and the post-dip indicated in the plot. Its x-axis refers to time, and its y-axis refers to the intensity of HRF. Usually an HRF lasts 20 to 30 seconds.

Many HRF studies use the basis function approach under the general linear model (GLM) framework [7, 4, 31, 19, 17, 27]. Other than the basis function approach, many time-domain approaches have been proposed in recent years. The deconvolution model for HRF estimation without any specific assumption on HRF was implemented in the fMRI software Analysis of Functional NeuroImages (AFNI) to estimate HRF [6, 29]. The time-domain deconvolution method, however, has the collinear problem, especially for the periodic experiment design. Some studies were carried out to compare the performance between three fixed HRFs (Gamma HRF, Glover HRF, and SPM canonical HRF) and the deconvolution method under the framework of the general linear model [21, 22]. A more flexible semiparametric approach based on finite impulse response (FIR) filters was proposed in modeling HRF [15]. Another semiparametric method that uses smoothing splines to estimate drift function and HRF while taking into account the covariance structure within the BOLD signal was developed [32, 33]. These semiparametric methods might assume an incorrect covariance structure and require expensive computing. More thorough comparisons were carried out based on SPM canonical HRF with or without its temporal or dispersion derivatives, FIR, semiparametric FIR, and inverse logit model [20]. In summary, these convenient "standard" methods lack the natural spatial variability of HRF, and the model-driven approach has collinearity problems.

The application of the HRF estimation from many methods is limited by the type of experiment design. It was pointed out that advances in single-event fMRI have allowed the extraction of the relative timing information of the hemodynamic response [9], and it is important to emphasize the correct form of the hemodynamic response in fMRI studies [24]. In block designs, the stimulus exists for a longer duration of time (typically 30 sec) compared to the event-related design, and it has been noted that the stimulus duration does not necessarily correspond to the duration of the neuronal activity [16]. A frequency-domain method is offered to detect activation for the periodic stimuli with self-calibrating [23]. The method of [1] corroborated these findings by proposing and implementing a nonparametric model for estimating univariate HRF using fast Fourier transform of event-related fMRI data. The frequency-domain methods simplify convolution evaluations for easier accommodation of temporal and spatial autocorrelation.

In this chapter, we propose a nonparametric frequency-domain approach to give a consistent estimator of HRFs. This approach is called the *transfer function estimate* (TFE) and views the Fourier transform of HRF as a transfer function in frequency domain.

1. We first extend the method of Bai [1] to a multivariate form in order to estimate multiple HRFs simultaneously using ordinary least square (OLS).
2. TFE detects brain activation while estimating HRF by providing the *F* map and also tests the linearity assumption inherited from the convolution model.
3. TFE is able to compare the difference among multiple HRFs in the experimental design.
4. TFE adapts to all kinds of experiment designs and does not depend on the pre-specified HRF length support.

Using multiple stimuli in one experimental session is often preferred in fMRI studies as it results in a stronger response signal compared to the single stimulus design. In terms of the biological processing of the human brain, the advantage from multiple stimuli is to avoid the refractory effect, which causes nonlinearity in the response over time [10]. Furthermore, subjects may get bored easily if only one stimulus is shown repeatedly in a session. In terms of the experimental design, multiple stimuli are helpful to create efficient designs in the limited scanner time that is available and tolerable to subjects. Thus, changing stimuli over successive trials is beneficial for the experimental design.

3.1.1 MODEL

Consider the BOLD response model given by

$$Y(t) = h_1 \otimes x_1(t) + h_2 \otimes x_2(t) + \cdots + h_n \otimes x_n(t) + \varepsilon(t), \quad t = 0, \ldots, T-1 \quad (3.1)$$

where $x_i(t)$ represents the ith stimulus function and $h_i(t)$ is the corresponding HRF, $i = 1, \ldots, n$. We assume that the error series, $\varepsilon(t)$, is a stationary random process with mean zero.

Model (3.1) can be written in a matrix form,

$$Y(t) = \sum_u \mathbf{h}(u) \mathbf{X}(t-u) + \varepsilon(t), \quad (3.2)$$

where $\mathbf{X}(t) = (x_1(t), x_2(t), \ldots, x_n(t))^\tau$ is a n-dimensional vector-valued series, and $\mathbf{h}(u) = (h_1(u), h_2(u), \ldots, h_n(u))$ is a $1 \times n$ filter. It is assumed that the HRF $\mathbf{h}(u)$ is 0 when $u < 0$ or $u > d$, where d is the length of HRF latency determined by underlying neural activity.

Set $\mathbf{H}(r) = \sum_t \mathbf{h}(t) \exp(-irt)$ and denote the finite Fourier transforms of Y and $\mathbf{X}$ by

$$\varphi_Y(r) \equiv \varphi_Y^{(T)}(r) = \sum_{t=0}^{T-1} Y(t) \exp(-irt),$$

$$\varphi_{\mathbf{X}}(r) \equiv \varphi_{\mathbf{X}}^{(T)}(r) = \sum_{t=0}^{T-1} \mathbf{X}(t) \exp(-irt), \quad r \in \mathbb{R}. \tag{3.3}$$

Here T is the duration or number of time points in an fMRI session. Let K be an integer with $2\pi K/T$ near radian frequency r. Suppose T is reasonably large. From the asymptotic property of the finite Fourier transform,

$$\varphi_Y\left(\frac{2\pi(K+k)}{T}\right) \doteq \mathbf{H}(r)\varphi_{\mathbf{X}}\left(\frac{2\pi(K+k)}{T}\right) + \varphi_\varepsilon\left(\frac{2\pi(K+k)}{T}\right) \tag{3.4}$$

for $k = 0, \pm 1, \ldots, \pm m$, say. By applying the Fourier transform, the convolution in the time-domain is transformed to the product in the frequency-domain. The product forms the linear relationship between $\varphi_Y(\cdot)$ and $\varphi_{\mathbf{X}}(\cdot)$, which makes the estimation simpler. Also, applying the Fourier transform is a good way to avoid estimating the autocorrelation in the time-domain. When we deal with Fourier coefficients instead of time points, these coefficients are asymptotically uncorrelated at different frequencies. According to this asymptotic attribute, when we consider the correlation structure in linear relation in the frequency-domain, variance at each frequency is much more important than covariance between frequencies, which allows us to consider WLS instead of generalized least square (GLS) in the next chapter.

3.1.2 ORDINARY LEAST SQUARE ESTIMATE

Suppose that $\mathbf{H}(r)$ is smooth. Relation (3.4) is seen to have the form of a multiple linear regression involving complex-valued variates. In the matrix form, the linear system (3.4) is

$$\Phi_Y(r) = \begin{bmatrix} \varphi_Y(\frac{2\pi}{T}(K-m)) \\ \varphi_Y(\frac{2\pi}{T}(K-m+1)) \\ \vdots \\ \varphi_Y(\frac{2\pi}{T}K) \\ \vdots \\ \varphi_Y(\frac{2\pi}{T}(K+m)) \end{bmatrix}, \Phi_{\mathbf{X}}(r) = \begin{bmatrix} \varphi_{\mathbf{X}}(\frac{2\pi}{T}(K-m))^\tau \\ \varphi_{\mathbf{X}}(\frac{2\pi}{T}(K-m+1))^\tau \\ \vdots \\ \varphi_{\mathbf{X}}(\frac{2\pi}{T}K)^\tau \\ \vdots \\ \varphi_{\mathbf{X}}(\frac{2\pi}{T}(K+m))^\tau \end{bmatrix}^\tau,$$

$$\Phi_\varepsilon(r) = \begin{bmatrix} \varphi_\varepsilon(\frac{2\pi}{T}(K-m)) \\ \varphi_\varepsilon(\frac{2\pi}{T}(K-m+1)) \\ \vdots \\ \varphi_\varepsilon(\frac{2\pi}{T}K) \\ \vdots \\ \varphi_\varepsilon(\frac{2\pi}{T}(K+m)) \end{bmatrix},$$

then we can compute the OLS estimate of $\mathbf{H}(\cdot)$ using

$$\hat{\mathbf{H}}(r) = (\Phi_\mathbf{X}(r)\overline{\Phi_\mathbf{X}(r)^\tau})^{-1}\Phi_\mathbf{X}(r)\overline{\Phi_Y(r)^\tau}, \tag{3.5}$$

where $\overline{A}$ is the conjugate of A.

More formally, define

$$\mathbf{I}_{Y\mathbf{X}}(r) = (2\pi T)^{-1}\varphi_Y(r)\overline{\varphi_\mathbf{X}(r)}^\tau, \tag{3.6}$$

$$\mathbf{I}_{\mathbf{XX}}(r) = (2\pi T)^{-1}\varphi_\mathbf{X}(r)\overline{\varphi_\mathbf{X}(r)}^\tau, \tag{3.7}$$

$$\hat{\mathbf{f}}_{Y\mathbf{X}}(r) = (2m+1)^{-1}\sum_{k=-m}^{m}\mathbf{I}_{Y\mathbf{X}}(\frac{2\pi(K+k)}{T}), \tag{3.8}$$

$$\hat{\mathbf{f}}_{\mathbf{XX}}(r) = (2m+1)^{-1}\sum_{k=-m}^{m}\mathbf{I}_{\mathbf{XX}}(\frac{2\pi(K+k)}{T}). \tag{3.9}$$

The function $\mathbf{I}_{\mathbf{XX}}(\cdot)$ is known as the *periodogram* in time series analysis. It is a *naive* estimate of the matrix of spectral density functions of the series $\mathbf{X}$. An improved version is given by (3.9) and is referred to as a window estimate.

Suppose that the $n \times n$ matrix $\hat{\mathbf{f}}_{\mathbf{XX}}(r)$ is non-singular. The OLS estimate (3.5) can be written as

$$\hat{\mathbf{H}}(r) = \hat{\mathbf{f}}_{Y\mathbf{X}}(r)\hat{\mathbf{f}}_{\mathbf{XX}}(r)^{-1}. \tag{3.10}$$

Let $s_{\varepsilon\varepsilon}(r)$ denote the spectral density function of the error series ε. Then its estimate is given by

$$\hat{f}_{\varepsilon\varepsilon}(r) = \frac{2m+1}{2m+1-r}[\hat{f}_{YY}(r) - \hat{\mathbf{f}}_{Y\mathbf{X}}(r)\hat{\mathbf{f}}_{\mathbf{XX}}(r)^{-1}\hat{\mathbf{f}}_{\mathbf{X}Y}(r)]. \tag{3.11}$$

Using (3.10), the OLS estimate of $\mathbf{h}(u)$ is given by

$$\hat{\mathbf{h}}(u) = \frac{1}{T}\sum_{t=0}^{T-1}\hat{\mathbf{H}}(\frac{2\pi t}{T})\exp(i2\pi tu/T). \tag{3.12}$$

3.1.2.1 Window Estimate

The window estimate has good sampling properties compared to the previous least square estimate. Actually, the naive estimates (3.8) and (3.9) are special cases of window estimate with the uniform window $W(\frac{2\pi k}{T}) = (2m+1)^{-1}$ as $k = 0, \pm 1, \ldots, \pm m$.

By observing (3.8) and (3.9), the window estimate can be written as

$$\hat{s}_{xy}(r) = \sum_{k=-m}^{m} W(\frac{2\pi k}{T})I_{xy}(\frac{2\pi}{T}(K+k)) = \sum_{k=K-m}^{K+m} W(r - \frac{2\pi k}{T})I_{xy}(\frac{2\pi k}{T}) \quad (3.13)$$

where $W(\cdot)$ is a non-negative function called the *weight* or *window function*. Since the estimation process is required to be symmetric, we extend the weight function periodically.

$$W(\alpha + 2\pi) = W(\alpha), \quad (3.14)$$

Note that $\mathbf{I}_{YX}(0) = 0$. In order to reflect the notion that the weight function should become more concentrated as the duration T tends to ∞, we introduce a bandwidth parameter b that depends on T such that $b \to 0$ as $T \to \infty$, then for sufficiently large T,

$$\int_0^{2\pi} b^{-1} W(b^{-1}\alpha) d\alpha = 1. \quad (3.15)$$

We therefore consider the following general window estimates:

$$\hat{s}_{YX}(r) = \sum_{k \neq 0} b^{-1} W(b^{-1}(r - \frac{2\pi k}{T})) \mathbf{I}_{YX}(\frac{2\pi k}{T}), \quad (3.16)$$

$$\hat{s}_{XX}(r) = \sum_{k \neq 0} b^{-1} W(b^{-1}(r - \frac{2\pi k}{T})) \mathbf{I}_{XX}(\frac{2\pi k}{T}), \quad (3.17)$$

and estimate $\mathbf{H}(r)$ by

$$\hat{\mathbf{H}}(r) = \hat{s}_{YX}(r)\hat{s}_{XX}(r)^{-1}. \quad (3.18)$$

Then we have

$$\hat{\mathbf{h}}(u) = \frac{1}{T} \sum_{t=0}^{T-1} \hat{\mathbf{H}}(\frac{2\pi t}{T}) \exp(i2\pi t u / T). \quad (3.19)$$

The estimate $\hat{\mathbf{h}}(u)$ has good sampling properties in the sense that it is an asymptotically consistent and efficient estimate of $\mathbf{h}(u)$ [3].

3.2 HYPOTHESIS TESTING

After introducing the TFE method, this section introduces the multivariate tests for fMRI analysis.

3.2.1 KEY CONCEPTS

First, we introduce two key concepts here to support the hypothesis testing [3].

Coherence is an important statistic that provides a measure of the strength of a linear time-invariant relation between the series $Y(t)$ and the series $\mathbf{X}(t)$, that is, it indicates whether there is a strongly linear relationship between the BOLD response and the stimuli. From a statistical view, we can test the linear time-invariant assumption for the convolution model; for the fMRI exploration, we can choose the voxels with significant large coherence where the BOLD series have a functional response to the stimulus, and then estimate the HRF in those voxels.

Coherence is defined as

$$|R_{Y\mathbf{X}}(r)|^2 = s_{Y\mathbf{X}}(r)s_{\mathbf{XX}}(r)^{-1}s_{\mathbf{X}Y}(r)/s_{YY}(r). \tag{3.20}$$

Coherence is seen as a form of correlation coefficient, bounded by 1 in absolute value. The closer to 1, the stronger the linear time-invariant relation between $Y(t)$ and $\mathbf{X}(t)$.

The second concept is *partial coherence*. If we look at the stimulus individually, it is interesting to consider the complex analogues of the partial correlations, or partial coherence. The estimated *partial cross-spectrum* of $Y(t)$ and $X_i(t)$ after removing the linear effects of $\mathbf{X}_j(t)$ is given by

$$s_{YX_i \cdot \mathbf{X}_j}(r) = s_{YX_i}(r) - s_{Y\mathbf{X}_j}(r)s_{\mathbf{X}_j X_j}(r)^{-1}s_{\mathbf{X}_j X_i}(r). \tag{3.21}$$

Usually the case of interest is the relationship between the response and a single stimulus after the other stimuli are accounted for, that is, X_i is the single stimulus of interest, and $\mathbf{X}_j$ is the other stimuli involved in the design paradigm.

The partial coherence of $Y(t)$ and $X_i(t)$ after removing the linear effects of $\mathbf{X_j}(t)$ is given by

$$\left|R_{YX_i \cdot \mathbf{X}_j}(r)\right|^2 = \frac{s_{YX_i \cdot \mathbf{X}_j}(r)^2}{s_{YY \cdot \mathbf{X}_j}(r)s_{X_i X_i \cdot \mathbf{X}_j}(r)}. \tag{3.22}$$

If $n = 2$, that is, if there are two kinds of stimulus in the experiment, it can be written as

$$\left|R_{YX_i \cdot X_j}(r)\right|^2 = \frac{\left|R_{YX_i}(r) - R_{YX_j}(r)R_{X_i X_j}(r)\right|^2}{[1 - |R_{YX_j}(r)|^2][1 - |R_{X_i X_j}(r)|^2]}. \tag{3.23}$$

Partial coherence is especially important when we focus on a specific stimulus. Not all stimuli are considered in equal measure. Stimuli such as the heartbeat and breathing, which cannot be avoided in any experiment involving humans, are of secondary concern. Sometimes for the experiment design, we are not likely to pay equal attention to all types of stimulus. There are some specific types of stimulus accounted for in the experiment goal. The other stimuli, such as heartbeat and breathing, are secondary concern that cannot be avoided in any kind of human brain experiment design. Furthermore, as each type of stimulus has its own characteristics, it is natural to perform an individual statistical analysis to see how each one affects the overall fMRI response.

3.2.2 TESTING THE LINEARITY

The linearity assumption functions as the essential basis of the convolution model. As we know, any nonlinearity in the fMRI data may be caused by the scanner system or human physical capabilities such as *refractory* effect. Refractory effects refer to the reductions in hemodynamic amplitude after several stimuli are presented. If refractory effects are present, then a linear model will overestimate the hemodynamic response to closely spaced stimuli, potentially reducing the effectiveness of experimental analyses. It is critical, therefore, to consider the evidence for and against the linearity of the fMRI hemodynamic response.

It is possible that the nonlinearity is overwhelmed during scanning. Consequently, it is crucial to make sure that the linearity assumption is acceptable. The advantage of our method is that we can first assess the validity of the linearity assumption before using the convolution model for analysis.

The value of coherence, between 0 and 1, reflects the strength of the linear relation between fMRI response and the stimuli. Under certain conditions, $\hat{R}_{Y\mathbf{X}}(r)$ is asymptotically normal with mean $R_{Y\mathbf{X}}(r)$ and variance proportional to constant $(1 - R_{Y\mathbf{X}}^2(r))/Tb$. Moreover, if $R_{Y\mathbf{X}} = 0$, then

$$F(r) = \frac{(c-n)|\hat{R}_{Y\mathbf{X}}(r)|^2}{n(1 - |\hat{R}_{Y\mathbf{X}}(r)^2|)} \sim F_{2n,2(c-n)} \tag{3.24}$$

where $c = bT/\gamma$ and $\gamma = \int \kappa^2$ with κ being the lag-window generator depending on the choice of window function. If the F statistic on coherence is significant, it is reasonable to accept the linearity assumption.

3.2.3   TESTING THE EFFECT FROM A SPECIFIC STIMULUS

For each brain area, stimuli have varying effects. For the motor cortex in the left hemisphere, right-hand motion causes much more neural activities than left-hand motion. Partial coherence is able to distinguish between right- and left-hand effects, determine whether left-hand motion evokes neural activity, and identify which motion has greater effect. In order to separate the right-hand effects from left-hand, to see whether the left-hand motion actually evokes the neuronal activity there, and to see which one has the larger effect, partial coherence is able to address these concerns through testing of hypotheses.

For partial coherence, if $R_{YX_i \cdot \mathbf{X}_j} = 0$, then

$$F(r) = \frac{c'|\hat{R}_{YX_i \cdot \mathbf{X}_j}(r)|^2}{1 - |\hat{R}_{YX_i \cdot \mathbf{X}_j}(r)^2|} \sim F_{2,2(c'-1)} \tag{3.25}$$

where $c' = bT/\gamma - n + 1$ and $\gamma = \int \kappa^2$ with κ being the lag-window generator.

3.2.4   DETECTING THE ACTIVATION

HRF in fMRI indicates the arising neural activity. If there is activation evoked by the stimulus, then the corresponding HRF cannot be ignored. If there is no HRF in a brain region, there is no ongoing neuronal activity. To detect activation in the brain region is to see whether there is underlying HRF. For our frequency method, we test $\mathbf{H}(r_0) = 0$ at stimulus-related frequency r_0.

This is carried out by means of analogs of the statistic (3.20). In the case $\mathbf{H}(r) = 0$,

$$\frac{(bT/\gamma)\hat{\mathbf{H}}(r)\hat{\mathbf{s}}_{\mathbf{XX}}(r)\overline{\hat{\mathbf{H}}(r)^\tau}}{n\hat{s}_{\varepsilon\varepsilon}(r)} \tag{3.26}$$

is distributed asymptotically as $\mathbf{F}_{2,2(bT/\gamma-n)}$.

3.2.5 TESTING THE DIFFERENCE BETWEEN HRF

The multivariate method simplifies the functional hypothesis testing by comparing the corresponding Fourier coefficients at frequency r in order to see whether there is any discrepancy between HRF curves corresponding to different stimuli. HRF curves are functions, but when we focus on the Fourier coefficients at frequency r, we look at a common hypothesis testing on points. To see the difference between the two HRFs, it is enough to consider the hypothesis that the two Fourier coefficients at task-related frequency r_0 are equivalent.

Recall that $\mathbf{H}(\cdot)$ is the Fourier transform of the HRF $\mathbf{h}$ given by

$$\mathbf{H}(r) = \sum_{t=0}^{T-1} \mathbf{h}(t)\exp(-irt), \qquad r \in \mathbb{R}. \tag{3.27}$$

Let $\mathbf{c}$ denote a vector of contrasts. For the hypothesis to compare HRF functions $\mathbf{c}^\tau\mathbf{h} = 0$, we have the equivalent hypothesis $\mathbf{c}^\tau\mathbf{H}(r) = 0$ by the form (3.27), where r is usually the task-related frequency r_0. From the OLS method, we have

$$\hat{\mathbf{H}}(r)^\tau \sim_a N_n^C(\mathbf{H}(r)^\tau, s_{\varepsilon\varepsilon}(r)\Sigma), \quad \mathbf{H}(r) \in \mathbb{C}^n, \tag{3.28}$$

where

$$\Sigma = \left\{ \begin{array}{ll} (bT/\gamma)^{-1}\hat{\mathbf{s}}_{\mathbf{XX}}(r)^{-1} & r \neq 0 \mod \pi \\ (bT/\gamma - 1)^{-1}\hat{\mathbf{s}}_{\mathbf{XX}}(r)^{-1} & r = 0 \mod \pi \end{array} \right.$$

and $N_n^C(\cdot,\cdot)$ is the complex multivariate normal distribution for the r-dimensional random vectors.

Definition 1 *The Complex Multivariate Normal Distribution. If $\Sigma = \Sigma_1 + i\Sigma_2$ is a complex-valued $m \times m$ matrix such that $\Sigma = \Sigma^\tau$ and $\mathbf{a}^\tau\Sigma a \geq 0$ for all $\mathbf{a} \in \mathbb{C}^m$, then we say that $\mathbf{Y} = \mathbf{Y}_1 + i\mathbf{Y}_2$ is a complex-valued multivariate normal random vector with mean $\mu = \mu_1 + i\mu_2$ and covariance matrix Σ if*

$$\left[\begin{array}{c} \mathbf{Y}_1 \\ \mathbf{Y}_2 \end{array} \right] \sim N\left(\left[\begin{array}{c} \mu_1 \\ \mu_2 \end{array} \right], \frac{1}{2} \left[\begin{array}{cc} \Sigma_1 & -\Sigma_2 \\ \Sigma_2 & \Sigma_1 \end{array} \right] \right). \tag{3.29}$$

We then write $\mathbf{Y} \sim N_m^C(\mu, \Sigma)$.

By the definition of complex normal distribution, then we have

$$\left[\begin{array}{c} \text{Re } \hat{\mathbf{H}}(r)^\tau \\ \text{Im } \hat{\mathbf{H}}(r)^\tau \end{array} \right] \sim N\left(\left[\begin{array}{c} \text{Re } \mathbf{H}(r)^\tau \\ \text{Im } \mathbf{H}(r)^\tau \end{array} \right], \frac{1}{2}s_{\varepsilon\varepsilon}(r) \left[\begin{array}{cc} \text{Re } \Sigma & -\text{Im } \Sigma \\ \text{Im } \Sigma & \text{Re } \Sigma \end{array} \right] \right). \tag{3.30}$$

We simply denote the multivariate normal as

$$\hat{\mathbf{H}}_v^\tau \sim N(\mathbf{H}_v, \frac{1}{2}s_{\varepsilon\varepsilon}(r)\Sigma_v) \tag{3.31}$$

where $\mathbf{H}_v$ is in a $2n$-dimensional real vector space, $\mathbf{H}_v = \begin{bmatrix} \mathrm{Re}\ \mathbf{H}(r)^\tau \\ \mathrm{Im}\ \mathbf{H}(r)^\tau \end{bmatrix}$ and Σ_v is a

$2n \times 2n$ real matrix, $\Sigma_v = \begin{bmatrix} \mathrm{Re}\ \Sigma & -\mathrm{Im}\ \Sigma \\ \mathrm{Im}\ \Sigma & \mathrm{Re}\ \Sigma \end{bmatrix}$.

The contrast between different HRF estimates can be represented by $\mathbf{c}^\tau\hat{\mathbf{H}}(r)^\tau$, where $\mathbf{c} = (c_1, c_2, \ldots, c_n)^\tau$ which satisfies that $\sum_{i=1}^{n} c_i = 0$. For the complex number $\mathbf{c}^\tau\hat{\mathbf{H}}(r)^\tau$, the hypothesis testing can be conducted by the definition of the complex normal distribution, which converts complex normal to multivariate normal distribution.

Under the hypothesis $\mathbf{c}^\tau\mathbf{H}(r) = 0$, $(\mathbf{c} \quad \mathbf{c}) \begin{bmatrix} \mathrm{Re}\ \hat{\mathbf{H}}(r)^\tau \\ \mathrm{Im}\ \hat{\mathbf{H}}(r)^\tau \end{bmatrix}$, denoted by $\mathbf{c}_v^\tau\hat{\mathbf{H}}_v$, where $\mathbf{c}_v^\tau = (c_1, \ldots, c_n, c_1, \ldots, c_n)$, is distributed asymptotically as

$$N\left(0, \frac{1}{2}s_{\varepsilon\varepsilon}(r)\mathbf{c}_v^\tau\Sigma_v\mathbf{c}_v\right). \tag{3.32}$$

At the same time, the distribution of $\hat{s}_{\varepsilon\varepsilon}(r)$ is approximated by an independent

$$\frac{s_{\varepsilon\varepsilon}(r)\chi^2_{2(bT/\gamma - n)}}{2(bT/\gamma - n)} \quad r \neq 0 \mod \pi. \tag{3.33}$$

Thus, the t statistic for the contrast between different HRF estimates is

$$\frac{\mathbf{c}_v^\tau\hat{\mathbf{H}}_v(r)^\tau}{\sqrt{\frac{2(bT/\gamma - n)}{bT/\gamma}\hat{s}_{\varepsilon\varepsilon}(r)\mathbf{c}_v^\tau\Sigma_v^{-1}\mathbf{c}_v}} \quad \sim \quad t_{2(bT/\gamma - n)} \tag{3.34}$$

when $r \neq 0 \mod \pi$.

The contrast is highly utilized in fMRI to point out the discrepancy of responses in different conditions. In the fMRI softwares SPM and FSL, we first specify "conditions," which are analogous to types of stimuli here. For example, if we have two types of stimuli from the right and left hands, there are two conditions: Right and Left. Then we need to set up the contrast of conditions according to our interest. For testing whether the right hand has greater effect than the left hand, the contrast should be Right>Left, equivalent to Right−Left> 0. So we state the contrast in a vector $\mathbf{c} = (1, -1)$, 1 for the condition Right and -1 for the condition Left. After settling the contrast, SPM and FSL will continue their general linear model, using parameters to conduct the t statistic.

The hypothesis of comparing the HRF similarity here equates the contrasts in SPM and FSL. We have two types of stimuli: Right and Left. Then we have respective HRF estimates for Right and Left. To test whether Right>Left, we specify $\mathbf{c} = (1, -1)^\tau$ in $\mathbf{c}^\tau\hat{\mathbf{H}}(r)^\tau$. As the result, the t statistic in (3.34) is used for testing their difference.

3.2.6 REMARKS

We will make a few remarks or comments on our method in contrast with the existing methods.

1. Some popular packages [11, 18] estimate the activation based on (3.1) and the double-gamma HEF using the GLM procedure. The model considered in SPM and FSL is given by

$$y(t) = \beta_1 h_1 \otimes x_1(t) + \beta_2 h_2 \otimes x_2(t) + \cdots + \beta_n h_n \otimes x_n(t) + z(t),$$
$$t = 0, \ldots, T-1, \tag{3.35}$$

where the HRFs $h_1, \ldots, h_n$ are pre-specified (such as the double-gamma function) and the time series $z(t)$ is a simple AR(1) model with two parameters to be estimated. Model (3.35) is a linear regression model with $\beta = (\beta_1, \ldots, \beta_n)$ as the regression parameters, and the design matrix is formed by the convolution of the stimuli $\mathbf{x}(t)$ and the double-gamma functions $\mathbf{h}(t)$. The estimates of $\beta_1, \ldots, \beta_n$ are obtained by the least squares (LS) methods which may be consistent (depending on the specification of the HRF) but may not be efficient. The latter is caused by the mis-specification of the AR(1) and the standard errors (se) of the LS estimates may be large. Another challenge is to conduct the voxel-wise significant test of activation.

Namely, at each voxel, the test is based on the t-statistic given by

$$t = \frac{\hat{\beta}}{\mathrm{se}(\hat{\beta})}$$

whose p-value may be over- or under-estimated depending on whether the standard error $\mathrm{se}(\hat{\beta})$ has been correctly estimated, which in turn depends on the specification of the model for the (noise) time series $z(t)$. This will result in a higher false-positive or false-negative rates in detecting activation. One may address the estimation of the model for $z(t)$, but this will be inefficient in fitting and testing a correct time series model at each voxel. The default setting in those packages usually starts with AR(1) at the expense of specification error.

This problem can be remedied using the TFE approach described above. Applying the discrete Fourier transform to (3.1) yields two important results at once. First, the TFE of HRF is a direct (without introducing the β's) and consistent estimator. Second, the FFT of the noise time series are uncorrelated (exactly independent in the Gaussian case), and consequently, the F-test (3.26) is consistent for testing activation. Note that our F-test is described in the frequency domain, which should not be confused with the model-based F-test in SPM or FSL described in the time-domain, which is easily accessible by the majority of statisticians. In the current setting, however, the Fourier transformation provides a direct and highly interpretable activation detection.

2. The TFE method was thought to be only applicable to block designs with some forms of periodicity. Hence the popular event-related designs have ruled out the use of it. This is certainly not the case as one can see from [3]. The current chapter demonstrates the usefulness of the TFE through simu-

lation and real data sets taken from the above package sites for the sake of comparison. More results can be found in [5].

3. The TFE also offers a check on the linearity assumption in (3.1) using the F-statistics based on the coherence (3.24). On the other hand, the time-domain method will further need to verify the error auto-covariance structure for this step, which can be inefficient for the whole brain.

4. A key issue to be addressed in using the TFE method is the smoothing parameter in estimating various auto- and cross-spectra. Some data-adaptive procedures such as cross-validation have been investigated in [5] and the HRF shape has made such a choice less challenging.

5. When the voxel time series are non-stationary, then both time- and frequency-domain methods will not work that well. In that situation, one can invoke the short time (or windowed) Fourier transform method (a special case of wavelet analysis) to estimate the HRF, this will be carried out in another paper.

6. An extensive review is given in one of the chapters of [5]. Here the main objective is to highlight some important contributions to the problem of HRF estimation using the Fourier transform. This type of deconvolution or finite impulse response (FIR) problems has been known in the area of signal processing. The various forms of the test statistics (3.24, 3.25), however, have not been appropriately addressed. Hence the aim of the current chapter is to illustrate the usefulness of the statistical inference tools offered by this important frequency-domain method.

3.3 SIMULATION

The simulation study was based on a multiple-stimuli experiment design and a simulated brain. The experiment in the section included two types of stimuli, called *left* and *right*. The simulated brain had 8×8 voxels, which was designed to have various brain functions in the left and right experiment design. The brain was divided into four regions: One only responded to left, one only responded to right, one can respond to both left and right, and the remaining one had no response in the experiment.

The fMRI data was simulated based on the convolution model (3.1): the convolution of the pre-specified left HRF $h_1(\cdot)$ and right HRF $h_2(\cdot)$, and the known experiment paradigm for left stimulus $x_1(t)$ and right stimulus $x_2(t)$. The response was given by

$$Y(t) = h_1 \otimes x_1(t) + h_2 \otimes x_2(t) + \varepsilon(t), \quad t = 0, \dots, T-1 \qquad (3.36)$$

with $T = 600$. The noise was generated from an *autoregressive-moving-average process* ARMA(2, 2):

$$\varepsilon(t) - 0.8897\varepsilon(t-1) + 0.4858\varepsilon(t-2) = z(t) - 0.2279z(t-1) + 0.2488z(t-2),$$

$$z(t) \sim N(0, 0.2^2).$$

The ARMA was chosen to test the strength of our method under other types of correlated structures, and the coefficients were selected to illustrate the performance of the procedure under moderate, serially correlated noise.

The illustrated experiment paradigm and the brain map are shown in Figure 3.2. This was an event-related design with left and right interchanging, where the green stands for the left, and the purple stands for the right. The left and right stimuli came periodically. The stimulus-related frequency for each was $1/40$, and the overall task-related frequency for the experiment was $1/20$. The brain map shows the brain function in each region (Figure 3.2b).

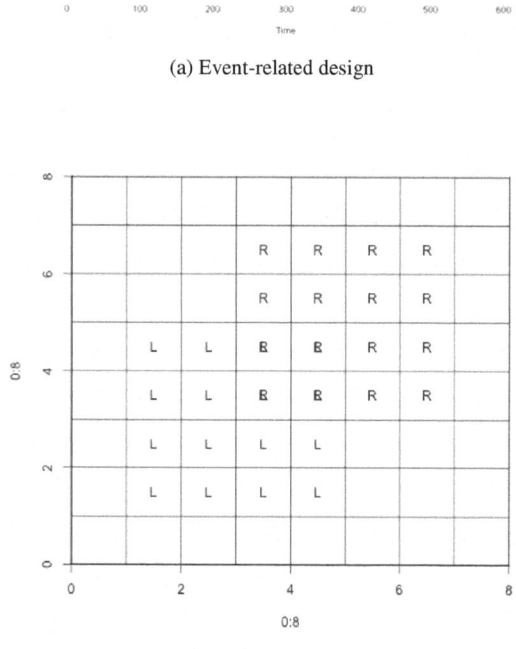

(a) Event-related design

(b) Simulated brain map

Figure 3.2: The simulated brain map in the simulation. (a) shows the experiment design of the simulation with two kinds of stimuli, which are finger taping on the right (shown in purple) and on the left (shown in green). (b) is the simulated brain map: the purple region only responds to the right-hand stimulus; the green region only responds to the left-hand stimulus; the brown region responds to both left and right; and the white region has only noise.

The first simulation was to detect the activation regions in the brain. We assumed both of the original HRFs for left and right were Glover's HRF. At the experiment frequency $1/20$, the coherence and F statistic map shows in Figure 3.3. The lighter the color is, the higher the values are. The high value of coherence in the responsive region implies a strong linear relation in the simulation. Also, the F statistic represents the strength of activation in the voxel. As expected, there were three activation regions: Left (L), Right (R), Left and Right (L&R), which was selected at $\alpha = 0.01$ level.

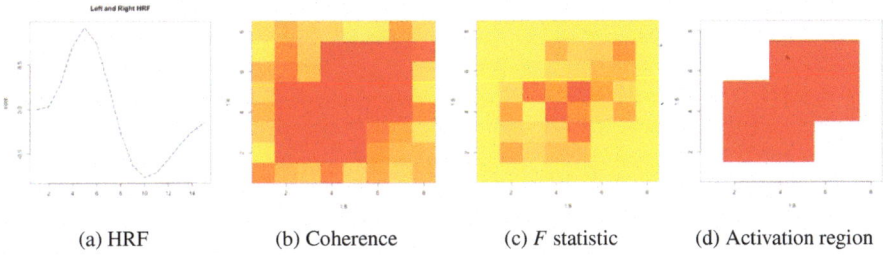

| (a) HRF | (b) Coherence | (c) F statistic | (d) Activation region |

Figure 3.3: Detecting the activation regions by TFE. The activation region is where the brain has a response to the experiment stimulus. (a) shows that the true HRFs for both left and right are the same. (b) shows the coherence value obtained in voxels (the red color means high intensity, and the yellow indicates low intensity). (c) shows the corresponding F statistic, called the F map. As shown in (d), both right and left activated regions (marked in red) are detected.

At the stimulus-related frequency $1/40$ level, we compared the similarity of left and right HRFs. The true left and right HRFs are the same Glover's HRF. There are only two HRF discrepancy regions: Left (L) and Right (R), where we regarded no response as zero HRF. The simulation result is displayed in Figure 3.4. The rejection region for L>R is the region L; the rejection region for L<R is the region R; the rejection region for L$\neq$R is L&R at level $\alpha = .05$. As we can see, if the voxel has the same response to different stimuli, and the result shows that there is no difference in the HRFs.

The second simulation was built on different HRFs. The left HRF kept Glover's HRF, and the right HRF reduced Glover's HRF to half. As we can see, the left and right HRFs had different amplitudes. The left was larger than the right one. At the experiment frequency $1/20$, the activation region is accurately spotted in Figure 3.5.

At the individual-stimulus-related frequency $1/40$, the difference between left and right HRF was detected, as shown in Figure 3.6. The rejection region for L>R contains the regions that respond to both L and R. The hypothesis testing of similar HRFs clearly separated different HRFs.

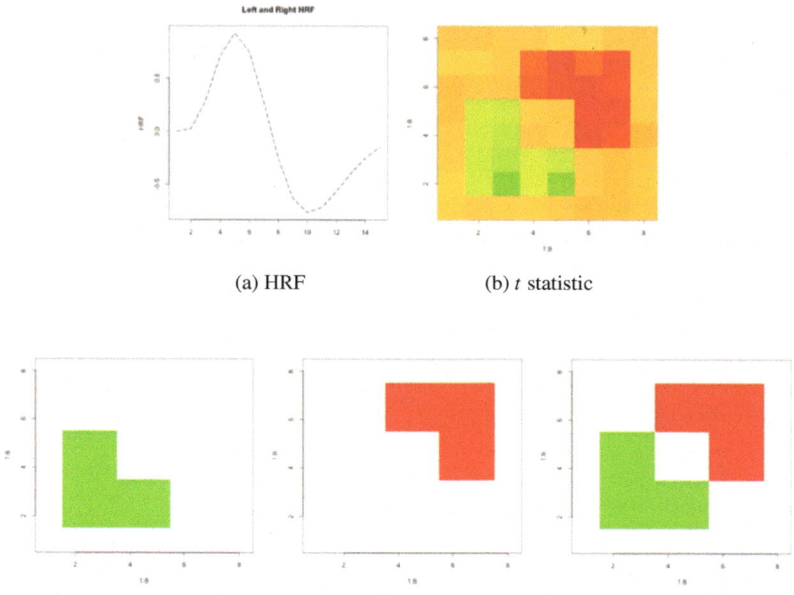

(a) HRF (b) *t* statistic

(c) Acceptance region: L>R (d) Acceptance region: L<R (e) Acceptance region: L≠R

Figure 3.4: Hypothesis testing with two identical HRFs in the simulated brain. (a) shows Glover's HRF for both left and right. (b) shows the overall *t* statistic over the brain map, where red color means high positive values, green color means negative values, and yellow means near 0. (c) shows the rejection region for the test: left≤right; (d) shows the rejection region for left≥right; (e) shows the rejection region for left=right.

3.4 REAL DATA ANALYSIS

3.4.1 AUDITORY DATA

In order to test whether the method of [1] is applicable to real data and able to detect fMRI activation, we applied the nonparametric method to the published auditory data set on the Statistical Parametric Mapping website (http://www.fil.ion. ucl.ac.uk/spm/data/auditory/). According to the information listed there, these whole brain BOLD/EPI images were acquired on a modified 2T Siemens MAGNETOM Vision system. Each acquisition consisted of 64 contiguous slices ($64 \times 64 \times 64, 3\,\text{mm} \times 3\,\text{mm} \times 3\,\text{mm}$ voxels). Acquisition took 6.05 seconds, with the scan to scan repeat time (TR) set arbitrarily to 7 seconds. During the experiment, 96 acquisitions were made in blocks of 6, which resulted in sixteen 42-second blocks. The blocks alternated between rest and the auditory stimulation. We included 8 trials in our dataset, with the first 6 images acquired in the first run discarded due to T1 effects. The data was preprocessed using SPM5, and included realignment, slice timing correction, coregistration, and spatial smoothing.

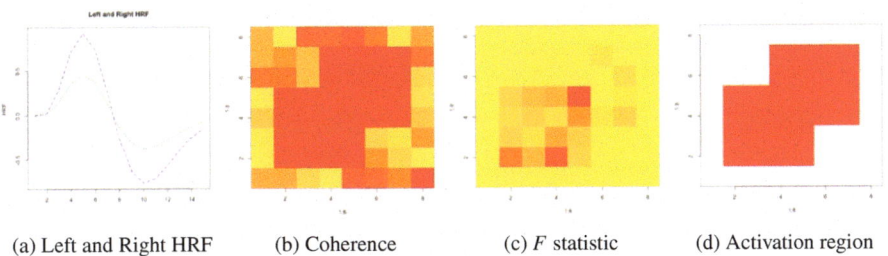

(a) Left and Right HRF (b) Coherence (c) F statistic (d) Activation region

Figure 3.5: Detecting the activation regions by TFE with non-identical HRFs. The activation region is where the brain has a response to the experiment stimulus. (a) shows the true HRFs for both left (green) and right (purple). (b) shows the coherence obtained in voxels (the red color means high intensity, and the yellow indicates low intensity). (c) shows the corresponding F statistic, F map. As shown in (d), both right and left activated regions (marked in red) are detected.

Figure 3.7a shows the time course data from the one voxel that had the greatest F-value in (3.26). The voxel time series depicted in Figure 3.7a has been detrended [23] because the trends may result in false-positive activations if they are not accounted for in the model. Since the voxel has a high F-value, its time series has a good relationship to the task, similar to the pattern we obtained in our second simulation. Figure 3.7b shows several HRF estimates from the 12 voxels with the highest F-values in the brain. The majority of the HRF estimates closely match the HRF shape, showing the increase in the signal that corresponds to the HRF peak and some even depicting the post-dip after the peak signal. The TR for this dataset is 7 seconds, which corresponds to the time interval between the acquisition of the data points. This leads to a very low temporal resolution with which to measure the hemodynamic response. The limitation of a large TR time for estimating the HRF results not only in the low temporal resolution, but also incorrect timing for the stimulus onset. For instance, if the TR is 7 seconds, we have 40-second blocks instead of 42-second blocks. If the stimulus functon $X(t)$ is 0-1 and indicates an onset every 7 seconds, then we will miss the exact onset at 40, 80, 120, 160, ... seconds. The strategy we used here is interpolation. We interpreted the preprocessed data on the second-based time series, and then applied TFE to see the HRF. As a result, we could see only the framework of the HRF and approximate its value. Despite this limitation, the resulting framework gave us evidence that our method does indeed capture the various HRFs in the voxels. In addition, it establishes that our HRF-based analysis can be applied to real data and may be improved with correspondingly refined temporal resolution.

In the experimental design of this study, there are 7 stimulus blocks in the time series data and a total of 90 acquisitions. As a result, the task-related frequency is $7/90 = 0.0778$. Using this information, we can apply our method in order to generate an F-statistic map to show the activation in the brain that is triggered by the stimuli (bottom row in Figure 3.8). For comparison, we also generated a T map using SPM5

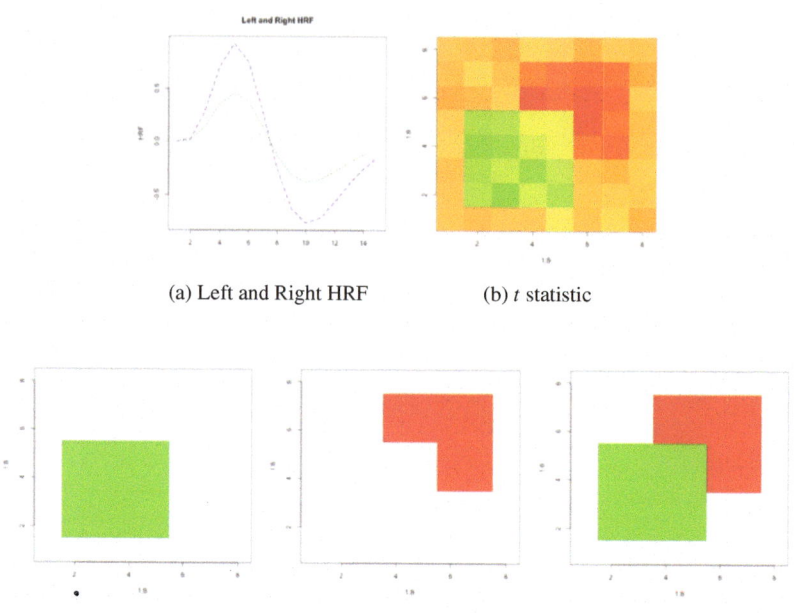

(a) Left and Right HRF   (b) t statistic

(c) Acceptance region: L>R   (d) Acceptance region: L<R   (e) Acceptance region: L≠R

Figure 3.6: Hypothesis testing with two non-identical HRFs in the simulated brain. (a) shows the Glover's HRF for left (green) and half Glover's HRF for right (purple). (b) shows the overall t statistic over the brain map, where red color means high positive values, green color means negative values, and yellow means near 0. (c) shows the rejection region for the test: left≤right; as the left HRF has much higher amplitude than the right one, the rejection region for the test left≤right is the two regions that respond to left-hand stimulus. (d) shows the rejection region for right≥left; (e) shows the rejection region for left=right.

(the SPM T map), which is shown in the upper row of Figure 3.8. The SPM T map is a contrast image that obtains the activation triggered by the stimulus after applying the canonical SPM HRF uniformly throughout the brain. As a result, it does not take into account any HRF variation that might occur in the different regions of the brain.

In both rows of Figure 3.8, increased activation is depicted by increasingly brighter colors, such that the bright yellow regions represents more activation. As expected from an auditory study, both the F map generated using our method and the SPM-generated T map display activation in the temporal lobe. Although the F map from our analysis shows increased activation almost exclusively in the temporal lobe, the contrast map generated using SPM also displays significant activation in other regions of the brain, including parietal and prefrontal cortical areas. In addition, the activation in the temporal lobe is more diffuse using SPM compared to that seen using our F method. We conclude that the map generated using our method

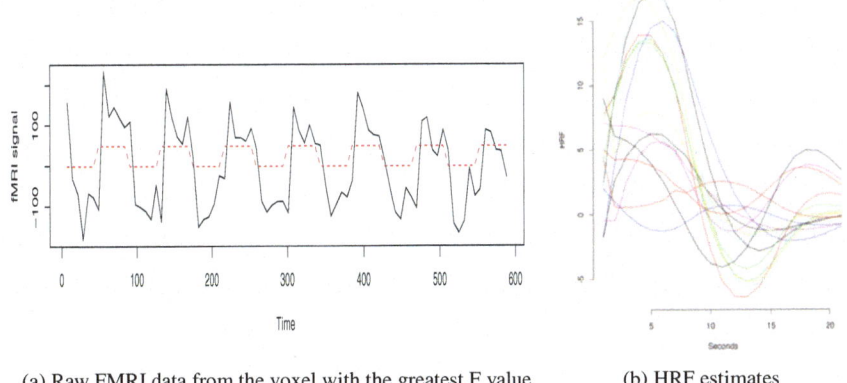

(a) Raw FMRI data from the voxel with the greatest F value   (b) HRF estimates

Figure 3.7: HRF estimation for auditory data (`http://www.fil.ion.ucl.ac.uk/spm/data/auditory/`). (a) is the experimental design paradigm (the red dashed line) for the auditory data. The solid line is the fMRI response from an activated voxel over time; (b) is the HRF estimates from the 12 highly activated voxels found by using TFE in the brain. Due to the large TR (7 seconds), there is a limitation on showing the HRF estimate in finer temporal resolution. In (b), we still can see the different shapes of HRF.

appears to display less noise, such that there is less activation in regions other than the primary auditory cortex. In addition, our method displayed a less diffuse activation area in the auditory region, which may be interpreted as either a more focused activation pattern or there may be some loss of sensitivity for detecting the actual activation. Despite this possible limitation associated with our method, it does have the additional benefit of being a test for the linearity assumption.

3.4.2 EVENT-RELATED VISUAL DATA

In order to further validate our method, we applied our TFE approach to another data set involving event-related visual data. The data set comes from one control subject who completed the event-related design in Figure 3.9, with pictures shown every 2 seconds. Four types of pictures were included: standard, neutral, scary, and a target circle. The data comes from Dr. Belger's Lab at UNC-CH. The standard pictures, regarded as the background instead of the stimulus, are presented most frequently during the scan session. Thus, the three remaining picture types are considered stimuli: Neutral, Scary, and Target. The entire scanning session included eight runs and each run lasted 4 minutes. The TR was 2 seconds. The region of interest (ROI) for this study is the right amygdala, which contains 38 voxels.

In order to see whether there is activation in the amygdala, we applied both SPM and TFE methods to the 38 voxels of the ROI. SPM generates a T statistic by using

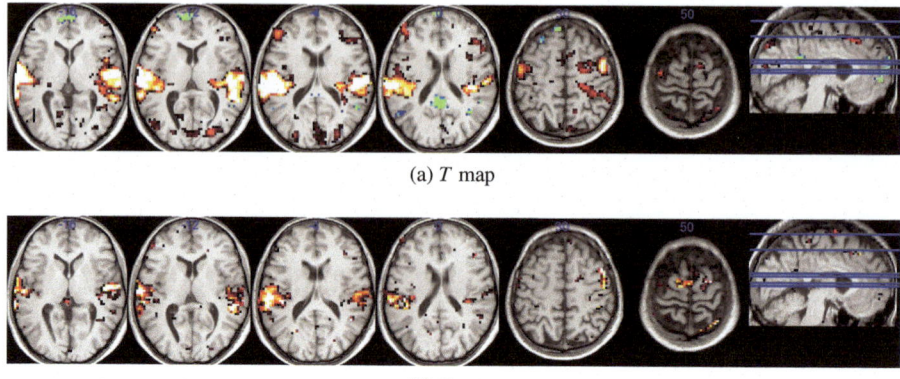

(a) T map

(b) F map

Figure 3.8: F map and T map of the activation by using TFE and SPM. The comparison is based on the auditory data (http://www.fil.ion.ucl.ac.uk/spm/data/auditory/). T maps contain blue and hot colors, which respectively indicate negative and positive factors. The F map generated by TFE (bottom row) appears to have less noise compared to the SPM-generated T map (upper row).

the coefficients in the general linear model to detect activation. Figure 3.10 shows that no activation was detected using the SPM method for any of the three picture stimuli. Even though the coefficients are different, the corresponding T statistic is not significant for detecting activation. Figure 3.11 shows that TFE detected some activation for the Neutral and Scary stimuli but no activation for Target. In contrast to the T statistic used by SPM, activation is detected by the F statistic in TFE by using task frequency information. For each single frequency, we used its fundamental frequency and did hypothesis testing on the frequency to see its activation within voxels. In the 38 voxels of the amygdala, 4 voxels (F values: 7.4, 4.7, 6.3, 4.7) are detected by TFE to be activated by the Neutral stimulus and 6 voxels (F values: 5.3, 9.1, 12.5, 7.5, 5.0, 7.5) by the Scary stimulus. It was confirmed by the researcher exploring the visual fields that the Scary stimulus evokes more activation than the Neutral stimulus and also that the Target stimulus has little effect in amygdala activation in control subjects. Based on the real data, we can see that TFE is more sensitive than SPM in detecting activation based on the frequency information.

The next step is to see the HRF estimates in the activation region. As we already have the activated voxels from TFE, we applied two methods, TFE and sFIR, to estimate the HRF in the activated voxels and region. Figure 3.12 shows the Neutral HRF estimates by both TFE and sFIR. Since they use the same time series, the HRF estimates can be compared side by side. Both of the HRF estimates have a similar shape, but the HRF estimates by TFE have higher amplitude. Figure 3.13 shows the Scary HRF estimates using both TFE and sFIR. TFE gives a smoother estimate in averaged time series than sFIR.

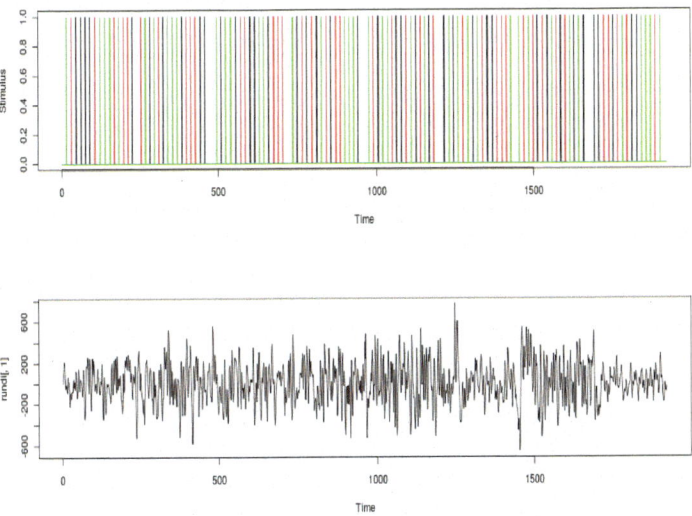

Figure 3.9: Event-related visual data design. The upper graph shows the experimental design with three kinds of events including Neutral (black), Scary (Red), and Target (blue). The bottom time series is an example of the fMRI data from one voxel.

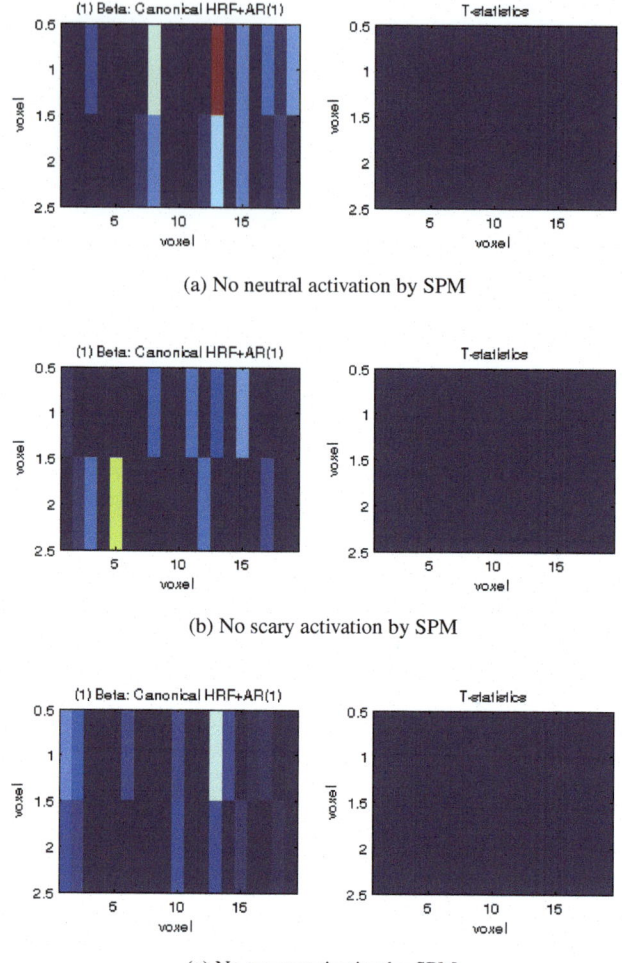

(a) No neutral activation by SPM

(b) No scary activation by SPM

(c) No target activation by SPM

Figure 3.10: The activation detected by SPM in event-related visual data. The value of coefficients in GLM is displayed in the $2 \times 19 = 38$ map on the left, and the activation detection is on the right. By thresholding at level 95% of t statistics in SPM, there is no activation detected for either of the three types of stimulus.

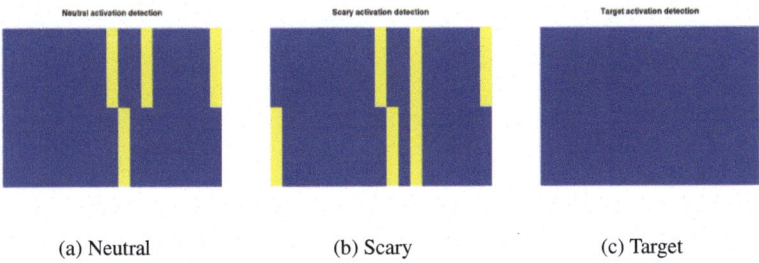

(a) Neutral   (b) Scary   (c) Target

Figure 3.11: Activation detected by TFE in event-related visual data (the same data as in Figure 3.10). There are activated voxels detected for the Neutral and Scary stimuli.

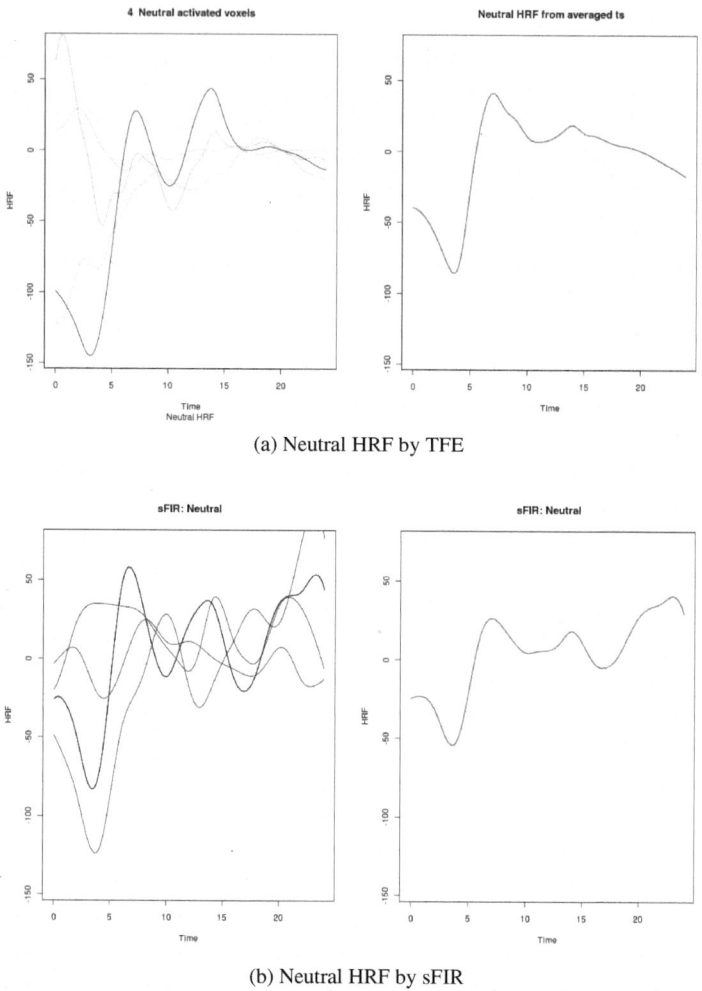

Figure 3.12: The HRF estimates for the four voxels activated by the Neutral stimulus. The upper row is from TFE and the bottom from sFIR. The left graph is the HRF estimates for each voxel, and the right is the HRF estimates by averaging the time series in the four activated voxels. For the Neutral stimulus, F-value=7.4, 4.7, 6.3, 4.7 in the four activated voxels. TFE gives a smoother estimate in averaged time series than sFIR.

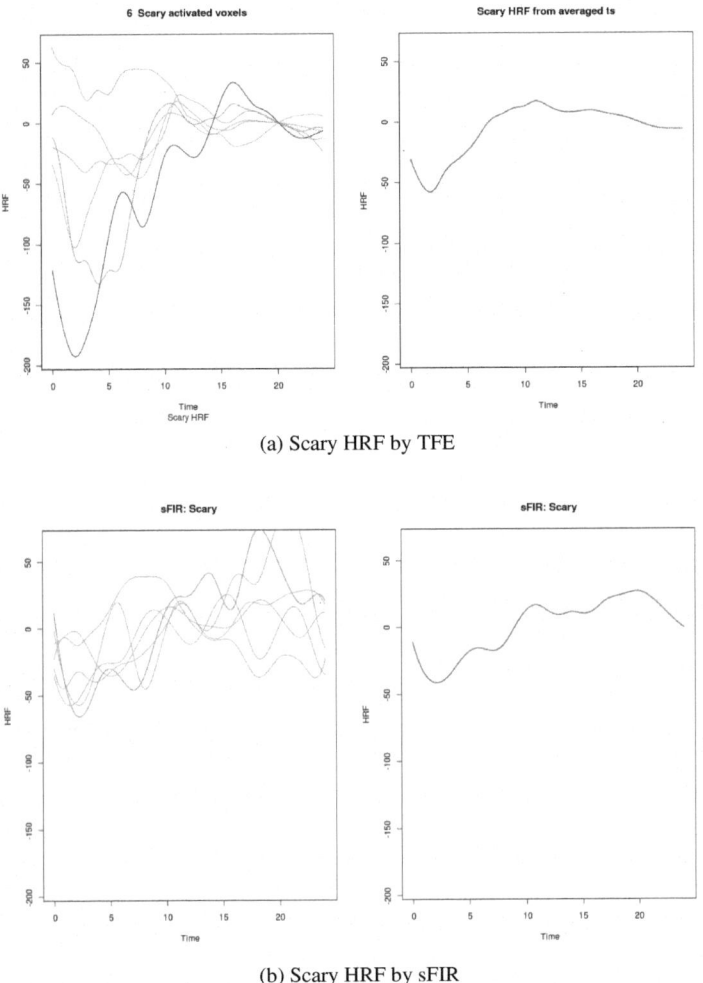

(a) Scary HRF by TFE

(b) Scary HRF by sFIR

Figure 3.13: The HRF estimates for the six voxels activated by the Scary stimulus. The upper row is from TFE and the bottom from sFIR. The left graph is the HRF estimates for each voxel, and the right is the HRF estimates by averaging the time series in the six activated voxels. For the Scary stimulus, F-value $=5.3, 9.1, 12.5, 7.5, 5.0, 7.5$ in the six activated voxels. TFE gives a smoother estimate in averaged time series than sFIR.

3.5 DISCUSSION

Based on the method of Bai et al. [1], we demonstrate that the nonparametric TFE approach successfully and sensitively can complete the entire fMRI data analysis procedure. In addition, it can be adapted to analyze various experimental designs, estimate HRFs in different subjects and brain regions, and detect activation sensitively in various regions of interest. Here we summarize the advantages to this approach in analyzing fMRI data.

The first benefit of the TFE method applies to the experimental design. TFE can be applied for any type of experimental design, including multiple stimuli designs, event-related designs, and block designs. As we demonstrated in the previous section, our nonparametric method can be applied successfully to the multiple stimulus experimental paradigm. From the standpoint of the HRF, different stimuli may cause different hemodynamic responses even in one specific region. The corresponding HRF estimates for each stimulus are estimated in our method, and we can carry out statistical testing to see whether they are equivalent to each other.

Our method also can be applied to block designs and some rapid event-related designs, which will be introduced in the next chapter. Most of the existing HRF estimation methods are applied only to event-related designs. With the adaptability of our method to various experimental designs, we extend the application of HRF estimating to rapid event-related designs and to block designs by adding an extra rest period. In fact, as long as there is a rest period during the design, our method is better in estimating the HRF.

The second benefit of the TFE method is the reduction in noise. Noise might come from multiple sources, such as the various types of scanners in use with their inherent systematic errors, background noise in the environment, and individual subject differences in the heartbeats and breathing. These noises lead to heterogeneity within the fMRI data. By using TFE, the heterogeneity is considered in the frequency domain, which simplifies the error structure estimation process. Such simplicity comes from the asymptotic independence of the spectrum in different Fourier frequencies when we transfer the time series analysis to the frequency domain. In addition, for efficiency, we use the WLS method to estimate the error spectrum. Unlike previous work [32, 3] based on the WLS method, which was implemented using a computationally costly high-dimensional-matrix operation, our method shows improved performance, since the dimension of our matrix operation depends only on the number of stimulus types in the experimental design.

The third benefit of our TFE method is the estimation of the HRF. TFE does not require the length of the HRF, which also is called the *latency* (width) of the HRF. As in most HRF modeling methods, the length of the HRF is entered as an a priori function at the start of the analysis. In practice, however, the latency of the HRF is unknown to researchers. If the length of HRF is assumed as known, such as in smooth FIR or the two-level method of [32], the final result may be very sensitive to the input length. For TFE, the latency of the HRF is not a factor that affects the estimates. Additionally, the TFE estimate gives us a rough idea about the latency of HRF by looking at how the estimates go to zero eventually over time.

One of the most important benefits of our method is that TFE is able to generate the brain activation map without using GLM. In fact, it simplified the analysis by reducing the number of steps from two to one. The typical fMRI analysis (SPM, FSL) requires two steps to customize the HRF in the analysis. The first step estimates the HRF and the second step applies the GLM to detect the activation. In TFE, however, activation detection is generalized by testing the hypothesis for the HRF estimates and does not require additional GLM analyses.

A unique feature of using TFE is being able to test the linearity assumption. As the linearity assumption is the foundation of the convolution model we used, our method is able to test its validity before estimation, which is definitely important for further analysis. As the linearity assumption is valid for the fMRI data after testing, we then are able to use our nonparametric method to perform the analysis, or any analysis tool based on the linearity assumption. If the linearity testing fails, nonlinearity dominates the fMRI data and the nonlinear estimation method might be used [10, 28, 13].

3.6 SOFTWARE: R

This section collects all the R codes used in this chapter. Some of the codes have been modified or adapted from Dr. J. Newton's TIMESLAB — an R library for time series analysis.

```
#### prep.R

source('BIV.R')
source('WLS.R')
source('fcn1.R')
source('MultiVM.R')

######## Stimulus type
## First
  gm <- function(t, a=1.0, c=5.5, b=5, d=0.9){
(a*(t-c)**b)*exp(-(t-c)/d) }
t <- 0:14
g1 <- gm(t, c=0)
g2 <- gm(t, a=0.4, c=0, b=12, d=0.7)
a1 <- max(g1); a2 <- max(g2)
beta <- ((g1/a1) - (g2/a2))
## Second
x <- seq(-2,-.6,.1)
beta1 <- .5*sin(2*pi*x)/x

## Third: another glover
g3 <- gm(t, a=0.9, c=.5, b=5, d=0.9)
g4 <- gm(t, a=0.4, c=1, b=12, d=0.7)
```

```
a3 <- max(g3); a4 <- max(g4)
beta2 <- g3/a3-g4/a4/2


## Cubic functions
ind <- 1:15
cubeta1 <- (ind-2)*(ind-11)*(ind-15)/200
cubeta2 <- (ind-1)*(ind-9)*(ind-15)/200


###########################
draw <- function(esthat, yrange, beta=beta, color, lty)
 {
 ndat <- dim(esthat)[2]
 plot(beta, ylim=yrange, typ='n', ylab="HRF", xlab="Time")
 for(i in 1:ndat) lines(esthat[, i], col="grey")
 lines(beta, lwd=2, col=color, lty=lty)
 lines(apply(esthat, 1, mean), col='red', lwd=2)
 quan <- apply(esthat, 1, quan025)
 lines(quan[1,], col='blue')
 lines(quan[2,], col='blue')
 }


drawpoly <- function(hhat, yrange, beta, color, lty)
{plot(beta, ylim=yrange, typ='n')
quan <- apply(hhat, 1, quan025)
x <- 1:15
xx <- c(x, rev(x))
yy <- c(quan[1,], rev(quan[2,]))
polygon(xx,yy,col="gray",border=NA,ylab="HRF",xlab="Time")
lines(beta, lwd=2, col=color, lty=lty)
lines(apply(hhat, 1, mean), col='red', lwd=2)}




# fig.R

##### hypothesis testing #####
##### Spatial Simulation #####
##### 8 x 8 three region: left, right, both left and right ###

sti1 <- rep(c(1, rep(0, 39)), 15)
sti2 <- rep(c(rep(0, 20), 1, rep(0, 19)), 15)

# fig1.R

plot(sti1, col="green4", lwd=2, lty=2, typ='l', xlab="Time",
```

```
        ylab="Design", ylim=c(0, 1.5))
lines(sti2, col="purple", lwd=2, lty=3) # Fig1a

source('prep.R')
ioptw <- 4; kn <- 14
source('R/MultiV.R')
source('a.R')


# fig2.R

# Fig2a
plot(beta, col="purple", lwd=2, lty=2, typ='l',
        ylab="HRF", xlab="")
lines(beta, col="green4", lwd=2, lty=3)
title("Left and Right HRF")

# Simulate BOLD signals

fz1 <- filter(c(rep(0, 15), sti1), beta, sides=1)[-(1:15)]
fz2 <- filter(c(rep(0, 15), sti2), beta, sides=1)[-(1:15)]

# Set up the responses

X <- cbind(sti1, sti2)
sig <- 0.1
n <- length(sti1)
ar <- c(0.8897, -0.4858)
ma <- c(-0.2279, 0.2488)

resp <- array(NA, dim=c(8, 8, n))

## noise
for(i in 1:8)
for(j in 1:8)
 resp[i, j, ] <- arima.sim(n=n, list(ar=ar,ma=ma), sd=sig)

##left
for(i in 2:5)
for(j in 2:5)
 resp[i, j, ] <- fz1 + resp[i, j, ]

##right
for(i in 4:7)
```

```
for(j in 4:7)
 resp[i, j, ] <- fz2 + resp[i, j, ]


##########
fre <- 1/20
source('cfa.R')  # cfa

image(1:8,1:8, -coh2, col=rainbow(n, start=0, end=1/6))  #Fig2b
image(1:8,1:8, -fstat, col=rainbow(n, start=0, end=1/6)) #Fig2c
image(1:8, 1:8, fstat>Fthres2, zlim=c(.01, 1), col=rainbow(n,
        start=1/6, end=0))


# fig3.R

# Fig3a
plot(beta, col="purple", lwd=2, lty=2, typ='l',
            ylab="HRF", xlab="")
lines(beta, col="green4", lwd=2, lty=3)
title("Left and Right HRF")

##########
fre <- 1/40
source('cfa.R')  # cfa

image(1:8, 1:8, tstat, col=rainbow(n,start=0,end=2/6)) #Fig3b
image(1:8, 1:8, tstat>thres.r, col=rainbow(n,start=0,end=2/6),
    zlim=c(0.1, 1))  # Fig3c
image(1:8, 1:8, tstat<thres.l, col=rainbow(n,start=2/6,end=0),
    zlim=c(0.1, 1))  # Fig3d
image(1:8, 1:8, 0.5*(tstat< -thres.b)+(tstat>thres.b),
        col=rainbow(n,
    start=0, end=2/6), zlim=c(0.5, 1))  # Fig3e


# fig4.R

# Fig4a
plot(beta, col="purple", lwd=2, lty=2, typ='l',
            ylab="HRF", xlab="")
lines(0.5*beta, col="green4", lwd=2, lty=3)
title("Left and Right HRF")
dev.off()
```

```
# Simulate BOLD signals

fz1 <- filter(c(rep(0, 15), sti1), beta, sides=1)[-(1:15)]
fz2 <- filter(c(rep(0, 15), sti2), 0.5*beta, sides=1)[-(1:15)]

# Set up the responses

X <- cbind(sti1, sti2)
sig <- 0.1
n <- length(sti1)
ar <- c(0.8897, -0.4858)
ma <- c(-0.2279, 0.2488)

resp <- array(NA, dim=c(8, 8, n))

## noise
for(i in 1:8)
for(j in 1:8)
 resp[i, j, ] <- arima.sim(n=n, list(ar=ar,ma=ma), sd=sig)

## Simon:ch03
##left
for(i in 2:5)
for(j in 2:5)
 resp[i, j, ] <- fz1 + resp[i, j, ]

##right
for(i in 4:7)
for(j in 4:7)
 resp[i, j, ] <- fz2 + resp[i, j, ]


#########
fre <- 1/20
source('cfa.R')   # cfa

image(1:8,1:8, -coh2, col=rainbow(n,start=0,end=1/6)) #Fig4b
image(1:8,1:8, -fstat, col=rainbow(n,start=0,end=1/6))#Fig4c
image(1:8, 1:8, fstat>Fthres2, zlim=c(.01,1), col=rainbow(n,
        start=1/6, end=0))

# fig5.R
```

```
# Fig5a
plot(beta, col="purple", lwd=2, lty=2, typ='l',
         ylab="HRF", xlab="")
lines(0.5*beta, col="green4", lwd=2, lty=3)
title("Left and Right HRF")

#########
fre <- 1/40
source('cfa.R')   # cfa


image(1:8, 1:8, tstat, col=rainbow(n,start=0,end=2/6)) #Fig5b
image(1:8, 1:8, tstat>thres.r, col=rainbow(n,start=0,end=2/6),
      zlim=c(0.1, 1))   # Fig5c
image(1:8, 1:8, tstat<thres.l, col=rainbow(n,start=2/6,end=0),
      zlim=c(0.1, 1))   # Fig5d
image(1:8, 1:8, 0.5*(tstat< -thres.b)+(tstat>thres.b),
col=rainbow(n, start=0, end=2/6), zlim=c(0.5, 1))   # Fig5e
```

Bibliography

1. P. Bai, X. Huang, and Y.K. Truong. Nonparametric estimation of hemodynamic response function: A frequency-domain approach. In J. Rojo, editor, *Optimality: The Third Erich L. Lehmann Symposium*, volume 57 of *IMS Lecture Notes Monograph Series*, pages 190–215. Institute of Mathematical Statistics, 2009.

2. G.M. Boynton, S.A. Engel, G.H. Glover, and D.J. Heeger. Linear systems analysis of functional magnetic resonance imaging in Human V1. *Journal of Neuroscience*, 16(13):4207–4221, 1996.

3. D.R. Brillinger. *Time Series*. Holden-Day, 1981.

4. E. Bullmore, M. Brammer, S.C.R. Williams, S. Rabe-Hesketh, N. Janot, A. David, J. Mellers, R. Howard, and P. Sham. Statistical methods of estimation and inference for functional MR image analysis. *Magnetic Resonance in Medicine*, 35(2):261–276, 1996.

5. W. Chen. On estimating hemodynamic response functions. PhD thesis, The University of North Carolina at Chapel Hill, NC, 2012.

6. R.W. Cox. AFNI: Software for analysis and visualization of functional magnetic resonance neuroimages. *Computers and Biomedical Research*, 29(3): 162–173, 1996.

7. K.J. Friston, C.D. Frith, R. Turner, and R.S.J. Frackowiak. Characterizing evoked hemodynamics with fMRI. *NeuroImage*, 2(2PA):157–165, 1995.

8. K.J. Friston, A.P. Holmes, K.J. Worsley, J.B. Poline, C.D. Frith, R.S.J. Frackowiak, et al. Statistical parametric maps in functional imaging: A general linear approach. *Hum Brain Mapp*, 2(4):189–210, 1995.

9. K.J. Friston, P. Fletcher, O. Josephs, A. Holmes, M.D. Rugg, and R. Turner. Event-related fMRI: Characterizing differential responses. *NeuroImage*, 7(1): 30–40, 1998.

10. K.J. Friston, O. Josephs, G. Rees, and R. Turner. Nonlinear event-related responses in fMRI. *Magnetic Resonance in Medicine*, 39:41–52, 1998.

11. K.J. Friston, J. Ashburner, S.J. Kiebel, T.E. Nichols, and W.D. Penny, editors. *Statistical parametric mapping: The analysis of functional brain images*. Academic Press, 2007. URL http://www.fil.ion.ucl.ac.uk/spm/.

12. C.R. Genovese. A Bayesian time-course model for functional magnetic resonance imaging data. *Journal of the American Statistical Association*, 95(451): 691–719, 2000.

13. G.H. Glover. Deconvolution of impulse response in event-related BOLD fMRI. *NeuroImage*, 9(4):416–429, 1999.

14. C. Gössl, L. Fahrmeir, and D.P. Auer. Bayesian modeling of the hemodynamic response function in BOLD fMRI. *NeuroImage*, 14(1):140–148, 2001.

15. C. Goutte, F.A. Nielsen, and K.H. Hansen. Modeling the hemodynamic response in fMRI using smooth FIR filters. *IEEE Transactions on Medical Imaging*, 19(12):1188–1201, 2000.

16. S.A. Huettel and G. McCarthy. Evidence for a refractory period in the hemodynamic response to visual stimuli as measured by MRI. *NeuroImage*, 11(5): 547–553, 2000.

17. J. Jacobs, C. Hawco, E. Kobayashi, R. Boor, P. LeVan, U. Stephani, M. Siniatchkin, and J. Gotman. Variability of the hemodynamic response as a function of age and frequency of epileptic discharge in children with epilepsy. *NeuroImage*, 40(2):601–614, 2008.

18. M. Jenkinson, C.F. Beckmann, T.E. Behrens, M.W. Woolrich, and S.M. Smith. Fsl. *NeuroImage*, 62:782–790, 2012. URL http://www.fmrib.ox.ac.uk/fsl/.

19. M.A. Lindquist and T.D. Wager. Validity and power in hemodynamic response modeling: A comparison study and a new approach. *Human Brain Mapping*, 28(8):764–784, 2007.

20. M.A. Lindquist, J. Meng Loh, L.Y. Atlas, and T.D. Wager. Modeling the hemodynamic response function in fMRI: Efficiency, bias and mis-modeling. *NeuroImage*, 45(1S1):187–198, 2009.

21. Y. Lu, A.P. Bagshaw, C. Grova, E. Kobayashi, F. Dubeau, and J. Gotman. Using voxel-specific hemodynamic response function in EEG-fMRI data analysis. *NeuroImage*, 32(1):238–247, 2006.

22. Y. Lu, C. Grova, E. Kobayashi, F. Dubeau, and J. Gotman. Using voxel-specific hemodynamic response function in EEG-fMRI data analysis: An estimation and detection model. *NeuroImage*, 34(1):195–203, 2007.

23. J.L. Marchini and B.D. Ripley. A new statistical approach to detecting significant activation in functional MRI. *NeuroImage*, 12(4):366–380, 2000.

24. R.S. Menon and S.G. Kim. Spatial and temporal limits in cognitive neuroimaging with fMRI. *Trends in Cognitive Sciences*, 3(6):207–216, 1999.

25. S. Ogawa, T.M. Lee, A.R. Kay, and D.W. Tank. Brain magnetic resonance imaging with contrast dependent on blood oxygenation. *Proceedings of the National Academy of Sciences*, 87(24):9868–9872, 1990.

26. A. Roche, S. Mériaux, M. Keller, and B. Thirion. Mixed-effect statistics for group analysis in fMRI: A nonparametric maximum likelihood approach. *NeuroImage*, 38(3):501–510, 2007.

27. J. Steffener, MH Tabert, and Y. Stern. Investigating hemodynamic response variability at the group level using basis functions. *NeuroImage*, 47:56–56, 2009.

28. A.L. Vazquez and D.C. Noll. Nonlinear aspects of the BOLD response in functional MRI. *NeuroImage*, 7(2):108–118, 1998.

29. B.D. Ward and L. Revision. Deconvolution analysis of fMRI time series data. *AFNI 3dDeconvolve Documentation*, Medical College of Wisconsin, 2000.

30. K.J. Worsley, S. Marrett, P. Neelin, A.C. Vandal, K.J. Friston, and A.C. Evans. A unified statistical approach for determining significant signals in images of cerebral activation. *Human Brain Mapping*, 458:73, 1996.

31. E. Zarahn. Testing for neural responses during temporal components of trials with BOLD fMRI. *NeuroImage*, 11(6):783–796, 2000.

32. C. Zhang, Y. Jiang, and T. Yu. A comparative study of one-level and two-level semiparametric estimation of hemodynamic response function for fMRI data. *Statistics in Medicine*, 26(21):3845–3861, 2007.

33. C. Zhang, Y. Lu, T. Johnstone, T. Oakes, and R.J. Davidson. Efficient modeling and inference for event-related fMRI data. *Computational Statistics & Data Analysis*, 52(10):4859–4871, 2008.

4 An Efficient Estimate of HRF

Wenjie Chen
AIG, New York

Haipeng Shen
University of Hong Kong, China

Young K. Truong
University of North Carolina at Chapel Hill

CONTENTS

Functional MRI is a measure of metabolic activity, instead of neural activity. It is well established that energy metabolism and neural activity are tightly coupled. The activity of the neuron requires energy from metabolism, which is provided sufficiently by blood flow in the brain. A small neuronal activity could cause a large increase in local energy demand. The energy comes from the consumption of glucose and oxygen in the blood. The oxygen is attached to hemoglobin molecules. The *oxygenated hemoglobin* (Hb) and *deoxygenated hemoglobin* (dHb) have different magnetic properties in the MR scanner. There are more MR signals when Hb is at a high level and fewer MR signal when dHb is at a high level. The changes of the dHb level can be captured in a strong static magnetic field from the MR scanner.

The change in the MR signal triggered by neural activity is known as the *hemodynamic response*. The HRF is the response to a brief, intense period of neural stimulation. The shape of the HRF varies according to the properties of the stimulus and, presumably, the underlying neuronal activity. The components of the typical HRF include a peak and a post-dip (undershoot) as shown in Figure 3.1. The peak is the maximum amplitude of the HRF, occurring typically about 4 to 6 seconds following a short-duration event. The undershoot is the decrease in MR signal amplitude below baseline due to the combination of reduced blood flow and increased blood volume.

Hemodynamic response varies from region to region in the brain and from subject to subject during the same experiment. Increasing the rate of neural firing increases HRF amplitude, whereas increasing the duration of neural activity increases HRF width (latency).

As hemodynamic response is the MR signal evoked by a short, single stimulus, it results from three factors during the neutral and metabolic activity: oxygen consumption, blood flow, and blood volume. It has three major components: initial dip, peak and undershoot.

Initial Dip: At the first 1 to 2 seconds after stimulus, an initial dip is reported by many studies. While neuronal activities start in a neuron region, the transient energy demand is satisfied by the oxygen extraction in local Hb. This results in an increase of dHb, and therefore the decrease of the MR signal, called *initial dip*.

Peak: After a short latency of initial dip, the blood flow comes in with increasing blood volume. The blood flow brings more oxygen than its needs in the neuron region, which results in the decrease of dHb and the increase of the MR signal. The increase takes about 5 seconds to reach a maximum value in the MR signal, called the *peak*.

Undershoot: After the peak, the blood flow decreased more rapidly than the blood volume to the baseline. During the period that the blood flow returns to baseline and the blood volume is still above the baseline, a greater amount of dHb is present, and therefore the MR signal is below the baseline for a prolonged period, called the *undershoot*.

The typical shape of HRF contains an initial dip, a rise to a peak, a fall to a baseline, and a prolonged post-stimulus undershoot. As the initial dip lasts only 1 to 2 seconds, it may not be detected without finer temporal resolution of MR signal. Different persons may have different HRF shapes including the timings of rise and fall, the amplitude of the peak, and the HRF latency. However, the connection from neuronal activity to energy metabolism is about the same. The BOLD signal is regarded as a detector of neural activity and the functions of the brain.

4.1   INTRODUCTION

4.1.1   EXPERIMENT DESIGN FOR DETECTING HRF

In this section, we introduce two typical experimental designs for fMRI.

There are two basic types of fMRI studies: *event-related design* and *block design*. An event-related design presents discrete, short-duration stimuli, called the *event*,

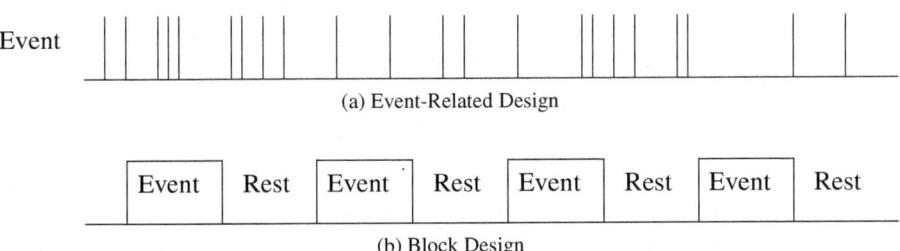

(a) Event-Related Design

(b) Block Design

Figure 4.1: Two major types of experimental designs. (a) illustrates a random event-related design. Each vertical line over time represents one event (a single stimulus). (b) illustrates a block design. Each block represents a period of stimulus presentation. There are a certain number of resting periods between them.

whose timing and order may be randomized. For instance, the event can be designed as a scary picture, a short sound, a gesture, etc. Figure 4.1a offers a graphic illustration of one event-related design, where the locations of the peaks correspond to the event times. A block design separates experimental conditions into distinct blocks, and each experimental condition is presented for an extended period of time. For example, see Figure 4.1b, where each block corresponds to the duration of the experiment stimulus.

For a simple example, there is only one kind of stimulus in the experiment. Figure 4.1a and 4.1b can be regarded as an illustration of the time series of the stimulus function.

4.1.2 GENERAL GUIDELINE FOR ESTIMATING HRF

A single stimulus can evoke a hemodynamic response in the brain that spans a 20-second or greater latency. For each trial, the duration of the BOLD response matches well with the subject's response time [37]. The shape and intensity of the hemodynamic response varies across brain regions and across individuals. Thus, when investigating the activity using fMRI, we must consider its spatial and temporal properties.

The HRF typically is closely linked to the event-related stimulus in fMRI, because the design paradigm allows the HRF to return to baseline or to recover after every trial. By characterizing the precise timing and waveform of the hemodynamic response, researchers can make inferences about the relative timing of neuronal activity, neuronal feedback processes, and sustained activity within a brain region.

The HRF modeling is usually carried out under the event-related design. It has been demonstrated that the areas of BOLD activity can be detected using even very short-duration stimuli and interstimulus intervals. In event-related designs, stimuli that generate short bursts of neural activity are known as *events* or *trials*. The different conditions are usually presented in random order rather than an alternating pattern. Event-related studies measure transient changes in brain activity associated with discrete stimuli. The pattern of changes over time becomes critical for experimental analysis. The characteristics of the event-related designs make the precise estimation of the timing and the waveform of a given HRF achievable. On the other hand, block designs are good for detecting brain activation regions, but are not widely used in estimating the HRF. Most existing HRF estimators have been developed under event-related design. We propose a procedure for estimating the HRF that works under both event-related and block designs.

A basic assumption for estimating HRF is *linearity*. In the event-related design, we consider the hemodynamic response to be evoked by a single, isolated stimulus; in the block design, the stimuli are presented in succession, and it is possible to assume that the same HRF is evoked for every stimulus, independently of the other stimuli presented. If the stimuli are sufficiently close together so that their hemodynamic responses overlap, then the measured total change in MRI signal will be the sum of the individual processes, known as a *linear system* [2, 10].

Modeling the HRF is essential to exploring the relationship between the experimental stimulus and the fMRI signal. If an estimated HRF does not accurately reflect the way the brain responds to stimulus, any follow-up statistical inference will unlikely be valid.

4.1.3 THE GENERAL LINEAR MODEL FRAMEWORK

As one of the most popular approaches, the general linear model (GLM) models the BOLD signal as a linear combination of several different component predictors. It is used to test whether the activity in a brain region is systematically related to any of those known input functions [28]. The data input for fMRI analysis includes the response $Y(t)$ of the voxel at time t, and also the stimulus function $s(t)$. Considering the relationship between the stimulus and the response, the GLM was first elucidated by [13]. It may be expressed as

$$\mathbf{Y} = \mathbf{X}\beta + \varepsilon \tag{4.1}$$

where $\mathbf{Y} = (Y(0), Y(1), \ldots, Y(T-1))^{\tau}$ is the vector of the time series data, $\mathbf{X}$ is the $T \times p$ design matrix, and ε is the error vector. Each column of $\mathbf{X}$ represents one of the experimental BOLD response predictors. Their corresponding parameters are represented by β as a $p \times 1$ vector. In most studies, the predictor is obtained by the convolution model (3.1) of the pre-specified HRF. $h(\cdot)$ (see Figure 3.1) and a single stimulus $s(\cdot)$ (Figures 4.1a, 4.1b). The GLM also adapts to the multi-stimuli design based on the linearity assumption. When there are n stimuli, there will be at least n columns of $\mathbf{X}$ with each column corresponding to one stimulus (sometimes multiple basis functions may be used to represent the HRF for a particular stimulus, so there is more than one column to represent the effect caused by one stimulus). The parameter vector β will be estimated in each voxel to weigh the effect of each experimental predictor.

As the HRF $h(\cdot)$ is involved as part of the predictor from the design matrix $\mathbf{X}$, an accurate hemodynamic response can improve the statistical power of the fMRI modeling analysis. Under the GLM framework, many studies looked into the design matrix $\mathbf{X}$ and tried to capture the characteristics of the hemodynamic response, and to come up with valid predictors. More description of the GLM framework and its implementation in various fMRI software can be found in [35]. Some studies [44, 5] also explore the error correlation structure to improve the efficiency of the estimation. Statistical inference after estimating HRF has been discussed in [30].

When multiple stimuli are presented in succession, the hemodynamic response is often assumed to be the summation of the individual responses generated by the each stimulus respectively [9]. Under certain conditions, the fMRI response has been found to be approximately linear [22, 2], and this has been the basis for most of the event-related fMRI analysis to date [3, 7, 14]. GLM adapts to multiple stimuli through the cooperation of multiple columns in the design matrix $\mathbf{X}$ based on the linearity assumption.

4.1.4   HRF MODELING

In early fMRI studies, researchers simply regarded the HRF $h(\cdot)$ as *a priori*, that is, the HRFs are assumed to be known, and then the predictor of BOLD response is easily obtained from the convolution model (3.1). Most studies in the GLM framework obtain predictors based on known HRFs, and plug them into the design matrix $\mathbf{X}$, so as to test whether the predictors are able to detect the activities. A variety of fixed HRFs were used in early studies, such as the Poisson function [11], the gamma function [7, 27, 15] and the Gaussian function [36, 26, 9]. After [19] intensively investigated the shape of the HRF, the typical HRF shape (Figure 3.1) is suggested in the form of a double gamma function. Since then, the double gamma function has been widely used in fMRI studies and popular fMRI data analysis packages, such as Statistical Parametric Mapping (SPM) [17], FMRIB Software Library (FSL) [24], and Analysis of Functional NeuroImages (AFNI) [8]. For example, the canonical HRF in SPM is double gamma function as the default choice for HRF.

Because the HRF varies across individuals and across brain regions, there exists no fixed "standard" form of HRF. It is crucial to allow flexibility in the exact HRF form while analyzing fMRI data; otherwise, even minor mismodeling of the HRF can result in severe power loss, and can inflate the false positive rate beyond the nominal level.

There is a list of literature on estimating HRF using fMRI data. Below we have organized the literature into two major categories: time-domain and frequency-domain methods.

4.1.4.1   Time-Domain Methods

Most HRF studies use a basis function approach under the GLM framework. To model the early and late components of an evoked hemodynamic response, two basis functions with the form of a product of a sine function and exponential functions has been considered [12]. There have been methods such as the sinusoidal orthogonal basis functions [4, 42], two sine basis functions [25], a superposition of three inverse logit functions [29], and Fourier basis sets [23, 38]. In general, the more basis functions are used in a linear model, the more flexible the model is in estimating the parameters of the GLM. However, flexibility relies on the cost of more free parameters, which means more error in estimating HRF, fewer degrees of freedom, and less statistical power due to potential collinearity.

Several Bayesian methods have also been developed in HRF modeling. In [18], the polynomial "bell" function is used to indicate the rise, fall, decay, and dip parts of the HRF. In [20], the method derived is based on physiological assumptions with the posterior estimated by numerical or Markov Chain Monte Carlo (MCMC) methods. In [16], an empirical Bayesian approach is taken to model the HRF with basis functions under constraints. In [41], a fully Bayesian approach is presented using the addition of four half-period cosines with six constrained parameters to account for the variation of HRF. All these Bayesian approaches require imposing restrictions on the HRF parameters.

Many other time-domain approaches have been proposed in recent years as discussed in Chapter 3. Some time-domain methods [43, 44] impose autocorrelation structure on estimates. Some [31] lacks the flexibility on shapes of HRF in different parts of brain. Some [39] have the flexibility but may cause collinearity during estimation. In [30], parametric methods of estimating HRF are reviewed: SPM canonical HRF, FIR, and semiparametric FIR, and also propose the inverse logit model by using a logit basis function.

4.1.4.2 Frequency-Domain Methods

Currently, some fMRI analyses use frequency-domain methods. [34] considered spatiotemporal changes in the cerebral hemodynamics and verified that the white-noise assumption is inadequate. A non-linear parametric model was used for brain-activation detection [27]. The model, including both temporal and spatial statistical inferences, recasts in the frequency-domain to simplify convolution evaluations and for easier accommodation of temporal and spatial autocorrelation. For periodic design, a frequency-domain method was proposed to detect activation on the fundamental (stimulus-related) frequency and harmonics of the stimulus design per voxel with the information from other frequencies to calibrate its statistic [33]. For HRF modeling, [32] proposes a nonparametric HRF estimate using fully Bayesian inference through MCMC at the price of a high computational demand. Another frequency-domain method is [40] based on the Fourier-wavelet regularized deconvolution technique. A nonparametric method for estimating the HRF using fast Fourier transformation of event-related fMRI data has been implemented [1].

4.1.4.3 Comparison of the Current Methods

Most existing analytical techniques for fMRI data need specific assumptions about the HRF. These assumptions may not be appropriate when the HRF varies from subject to subject or from region to region, especially for the pre-specified HRF methods. Sometimes the experiment contains more than one type of stimulus, and the application for the parametric methods may not be adapted for the multiple stimuli. Additionally, it is unlikely that the fMRI data is homogeneous from the whole scanning session, so the error correlation structure should be considered in the analysis, the estimation of which could be another difficulty in fMRI modeling.

Based on a periodic stimulus design, the frequency-domain approaches simplify the analysis by determining a few parameters related to the stimulus frequency information without considering the correlation structure in the noise series [27, 33]. These frequency approaches rely on the pre-defined HRF. They lack the ability to specify the HRF variation or adapt to the different types of stimulus design (Figures 4.1a, 4.1b).

A more flexible nonparametric way to estimate the HRF in the single event-related experimental design was implemented by modeling the stimulus sequences using a stochastic point process [1]. This method accounts for the variability of HRFs across regions of the brain. The method was for a single event-related design, and in Chapter 3 we extended this method to the block design and the design that allows the multiple, rapid events throughout an experiment, such as the rapid event-related design. Most importantly in Chapter 3, we developed the multivariate estimation method called the *Transfer Function Estimate* (TFE) based on the nonparametric model, which broadened the method's application on all kinds of experiment design under the assumption of linearity. By using hypothesis testing under TFE, the method was easily extended to detect brain activation and examine linearity. Easy application and activation detection are its two important features.

Another feature TFE provides is that it can give an asymptotically consistent and efficient estimate for HRF. Its asymptotic consistency can be easily derived from the ordinary least square (OLS) estimates introduced in Chapter 3. As for efficiency, by introducing weighted least square (WLS) estimates in the TFE, we can derive its asymptotic efficiency. Furthermore, we showed its consistency and efficiency compared with other methods in two simulations of this chapter. Some of the TFE's asymptotic properties can be found in [6].

4.2 TFE METHOD: WLS ESTIMATE

We are going to introduce a WLS estimate for the TFE in this section.

As shown in (3.5), we applied OLS to obtain the estimate of $\mathbf{H}(\cdot)$. The estimate (3.10) will be near an optimal estimate if $s_{\varepsilon\varepsilon}(r)$ is uniform. This is perhaps not likely in fMRI data, since the BOLD response is not homogeneous throughout the whole session in the experiment. The efficient estimate is given by WLS, where we used, in practice and in theory, the estimate $\hat{s}_{\varepsilon\varepsilon}(r)$ obtained by the use of the same formulas given in (3.11) as the weight in the uniform estimate

$$\hat{f}_{\varepsilon\varepsilon}(r) = \frac{2m+1}{2m+1-r}[\hat{f}_{YY}(r) - \hat{\mathbf{f}}_{YX}(r)\hat{\mathbf{f}}_{XX}(r)^{-1}\hat{\mathbf{f}}_{XY}(r)]. \tag{4.2}$$

The WLS estimate of the linear system (3.4) is

$$\hat{\mathbf{H}}(r) = (\varphi_{\mathbf{X}}^{\tau}(r)\Sigma(r)\overline{\varphi_{\mathbf{X}}(r)})^{-1}\varphi_{\mathbf{X}}^{\tau}(r)\Sigma(r)\overline{\varphi_{Y}(r)}, \tag{4.3}$$

where

$$\Sigma(r) = \text{diag}\{\hat{f}_{\varepsilon\varepsilon}^{-1}(\frac{2\pi}{T}(K-m)), \hat{f}_{\varepsilon\varepsilon}^{-1}(\frac{2\pi}{T}(K-m+1)), \ldots, \hat{f}_{\varepsilon\varepsilon}^{-1}(\frac{2\pi}{T}(K+m))\}.$$

Then (3.8) and (3.9) are updated as

$$\hat{\mathbf{f}}_{Y\mathbf{X}}(r) = (2m+1)^{-1}\sum_{k=-m}^{m}\hat{f}_{\varepsilon\varepsilon}^{-1}(\frac{2\pi(K+k)}{T})\mathbf{I}_{Y\mathbf{X}}(\frac{2\pi(K+k)}{T}), \tag{4.4}$$

$$\hat{\mathbf{f}}_{\mathbf{XX}}(r) = (2m+1)^{-1}\sum_{k=-m}^{m}\hat{f}_{\varepsilon\varepsilon}^{-1}(\frac{2\pi(K+k)}{T})\mathbf{I}_{\mathbf{XX}}(\frac{2\pi(K+k)}{T}). \tag{4.5}$$

For the general window estimate, the weight in WLS is

$$\hat{s}_{\varepsilon\varepsilon}(r) = \frac{bT/\gamma}{bT/\gamma - r}[\hat{s}_{YY}(r) - \hat{s}_{Y\mathbf{X}}(r)\hat{s}_{\mathbf{XX}}(r)^{-1}\hat{s}_{\mathbf{X}Y}(r)]. \tag{4.6}$$

Then the weighted spectrum is defined as

$$\tilde{\mathbf{s}}_{Y\mathbf{X}}(r) = \sum_{k\neq 0}b^{-1}W(b^{-1}(r-\frac{2\pi k}{T}))\hat{s}_{\varepsilon\varepsilon}^{-1}(\frac{2\pi k}{T})\hat{\mathbf{s}}_{Y\mathbf{X}}(\frac{2\pi k}{T}), \tag{4.7}$$

$$\tilde{\mathbf{s}}_{\mathbf{XX}}(r) = \sum_{k\neq 0}b^{-1}W(b^{-1}(r-\frac{2\pi k}{T}))\hat{s}_{\varepsilon\varepsilon}^{-1}(\frac{2\pi k}{T})\hat{\mathbf{s}}_{\mathbf{XX}}(\frac{2\pi k}{T}). \tag{4.8}$$

The transfer function estimate is generally given by

$$\tilde{\mathbf{H}}(r) = \tilde{\mathbf{s}}_{Y\mathbf{X}}(r)\tilde{\mathbf{s}}_{\mathbf{XX}}(r)^{-1}. \tag{4.9}$$

To estimate HRF, we consider Equation (3.12) as

$$\hat{\mathbf{h}}(u) = \frac{1}{T}\sum_{t=0}^{T-1}\hat{\mathbf{H}}(\frac{2\pi t}{T})\exp(i2\pi tu/T). \tag{4.10}$$

Thus, $\hat{h}$ is the WLS HRF estimate from the TFE.

4.3 SIMULATION

In this section, we will describe the application of the methodology described in Section 4.2 to two simulated datasets. Assuming a linear time-invariant system in HRF [11], it is natural to simulate the fMRI response as

$$Y(t) = \sum_{j}h(t-\tau_j) + \varepsilon(t), \tag{4.11}$$

where $Y(t)$ is the simulated response, $h(\cdot)$ is the unknown HRF, and $\varepsilon(t)$ is a stationary, zero mean, noise series. We know the experimental design, i.e., the stimulus timing, which is denoted by $X(t)$. $X(t)$ is a series of 0s and 1s, where 0 represents a stimulus-free point and 1 denotes that a stimulus is presenting at that time. Consequently, we now can write

$$Y(t) = h \otimes X(t) + \varepsilon(t). \tag{4.12}$$

4.3.1 SIMULATION 1: WLS

The WLS estimate is the extension of OLS and is preferred. To have a clear idea about the advantage of the multivariate method and WLS, we compared the simulation results of univariate and multivariate, OLS and WLS in Figure 4.2. And we discuss the comparison of the following five methods: *Naive, Univariate* method with OLS/WLS, and *Multivariate* method with OLS/WLS.

When we conduct the statistical analysis for a fixed experiment design with multiple types of stimuli, in order to estimate HRF we usually have certain assumptions. One simple example of the assumption is that HRFs are same for all types of stimuli, that is, the brain has the same response to any kind of stimuli. We call this assumption *naive*. The brain has, however, various responses to different types of stimuli. To separate the stimuli and estimate HRFs for each, we consider each type of stimuli separately, that is, we use one stimulus function at a time in the model without considering the other existing stimuli. This method is called *univariate* since just one of the stimuli is used each time. Moreover, the ideal case is that we consider the multivariate analysis including all the stimuli, and estimate the HRFs for each of them simultaneously. We call this the *multivariate* method here.

By counting the OLS and WLS estimates from the above three assumptions, the five methods are listed below.

(i)   Naive method (Naive)
(ii) Univariate method with OLS (OLSu)
(iii) Univariate method with WLS (WLSu)
(iv) Multivariate method with OLS (OLSm)
(v)   Multivariate method with WLS (OLSm)

It should be noted that the naive method used OLS estimation.

The fMRI response is simulated as the sum of the convolutions of the HRF $h_i(\cdot)$ and the stimulus function $x_i(\cdot)$.

$$Y(t) = h_1 \otimes x_1(t) + h_2 \otimes x_2(t) + \varepsilon(t), t = 0, \ldots, T-1 \qquad (4.13)$$

The noise $\varepsilon(\cdot)$ is heterogeneous. It follows a normal distribution with parameters $(0, \sigma^2)$, where σ follows a uniform distribution.

Figure 4.3 shows the simulation of the block design. We used two types of stimuli (the purple and green lines in Figure 4.3a), and two corresponding HRFs are different. The two kinds of HRFs have the same smoothness since they are cubic functions: $(x-2)(x-11)(x-15)$ and $(x-1)(x-9)(x-15)$. Two hundred simulations were generated with the noise $N(0, \sigma^2)$, where $\sigma \sim U(0, 0.5)$, as denoted above. Figure 4.3b shows the HRF estimates from the five methods: the upper row is the HRF estimates for the purple stimulus, and the bottom row for the green. In each result, the gray lines are the 200 HRF estimates from the 200 simulations. The black curve is the true HRF, and the bold red dashed line is the average of the 200 HRF estimates. The gray band is between the 2.5% and 97.5% quantiles of the 200 estimates.

In Figure 4.3b, the naive method (the 1st column) gave biased estimates, since it incorrectly assumes that the two types of stimuli have the same HRF estimates.

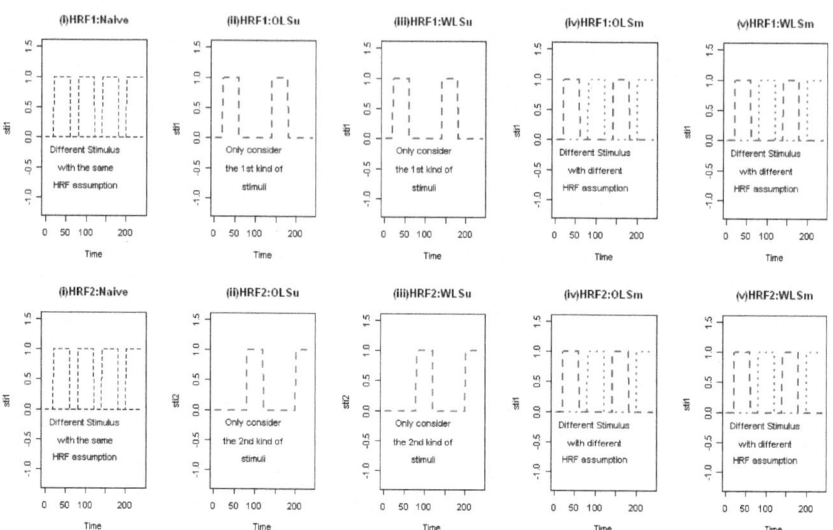

Figure 4.2: The experimental design for the five-method comparison. This is the plot arrangement for Figures 4.3–4.4. The five columns indicate five different methods: Naive, OLSu, WLSu, OLSm, WLSm, respectively. The two rows are for two types of HRFs, respectively.

The univariate method (the 2nd and 3rd columns) gave two different HRF estimates with lower variation. However, the bias is very large. The univariate method hardly distinguished the two HRFs. The multivariate method (the 4th and 5th columns) examined the two types of HRFs at the same time, so the bias has been dramatically reduced in both HRF estimates. The average of 200 estimates is consistent with the true HRF. The variance, however, increased from the univariate method due to the additional covariates in the model, which leads to more variable estimates. Lower variation in estimation could be approached by increasing the number of block trials.

The improvement from OLS to WLS is obvious in Figure 4.3. Comparing the 2nd and 3rd columns, or the 4th and 5th columns, the WLS reduces the variety in estimation in the heterogeneous noise simulation. The multivariate method with WLS estimation is robust and consistent.

The estimation result in Figure 4.3 demonstrates that the multivariate method gives admirable estimates even though the HRFs from different types of stimuli are very similar to each other. What if the HRFs are very different? One extreme and possible case is that there is no response to one of the stimuli, that is, this HRF is zero. The simulation is shown in Figure 4.4. When we look at the zero HRF estimates, the multivariate method shows its robustness. It concludes that the average of the estimates is almost zero, while the univariate method shows there is still some small perturbation for the zero HRFs.

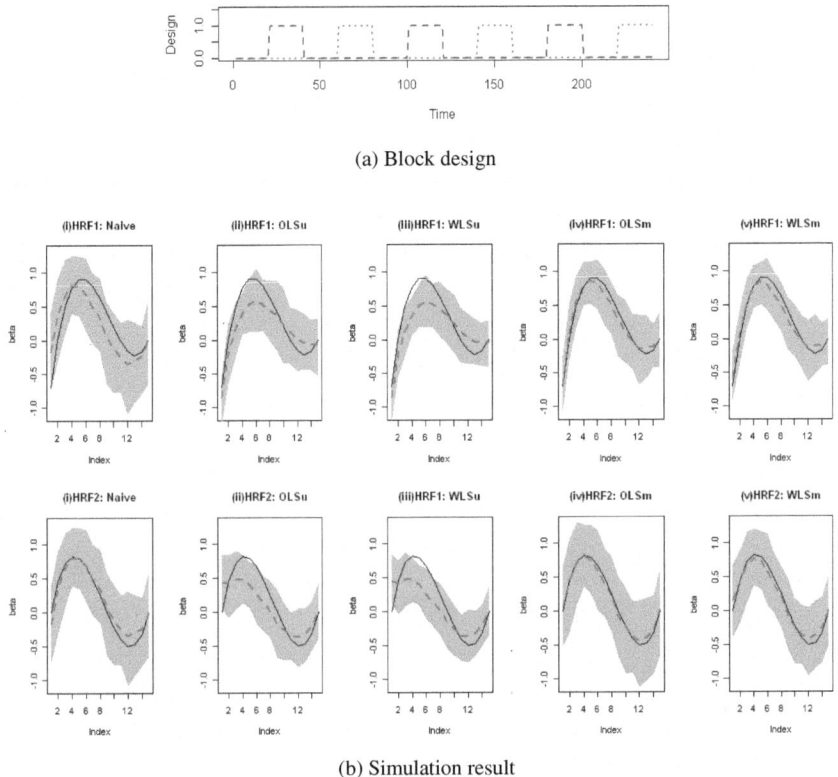

(a) Block design

(b) Simulation result

Figure 4.3: HRF estimates from typical block design. The multivariate method using WLS has substantially reduced the bias associated with the univariate method in both HRF estimate.

Although with a fixed number of trials, the more variable estimate is generated by the multivariate method, the conducted hypothesis testing from the multivariate method is more efficient. If we wish to compare the HRFs (that is, the null hypothesis is $H_1(r_0) = H_2(r_0)$ where r_0 is the task-related radian frequency), the multivariate method will reduce the variance for estimating the t statistic. After hypothesis testing on activation, if there is no significant evidence about HRF existing for one stimulus, then we should reduce the number of variables in the model in order to have more accurate HRF estimates for the other stimuli.

4.3.2   SIMULATION 2: COMPARISON

In this section we compare the TFE with other existing HRF methods such as 3dDe-convolution (AFNI) [39], smooth Finite Impulse Response (sFIR) [24], Inverse Logit (IL), and SPM Canonical HRF [17].

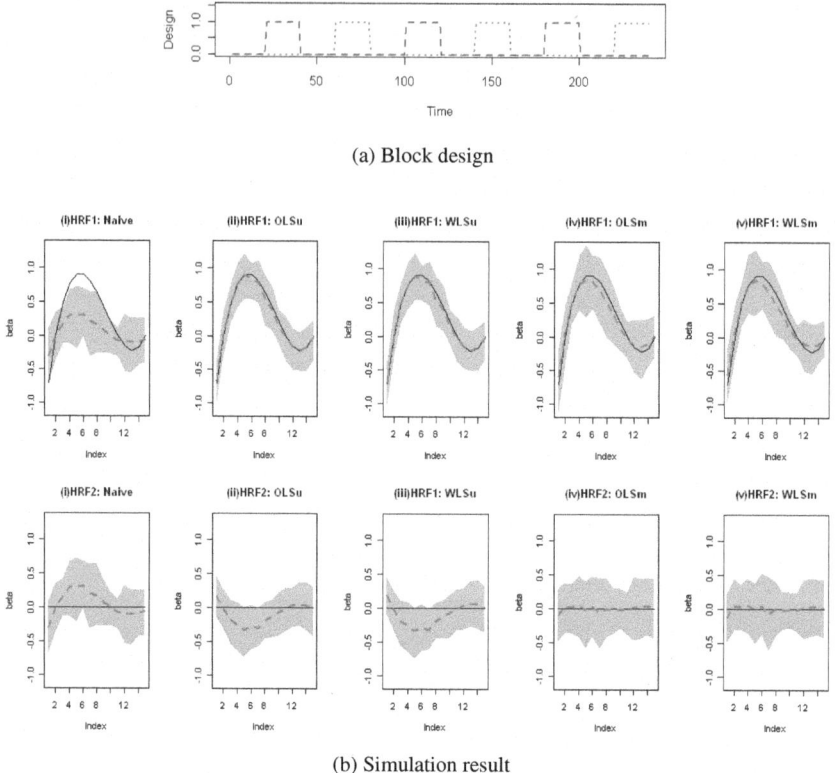

(a) Block design

(b) Simulation result

Figure 4.4: HRF estimates from the simulation that only show the response to one of the stimuli in the block design. The multivariate method demonstrates its robustness by giving consistent estimate.

The face fMRI data set is an event-related design in [21], where famous and non-famous faces were presented twice against a checkerboard baseline. There are thus four event types of interest: the first and second presentations of non-famous and famous faces, which we denote as N1, N2, F1, and F2, respectively. The experimental stimuli and timings of events are shown in Figure 4.5a. According to the description of the study [21], the subject was required to push a button when he/she saw the second presentation of any face. Thus, the first and second presentations of a particular face are different types of stimuli. There were 52 faces presented in the experiment design, 26 famous and 26 non-famous. Also, the famous and non-famous faces may evoke different responses in the subject's brain. Considering the two factors, fame and repetition, in total we had four kinds of stimuli during the whole experiment session.

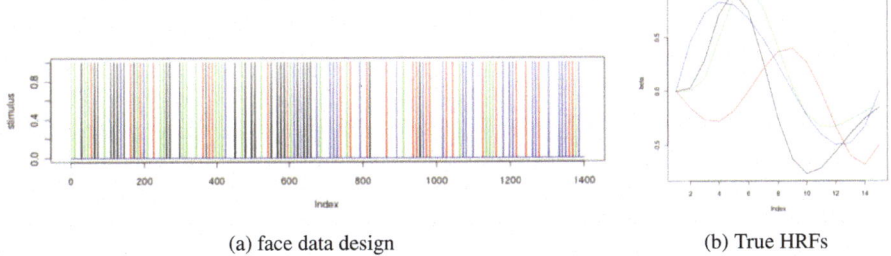

(a) face data design

(b) True HRFs

Figure 4.5: Face data design. There are four event types of interest; the first and second presentations of non-famous and famous faces, which we denote as N1 (in black), N2 (in red), F1 (in green) and F2 (in blue), respectively. (a) shows the timing of the events in the experiment, and (b) shows the respective True HRFs, with which we will compare the estimate results in Figure 4.6.

For the simulation here, we used the face data design but simulated the fMRI data by using the convolution model of the BOLD signal

$$Y(t) = \sum_u s_1(t-u)h_1(u) + \sum_u s_2(t-u)h_2(u) + \sum_u s_3(t-u)h_3(u) + \sum_u s_4(t-u)h_4(u).$$

The four summations responded to the four types of stimuli. Each type of stimuli had its own unique HRF such as $h_1(\cdot), h_2(\cdot), h_3(\cdot), h_4(\cdot)$. The simulation study was conducted by convolving the above stimulus function (experiment design) with HRFs. Since there are four event types, four different HRFs were put to use for each, as shown in Figure 4.5b. The white noise was added over time with SNR=3. As shown in Figure 4.5b, the four true HRFs have different peak timings and amplitudes. The simulation here was to see whether the HRF modeling method could distinguish them and give an unbiased estimate.

Using the simulation data, we applied five HRF modeling methods and plotted their estimates. The four methods in HRF modeling are *transfer function estimate* (TFE), *3dDeconvolution* (AFNI), *smooth finite impulse response* (sFIR), *inverse logit* (IL), and *SPM canonical HRF* (SPM). For the consistency study, we simulated 200 face data sets, applied the four methods for each set, and drew the 200 estimates in one plot.

TFE estimates (Figure 4.6a) show consistency in estimating the HRF by looking at the 200 estimates. The 200 estimates do not show large variation, but we see the zig-zag on the upper and lower boundaries. To improve the individual TFE estimates, we smoothed each HRF estimate from the TFE through LOESS, which is a classic local regression method. The LOESS-smoothed TFE estimates are shown in Figure 4.6b. Compared to Figure 4.6a, the individual HRF did show a smoother estimate, and the consistency is maintained.

Figure 4.6c illustrates the performance of 3dDeconvolution methods in the neuroimaging software AFNI. It is an ordinary linear regression method to estimate HRF by specifying the length of the HRF. Compared to the TFE, it has more variation in the HRF estimates. Additionally, it has bias in the average estimate.

SFIR is similar to AFNI with a penalty on the smoothness. Figure 4.6d illustrates the performance of sFIR. It captured each shape of the four HRFs but might not be able to distinguish them very well. For each HRF, its estimates are biased. The reason that sFIR could not capture the exact shape of each HRF might be due to the comparably large number of types of stimulus involved in the simulation and/or its fixed smooth penalty for any kind of HRF.

Figures 4.6e and 4.6f are the results from IL and SPM. Both of them are basis function methods, which means their results are the linear combinations of their basis functions. For instance, IL estimates belong to the Inverse Logit family, and SPM estimates, to the Gamma family. As the true HRF was not included in any of the two families, their results were biased.

Based on the simulation results of the five methods, the TFE was the method that resulted in a more unbiased and efficient estimate. If the data is too noisy, smoothing the individual HRF is the strategy to use for a more efficient and unbiased estimate.

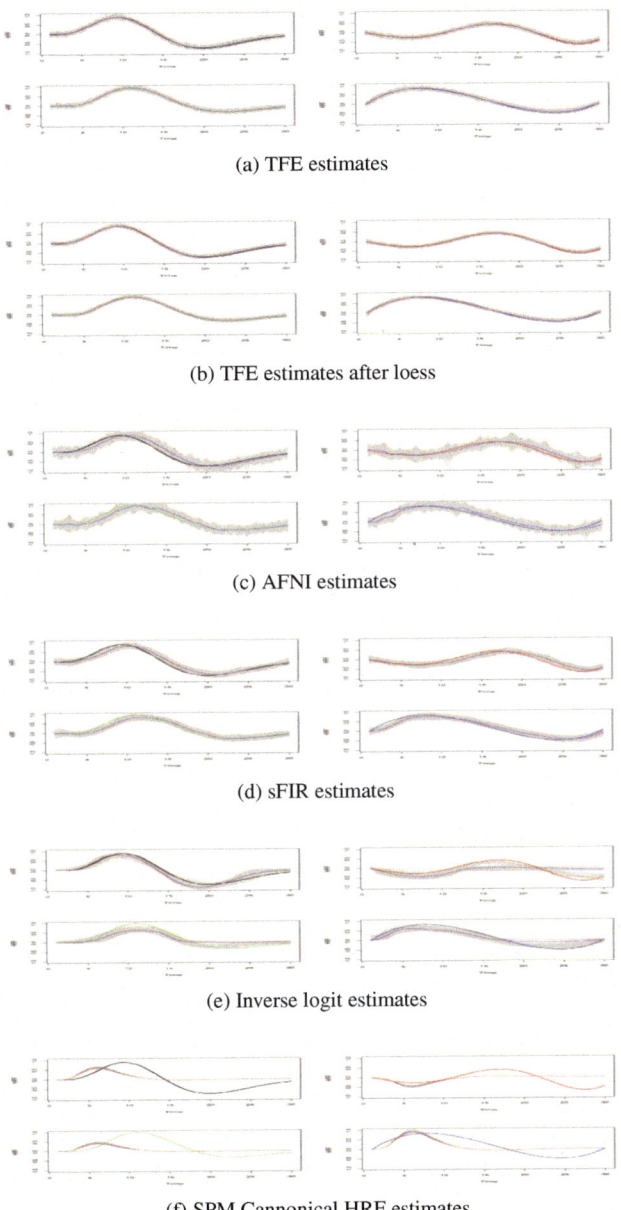

(a) TFE estimates

(b) TFE estimates after loess

(c) AFNI estimates

(d) sFIR estimates

(e) Inverse logit estimates

(f) SPM Cannonical HRF estimates

Figure 4.6: HRF estimates comparison of 200 face data simulations for the experiment shown in Figure 4.5. The 200 TFE estimates are gray lines, and the average of them is purple. The true HRF of four different events are shown in color (black, red, green, and blue in respective plots). There are two thin orange lines close to the upper and lower boundaries indicating the 95% confidence band. You may not see the average line clearly if the average fits the true HRF very well.

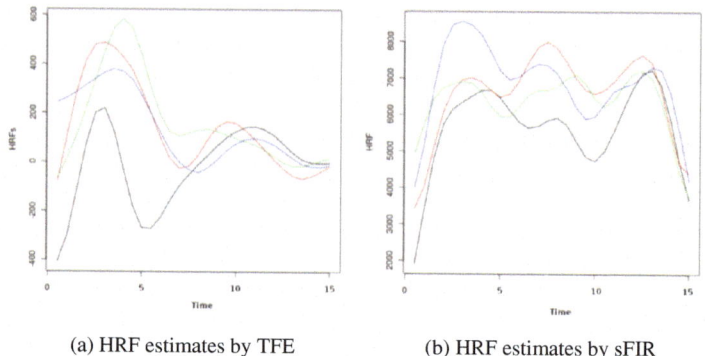

(a) HRF estimates by TFE   (b) HRF estimates by sFIR

Figure 4.7: HRF estimates in one voxel. The voxel is selected by using the highest *t* statistic generated by SPM among the whole brain. The TFE gives a much better estimation in the tail of the HRF than sFIR.

4.4   REAL DATA ANALYSIS

The simulation study on face data has been presented in the simulation section with the comparison among several methods, which gave a very positive demonstration of the performance of the TFE. Here we present the real data analysis from face data sets [21] by using the TFE.

As described in simulation, it is an event-related experiment design (Figure 4.5a) with four kinds of stimulus: the first and second presentation of famous and non-famous faces (F1, F2, N1, N2). So we regard F1, F2, N1, N2 as four types of stimuli with four HRFs to estimate in the experiment.

Here, the HRF estimate from the face data is conducted in the most activated voxel. The voxel is selected by using the highest *t* statistic generated in SPM, which denotes the most activated voxel in the brain. Since it has four types of stimulus, we have four HRFs corresponding to F1, F2, N1, N2 to estimate in this voxel. Additionally, we applied two methods, TFE and sFIR, to the voxel time series. Figure 4.7 is the comparison result. The HRF estimates from the TFE (Figure 4.7a) give a nice shape of HRF that captures the initial peak very well. The HRF estimates from sFIR (Figure 4.7b) have several peaks in the estimates. The possible reason may be due to the aliased HRF. When the events appear in a regular pattern, such as when the first event and second event always present in the same inter-trial intervals (ITI), the single HRF estimate may include the second event information, which makes the multiple peaks in the result.

4.5   SOFTWARE: R

```
# BIV.R

auspe <- function(x, ioptw=4, M=15, logs=F, ci=T)
#-----------------------------------------------------------
#
#  Function to estimate and plot the auto spectral density
#  of a time series x using window ioptw and M
#
#  Arguments:
#    x1, x2:    time series
#    ioptw:     window option 1 ... 8
#    M:         scale parameter for window
#    logs:      flag for logarithmic scale
#    ci:        confidence intervals
#
#  Value: None
#
#-----------------------------------------------------------
{
  x.spe <- spe(x, ioptw, M)
  f11 <- x.spe$f
  crv <- x.spe$c
  n <- length(f11)

  if ( logs )
  {
    f11 <- log(f11)
    fl <- f11-crv
    fu <- f11+crv

    if (ci) ci.plot(f11,fl,fu,y.lab="Log-spectrum")
     else sp.plot(f11,y.lab="Log-spectrum")
  }
  else
  {
    crv <- exp(crv)
    fl <- f11/crv
    fu <- f11*crv

    if (ci) ci.plot(f11,fl,fu,y.lab="Spectrum")
     else sp.plot(f11,y.lab="Spectrum")
  }
```

```
}

sp.plot <- function(f, y.lab="Spectrum", p.type="l")
# ---------------------------------------------------------
#
#    Function for plotting spectral quantities
#
# ---------------------------------------------------------
{
    q <- length(f)
    plot(c(0:(q-1))/(2*(q-1)),f,xlab="frequency",
        ylab=y.lab,type=p.type)
}

ci.plot <- function(f,fl,fu,y.lab="Spectrum")
# ---------------------------------------------------------
#
#    Function for plotting spectral quantities and CI
#
# ---------------------------------------------------------
{
  n <- length(f)
  plot(c(0,.5), c(min(fl), max(fu)), xlab = "frequency",
      ylab = y.lab, type = "n")
  lines(c(0:(n - 1))/(2 * (n - 1)), f)
  lines(c(0:(n - 1))/(2 * (n - 1)), fl, lty=2)
  lines(c(0:(n - 1))/(2 * (n - 1)), fu, lty=2)
}


coher <- function(x1,x2,ioptw,M)
#-----------------------------------------------------------
#
#  Function to estimate the spectra, coherence and phase
#  of bivariate series x1 and x2 using window ioptw and M
#
#  Arguments:
#    x1, x2:    time series
#    ioptw:     window option 1 ... 8
#    M:         scale parameter for window
#
#  Value: None
#
#-----------------------------------------------------------
```

```
{
 f11 <- spe(x1,ioptw,M)$f
 f22 <- spe(x2,ioptw,M)$f
 f12 <- xspe(x1,x2,ioptw,M)
 c12 <- f12$re
 q12 <- f12$im

 amp <- polar(c12,q12)$amp
 phi <- polar(c12,q12)$phase
 coh <- amp^2/(f11*f22)

 par(mfrow=c(2,2))

 sp.plot(f11, y.lab = "Spectrum f11")
 sp.plot(f22, y.lab = "Spectrum f22")

 ilam2 <- c(2,2./3.,.795,.539,.586,1,1.2,1.66)
 n <- length(x1)

 w12 <- sqrt(coh)
 se <- sqrt(abs(1-coh)*M*ilam2[ioptw]/(2*n))
 lo <- w12-2*se
 up <- w12+2*se
 ci.plot(w12, lo, up, y.lab="Coherence")

 se <- amp*sqrt(abs((1/coh)-1)*M*ilam2[ioptw]/(2*n))
 lo <- phi-2*se
 up <- phi+2*se
 ci.plot(phi, lo, up, y.lab="Phase")

 par(mfrow=c(1,1))
 invisible()
}

cohtest <- function(x1,x2,ioptw,M)
#------------------------------------------------------------
#
#  Function to estimate the spectra, coherence and phase
#  of bivariate series x1 and x2 using window ioptw and M
#
#  Arguments:
#    x1, x2:    time series
#    ioptw:     window option 1 ... 8
#    M:         scale parameter for window
```

```
#
#   Value: None
#
#-------------------------------------------------------------
{
    f11 <- spe(x1,ioptw,M)$f
    f22 <- spe(x2,ioptw,M)$f
    f12 <- xspe(x1,x2,ioptw,M)
    c12 <- f12$re
    q12 <- f12$im

    amp <- polar(c12,q12)$amp
    phi <- polar(c12,q12)$phase
    coh <- amp^2/(f11*f22)

#   n <- length(f11)
    n <- length(x1)
    par(mfrow=c(2,2))

    sp.plot(f11, y.lab = "Spectrum f11")
    sp.plot(f22, y.lab = "Spectrum f22")

    ilam2 <- c(2,2./3.,.795,.539,.586,1,1.2,1.66)

    const <- (n - M*ilam2[ioptw])/(M*ilam2[ioptw])
    fstat <- const*coh/(1-coh)
    nc <- 2*const
    ft <- qf(.95,2,nc)
    sp.plot(fstat, y.lab = "F-stat")
    lines(c(0:(n - 1))/(2 * (n - 1)), rep(ft,n),lty=2)

    sp.plot(phi, y.lab = "Phase")
    par(mfrow=c(1,1))
    invisible()
return(list(fstat=fstat, cons=const, f11=f11, f22=f22, coh=coh));
}


crossp <- function(x1,x2,ioptw,M, ci=T)
#-------------------------------------------------------------
#
#   Function to estimate, plot the co and quad spectra
#   of bivariate series x1 and x2 using window ioptw and M
#
#   Arguments:
```

```
#    x1, x2:     time series
#    ioptw:      window option 1 ... 8
#    M:          scale parameter for window
#
#   Value: None
#
#    TIMESLAB (p.338)
#-------------------------------------------------------
{
  f11 <- spe(x1,ioptw,M)$f
  f22 <- spe(x2,ioptw,M)$f
  f12 <- xspe(x1,x2,ioptw,M)
  crv <- f12$c
  c12 <- f12$re
  q12 <- -f12$im

# Compute the avar

  v1 <- f11*f22
  v2 <- c12^2-q12^2
  sd <- matrix(c(sqrt((v1+v2)/2),sqrt((v1-v2)/2)),ncol=2)

  f <- matrix(c(c12,q12),ncol=2)
  n <- length(f[,1])

  par(mfrow=c(1,2))
  for (j in (1:2))
  {
      s <- f[,j]
      se <- sd[,j]*crv
      cil <- s - se
      ciu <- s + se
      if (j==1)
        {
          y.lab <- "Co-spectrum"
        }
      else
        {
          y.lab <- "Quad-spectrum"
        }
      if (ci) ci.plot(s, cil, ciu, y.lab = y.lab)
        else sp.plot(s,y.lab=y.lab)
  }
  par(mfrow=c(1,1))
```

```
   invisible()
}


divpoly <- function(a,b,n)
# ------------------------------------------------
#
#  Find the first n coefficients of
#        (1+sum_1^q b(j)z^j) / (1+sum_1^p a(i)z^i)
#
# ------------------------------------------------
{
  p <- length(a)
  q <- length(b)
  if (p+q == 0) return(rep(0,p+q))
  g <- c(b,rep(0,n-q))
  if (p > 0)
  {
    for (i in 1:n)
    {
      c0 <- g[i]
      for (j in 1:min(i,p))
      {
        c1 <- 1
        if (i-j>0) c1 <- g[i-j]
        c0 <- c0 - a[j]*c1
      }
      g[i] <- c0
    }
  }
  g
}


estfilt <- function(x1,x2,ioptw,M,kn)
#------------------------------------------------------------
#
#  Function to estimate the best filter coefficients
#  of bivariate series x1 and x2 using window ioptw and M.
#  It uses the inverse Fourier transform on A=f21/f11.
#
#  Arguments:
#    x1, x2:     time series
#    ioptw:      window option 1 ... 8
#    M:          scale parameter for window
#    kn+1:        number of beta's
```

```
#
#  Value: beta[1:(kn+1)] (coefficients of the transfer function)
#
#  See Theorem 4.1.6 of TIMESLAB (p.321)
#
#------------------------------------------------------------
{
  f11 <- spe(x1,ioptw,M)$f
  f22 <- spe(x2,ioptw,M)$f
  f21 <- xspe(x2,x1,ioptw,M)
  c21 <- f21$re
  q21 <- f21$im

  c211 <- c21/f11
  q211 <- q21/f11

  q <- length(c21)
  c211 <- c(c211,c211[(q-1):2])      # extend to [0,1] via sym
  q211 <- c(q211,-q211[(q-1):2])     # extend to [0,1] via antisym

   z <- c211 + q211*1i               # Form complex array for FFT

  beta <- Re(fft(z,inv=T))/(2*q-2)
  beta[1:(kn+1)]
}

filtest <- function(x1,x2,ioptw,M,kn)
#------------------------------------------------------------
#
#  Function to estimate the transfer function of a filter
#  of bivariate series x1 and x2 using estfilt and prewhitening
#
#  Arguments:
#    x1, x2:    time series
#    ioptw:     window option 1 ... 8
#    M:         scale parameter for window
#    kn+1:       number of beta's
#
#  Value: coef  (coefficients of the transfer function)
#
#  See TIMESLAB (p.344)
#
#------------------------------------------------------------
{
```

```
   z <- prewhite(x1,x2)
   x1 <- z$x
   x2 <- z$y
   ax <- -z$ax
   ay <- -z$ay

   beta <- estfilt(x1,x2,ioptw,M,kn)

   beta0 <- beta[1]
   beta <- beta/beta0

   kn1 <- kn+1
   gama <- multpoly(beta[2:kn1],ax)
   betay <- divpoly(a=ay,b=c(),40)
   gama <- multpoly(betay,gama)
   gama <- gama*beta0

   list(coef=gama[1:kn], b0=beta0)
}


indtest <- function(x1, x2, nlag=20, alpha=.05)
# ----------------------------------------------------
#
#   To test the independence of two time series
#   using cross-correlations
#
#   Argument: x1, x2, nlag, alpha
#
#   Value:     upper confidence limit
#
# ----------------------------------------------------
{
  z <- prewhite(x1,x2)
  zz <- matrix(c(z$x,z$y),ncol=2)
  zacf <- acf(zz, lag.max=nlag)              # plot acf

  n <- length(z$x)        # for CI
  qnorm( (1+(1-alpha)^(1/(2*nlag+1)))/2) / (n^.5)  # upper CI
}


multpoly <- function(a,b)
# ----------------------------------------------------
#
#   Multiply 1+sum_1^p a(i)z^i and 1+sum_1^q b(j)z^j
```

```
#
# -------------------------------------------------
{
  p <- length(a)
  q <- length(b)
  if (p==0) g = b
  else if (q==0) g=a
  else {
  g <- rep(0,p+q)
  g <- g + c(a,rep(0,q))
  g <- g + c(b,rep(0,p))
  for (i in 1:p)
  {
    for (j in 1:q) g[i+j] <- g[i+j] + a[i]*b[j]
  }
}
  g
}

polar <- function(zr,zi)
#-------------------------------------------------------
#
#  Function for converting to polar representation
#
#-------------------------------------------------------
{
 return(list(amp=sqrt(zr^2+zi^2), phase=atan2(zi,zr)))
}

prewhite <- function(x,y)
# -------------------------------------------------------
#
# Use AR filters to prewhiten and align two time series
#
# -------------------------------------------------------
{
  n <- length(x)
  x <- x - mean(x)
  y <- y - mean(y)

  x.ar <- ar(x)
  y.ar <- ar(y)
  px <- x.ar$order
  py <- y.ar$order
```

```
  if (px == 0) xx <- x
  if (px > 0) {
    xx <- x.ar$resid
    xx <- xx[(px+1):n]
  }

  if (py == 0) yy <- y
  if (py > 0) {
    yy <- y.ar$resid
    yy <- yy[(py+1):n]
  }

  nx <- n - px
  ny <- n - py
  nn <- nx

  if (nx > ny) {
    nn <- ny
    n1 <- nx-ny
    n1 <- n1 + 1
    xx <- xx[n1:nx]
  }

  if (nx < ny) {
    nn <- nx
    n1 <- ny-nx
    n1 <- n1 + 1
    yy <- yy[n1:ny]
  }

  list(n=nn, x=xx, y=yy, ax=x.ar$ar, ay=y.ar$ar)
}

spe <- function(x, ioptw, M, alpha=.05)
#------------------------------------------------------------
#
#   Function to find auto-spectra using scale parameter M,
#   window number ioptw:
#
#   1   Truncated
#   2   Bartlett
#   3   Tukey
#   4   Parzen
```

```
#     5    Bohman
#     6    Daniell
#     7    Bartlett-Priestley
#     8    Parzen-Cogburn-Davis
#
#     The spectral estimator and the constant c used in confidence
#     intervals are returned in a list.
#
#     TIMESLAB (p.185)
#
#-------------------------------------------------------------
{
    n <- length(x)
    if(M >= n || ioptw < 1 || ioptw > 8
        || alpha <= 0. || alpha >= 1.)
    stop("Illegal Input to spe()")

    x.acf <- acf(x,lag.max=n-1,type="cov",plot=F)
    z <- x.acf$acf[,1,1] # c11

    z1 <- rep(0,n)

    if(ioptw == 1) z1[1:(M+1)] <- rep(1, M+1)

    else if(ioptw == 2)  z1[1:(M+1)] <- 1 - (c(0:M) / M)

    else if(ioptw == 3)  z1[1:(M+1)] <- .54+.46*cos(pi*c(0:M)/M)

    else if(ioptw == 4) {
        u  <- c(1:M)/M
        u1 <- u[u<=.5]
        u2 <- u[u>.5]
        z1[1:(M+1)]  <- c(1,1-6*u1^2+6*u1^3,2*(1-u2)^3)
    }

    else if(ioptw == 5) {
        u   <- c(0:M) / M
        piu <- pi * u
        z1[1:(M+1)]  <- (1 - u) * cos(piu) + sin(piu) / pi
    }

    else if(ioptw == 6) {
        piu <- pi * (c(1:(n-1)) / M)
        z1[1:n] <- c(1,sin(piu)/piu)
```

```
    }

    else if(ioptw == 7) {
       piu <- pi * (c(1:(n-1)) / M)
       z1[1:n] <- c(1,3*((sin(piu)/piu) - cos(piu)) / (piu*piu))
    }

    else if(ioptw == 8) {
       u <- c(0:(n-1))/M
       z1[1:n] <- 1 /(1+u^4)
    }

    z <- c(z1[1] * z[1], 2 * z1[2:n] * z[2:n])

    z <- Re(fft(z))
    z <- z[1:((n / 2) + 1)]

    ilam2 <- c(2,2./3.,.795,.539,.586,1,1.2,1.66)
    fac <- qnorm(1-(alpha/2)) * sqrt(M*ilam2[ioptw]/n)

    return(list(f=z,c=fac))

}

xspe <- function(x1, x2, ioptw, M, alpha=.05)
#-------------------------------------------------------------
#
#    Function to find  cross-spectra using scale parameter M,
#    window number ioptw:
#
#    1    Truncated
#    2    Bartlett
#    3    Tukey
#    4    Parzen
#    5    Bohman
#    6    Daniell
#    7    Bartlett-Priestley
#    8    Parzen-Cogburn-Davis
#
#    The spectral estimator and the constant c used in confidence
#    intervals are returned in a list.
#
#    TIMESLAB (p.335)
#
```

```
#-----------------------------------------------------------------
{
   if (length(x1) != length(x2))
     stop("Series with unequal length!")
   n <- length(x1)

   if(M >= n || ioptw < 1 || ioptw > 8
      || alpha <= 0. || alpha >= 1.)
   stop("Illegal Input to xspe()")

 x <- matrix(c(x1,x2),n,2)
   x.acf <- acf(x,lag.max=n-1,type="cov",plot=F)
c12 <- x.acf$acf[,1,2]
c21 <- x.acf$acf[,2,1]

   z1 <- rep(0,n)

   if(ioptw == 1) z1[1:(M+1)] <- rep(1, M+1)

   else if(ioptw == 2)  z1[1:(M+1)] <- 1 - (c(0:M) / M)

   else if(ioptw == 3)  z1[1:(M+1)] <- .54+.46*cos(pi*c(0:M)/M)

   else if(ioptw == 4) {
      u  <- c(1:M)/M
      u1 <- u[u<=.5]
      u2 <- u[u>.5]
      z1[1:(M+1)]  <- c(1,1-6*u1^2+6*u1^3,2*(1-u2)^3)
   }

   else if(ioptw == 5) {
      u  <- c(0:M) / M
      piu <- pi * u
      z1[1:(M+1)]  <- (1 - u) * cos(piu) + sin(piu) / pi
   }

   else if(ioptw == 6) {
      piu <- pi * (c(1:(n-1)) / M)
      z1[1:n] <- c(1,sin(piu)/piu).
   }

   else if(ioptw == 7) {
      piu <- pi * (c(1:(n-1)) / M)
      z1[1:n] <- c(1,3*((sin(piu)/piu) - cos(piu)) / (piu*piu))
```

```
    }

    else if(ioptw == 8) {
       u <- c(0:(n-1))/M
       z1[1:n] <- 1 /(1+u^4)
    }

    z <- fft(z1[1:n]*c12[1:n])+Conj(fft(c(0,z1[2:n]*c21[2:n])))
    z <- z[1:((n / 2) + 1)]

    ilam2 <- c(2,2./3.,.795,.539,.586,1,1.2,1.66)
    fac <- qnorm(1-(alpha/2)) * sqrt(M*ilam2[ioptw]/n)

    return(list(re=Re(z), im=Im(z), c=fac))

}


# WLS.R


filterr <- function(x1,x2,ioptw,M)
#----------------------------------------------------------
#
#  Function to estimate fee: the spectrum of error after OLS
#  of bivariate series x1 and x2 using window ioptw and M.
#
#  Arguments:
#    x1, x2:    time series
#    ioptw:     window option 1 ... 8
#    M:         scale parameter for window
#
#
#  See Theorem 4.1.6 of TIMESLAB (p.321)
#
#----------------------------------------------------------
{
  f11 <- spe(x1,ioptw,M)$f
  f22 <- spe(x2,ioptw,M)$f
  f21 <- xspe(x2,x1,ioptw,M)
  c21 <- f21$re
  q21 <- f21$im
  q <- length(f11)
```

```
  er <- f22-(c21^2+q21^2)/f11

  er <- c(er,er[(q-1):2])

  return(er)
}
```

############### General Methods ###############

```
h.wls.uniform <- function(x1,x2,ioptw,M,K)
{
  n <- length(x1)
  f11 <- spe(x1,ioptw,M)$f
  f22 <- spe(x2,ioptw,M)$f
  f21 <- xspe(x2,x1,ioptw,M)
  c21 <- f21$re
  q21 <- f21$im

  er <- f22-(c21^2+q21^2)/f11

  q <- length(er)
  # even: extend to [0,1] via sym
  fee <- c(er,er[(q-1):2])

  # even: extend to [0,1] via sym
  f11 <- c(f11,f11[(q-1):2])
  f11 <- f11/fee
  f11r <- Re(f11)
  f11i <- Im(f11)

  c21 <- c(c21,c21[(q-1):2])
  # even: extend to [0,1] via antisym
  q21 <- c(q21,-q21[(q-1):2])
  f21 <- c21 + q21*1i
  f21 <- f21/fee
  f21r <- Re(f21)
  f21i <- Im(f21)

  f11r <- c(f11r[(n-K+1):n], f11r, f11r[1:K])
  f110r <- 1/(2*K+1)*filter(f11r, rep(1, 2*K+1))[(K+1):(K+n)]
  f11i <- c(f11i[(n-K+1):n], f11i, f11i[1:K])
```

```
  f110i <- 1/(2*K+1)*filter(f11i, rep(1, 2*K+1))[(K+1):(K+n)]
  f110 <- f110r+f110i*1i

  f21r <- c(f21r[(n-K+1):n], f21r, f21r[1:K])
  f210r <- 1/(2*K+1)*filter(f21r, rep(1, 2*K+1))[(K+1):(K+n)]
  f21i <- c(f21i[(n-K+1):n], f21i, f21i[1:K])
  f210i <- 1/(2*K+1)*filter(f21i, rep(1, 2*K+1))[(K+1):(K+n)]
  f210 <- f210r+f210i*1i

  f211 <- f210/f110

  beta <- Re(fft(f211, inv=T))/(2*q-2)
  beta[1:(kn+1)]

}


h.wls <- function(x1,x2,ioptw, M1, M, kn)
{
  f11 <- spe(x1,ioptw,M1)$f
  f22 <- spe(x2,ioptw,M1)$f
  f21 <- xspe(x2,x1,ioptw,M1)
  c21 <- f21$re
  q21 <- f21$im
  q <- length(c21)

  er <- f22-(c21^2+q21^2)/f11

  fee <- c(er,er[(q-1):2])

  # even: extend to [0,1] via sym
  f11 <- c(f11,f11[(q-1):2])
  f11 <- f11/fee
  f110 <- smooth.fre(f11, ioptw, M)

  c21 <- c(c21,c21[(q-1):2])
  # even: extend to [0,1] via antisym
  q21 <- c(q21,-q21[(q-1):2])
  f21 <- c21 + q21*1i
  f21 <- f21/fee
  f210 <- smooth.fre(f21, ioptw, M)

  f211 <- f210/f110
```

```
  n <- length(f211)
  beta <- Re(fft(f211, inv=T))/n
  beta[1:(kn+1)]

}


smooth.fre <- function(fxy, ioptw, M)
######################################
#
# fxy is the spectrum, even
#
######################################
{
   n <- length(fxy)

   z1 <- rep(0,n)
   z2 <- 1:(n-1)
   z2 <- z2/n
   z3 <- 0:(n-1)
   z3 <- z3/n

   if(ioptw == 1) z1 <- c(2*M, sin(2*pi*M*z2)/pi/z2)

   else if(ioptw == 2)  z1 <- c(M, sin(pi*M*z2)^2/pi^2/M/z2^2)

   else if(ioptw == 3)
       z1 <- c(M, sin(2*pi*M*z2)/pi/z2/2/(1-4*z2^2*M^2))

   else if(ioptw == 4) {
      z1 <- c(3*M/4, 3*M/4*(sin(pi*M*z2/2)/(pi*M*z2/2))^4)
   }

   else if(ioptw == 5) {
      z1 <- 4*M/pi^2/(1-4*M^2*z3^2)^2*(1+cos(2*pi*M*z3))
   }

   else if(ioptw == 6) {
      if(2*M > n) stop("M should be smaller than n/2!")
      z1[z3 <= 1/2/M] <- M
      z1[(1-z3) <= 1/2/M] <- M
   }

   else if(ioptw == 7) {
```

```
      if(2*M > n) stop("M should be smaller than n/2!")
      z4 <- z3[z3 <= 1/2/M]
      z1[z3 <= 1/2/M] <- 3*M/2*(1-(2*M*z4)^2)
      z4 <- z3[(1-z3)<= 1/2/M]
      z1[(1-z3)<= 1/2/M] <- 3*M/2*(1-(2*M*z4)^2)
   }

   else if(ioptw == 8) {
      z4 <- z3[1: ((n/2)+1)]
      z1[1:((n/2)+1)] <- pi*M*exp(-abs(2*pi*M*z4)/sqrt(2))
                          *sin(abs(2*pi*M*z4)/sqrt(2)+pi/4)
      z4 <- z3[(n/2+2):n]-1
      z1[(n/2+2):n] <- pi*M*exp(-abs(2*pi*M*z4)/sqrt(2))
                          *sin(abs(2*pi*M*z4)/sqrt(2)+pi/4)
   }

   rfxy <- Re(fxy)
   ifxy <- Im(fxy)

   rs <- convolve(rfxy, z1, type="c")
   ri <- convolve(ifxy, z1, type="c")

   return(rs+ri*1i)
}




################## HRF plot ###################

hwest <- function(x, Y, ioptw, M, kn) ##Y is matrix
{
 sn <- dim(Y)[2]
 bhat <- NULL
   for(i in 1:sn)
     {
      bh <- h.wls(x, Y[, i], ioptw, M, kn)
      bhat <- cbind(bhat, bh)
     }
 return(bhat)
}

hwplot <- function(x, Y, ioptw, M, kn, ylim=FALSE)
```

```
{
 sn <- dim(Y)[2]
 bhat <- hwest(x, Y, ioptw, M, kn)
 if(ylim[1]==FALSE)
     plot(beta, ylim=c(min(c(-1,bhat)), max(c(1,bhat))), typ='n')
 else plot(beta, ylim=ylim, typ='n')
 for(i in 1:sn) lines(bhat[, i], col="grey")
 lines(beta, lwd=2)
 lines(apply(bhat, 1, mean), col='red', lwd=2)
 quan <- apply(bhat, 1, quan025)
 lines(quan[1,], col='blue')
 lines(quan[2,], col='blue')
 title(paste('Estimated HRF WLS',': M=', M, '; Sig=', sig))
}

# fcn1.R

###################################################################
###Functions. Three parts: simulation, estimation, and plot.
###################################################################



library(calibrate)



################# Simulation ################
### blocks design ################

##complete periodic design Without padding
eres1 <- function(sig=0, blocksize=2, rest=16, n=16, plot=TRUE)
## sig is to add noise (sd); rest is the time span for rest
## blocksize is the length for stimuli
## n is the repeats
## plot=TRUE draws the simulation pattern while simulating
{
  gm <- function(t, a=1.0, c=5.5, b=5, d=0.9){
(a*(t-c)**b)*exp(-(t-c)/d) }
t <- 0:14
g1 <- gm(t, c=0)
g2 <- gm(t, a=0.4, c=0, b=12, d=0.7)
a1 <- max(g1); a2 <- max(g2)
beta <- ((g1/a1) - (g2/a2))
```

```
k <- length(beta)
z <- c(rep(0, k), rep(rep(1:0, c(blocksize, rest)), n))
fz <- filter(z, beta, sides=1)
l <- length(fz)
fz <- fz[(k+1):l]
fz <- fz+sig*rnorm(fz)
z <-   z[(k+1):l]
if(plot==TRUE)
{ts.plot(fz)
lines(z, col=2, lty=2)}
return(list(z=z, fz=fz))
}



##With padding in the tails
eres2 <- function(sig=0, blocksize=2, rest=16, n=16,
                     tail=14, plot=TRUE)
## sig is to add noise; rest is the time span for rest in period
## blocksize is the length for stimuli
## n is the repeats
## plot=TRUE draws the simulation pattern while simulating
## Tail is the length of padding
{
  gm <- function(t, a=1.0, c=5.5, b=5, d=0.9){
(a*(t-c)**b)*exp(-(t-c)/d) }
t <- 0:14
g1 <- gm(t, c=0)
g2 <- gm(t, a=0.4, c=0, b=12, d=0.7)
a1 <- max(g1); a2 <- max(g2)
beta <- ((g1/a1) - (g2/a2))
k <- length(beta)
z <- c(rep(0, k), rep(rep(1:0, c(blocksize, rest)), n),
           rep(0, tail))
fz <- filter(z, beta, sides=1)
l <- length(fz)
fz <- fz[(k+1):l]
fz <- fz+sig*rnorm(fz)
z <-   z[(k+1):l]
if(plot==TRUE)
{ts.plot(fz)
lines(z, col=2, lty=2)}
return(list(z=z, fz=fz))
}
```

```
##Choose adding locations and length
eres3 <- function(sig=0, blocksize=1, rest=16, n=16, restN=1,
            vocation=17, plot=TRUE)
## sig is to add noise; rest is the time span for rest;
## blocksize is the length for stimuli
## n is the repeats
## plot=TRUE draws the simulation pattern while simulating
## restN is the location of padding
## vocation specifies the length of the vocation period
{
  gm <- function(t, a=1.0, c=5.5, b=5, d=0.9){
(a*(t-c)**b)*exp(-(t-c)/d) }
t <- 0:14
g1 <- gm(t, c=0)
g2 <- gm(t, a=0.4, c=0, b=12, d=0.7)
a1 <- max(g1); a2 <- max(g2)
beta <- ((g1/a1) - (g2/a2))
k <- length(beta)
z <- c(rep(0, k), rep(rep(1:0, c(blocksize, rest)), restN),
        rep(0, vocation), rep(rep(1:0, c(blocksize, rest)),
        n-restN))
fz <- filter(z, beta, sides=1)
l <- length(fz)
fz <- fz[(k+1):l]
fz <- fz+sig*rnorm(fz)
z <-  z[(k+1):l]
if(plot==1)
{ts.plot(fz)
lines(z, col=2, lty=2)}
return(list(z=z, fz=fz))
}



# Examples of HRF
# 1. part of sin(x)/x function
# x <- seq(-2,-.6,.1)
# beta <- .5*sin(2*pi*x)/x
# ts.plot(.5*sin(2*pi*x)/x)
# 2. Buckner
# delta <- 2.5; tau <- 1.25; sigma <- 0; ng <- 2
# # delta <- 2.05; tau <- 1.08; sigma <- 0.92; ng <- 3
```

```
# hrf <- function(x) ((x-delta)/tau)^ng*exp(-(x-sigma)/tau)
# beta <- hrf(seq(0,14)) # see Friston
# 3. splines
# beta <- c(-.25*seq(0,4)/4,1.25*seq(5,8)/4-1.5,-1.25*seq(9,14)/6
#                  +(1+1.25*8/6),
# .25*seq(15,16)/2-2)
# 4. simple beta
# beta <- c(.7,.1)
#
# 5. Glover's HRF
gm <- function(t, a=1.0, c=5.5, b=5, d=0.9){
(a*(t-c)**b)*exp(-(t-c)/d) }
glover <- function(){
t <- 0:14
g1 <- gm(t, c=0)
g2 <- gm(t, a=0.4, c=0, b=12, d=0.7)
a1 <- max(g1); a2 <- max(g2)
beta <- ((g1/a1) - (g2/a2)); return(beta)}




#### different HRFs
eres_beta <- function(sig=0, blocksize=1, rest=16, n=16, restN=1,
               vocation=15, plot=TRUE, beta)
## sig is to add noise; rest is the time span for rest
## beta is the HRF
{
k <- length(beta)
z <- c(rep(0, k), rep(rep(1:0, c(blocksize, rest)), restN),
           rep(0, vocation), rep(rep(1:0, c(blocksize, rest)),
           n-restN))
fz <- filter(z, beta, sides=1)
l <- length(fz)
fz <- fz[(k+1):l]
fz <- fz+sig*rnorm(fz)
z <-  z[(k+1):l]
if(plot==1)
{ts.plot(fz)
lines(z, col=2, lty=2)}
return(list(z=z, fz=fz))
}
```

```
####Event-related Design simulation
 ergroup <-  function(sig, rest, n)
{
  gm <- function(t, a=1.0, c=5.5, b=5, d=0.9){
(a*(t-c)**b)*exp(-(t-c)/d) }
t <- 0:14
g1 <- gm(t, c=0)
g2 <- gm(t, a=0.4, c=0, b=12, d=0.7)
a1 <- max(g1); a2 <- max(g2)
beta <- ((g1/a1) - (g2/a2))
k <- length(beta)
   ind <- 0
   r0 <- rep(0,rest+1)
   r1 <- rep(0:1,c(rest,1))
   p <- runif(1)
   if (p < .5) {x1 <- r0; begin <- 1} else {x1 <- r1; begin <- 0}
   for (i in 2:n) {
p <- runif(1)
if ( p<.5) {
x1 <- c(x1, r0)
ind <- ind+1
} else {
x1 <- c(x1, r1)
}
}
   x1 <- c(x1, rep(0,k))
   z <- x1
fz <- filter(z, beta, sides=1)
fz <- fz[(k+1):(n*(rest+1)+k)]
fz <- fz+sig*rnorm(n*(rest+1))
z <-  z[(k+1):(n*(rest+1)+k)]
par(mfrow=c(2, 1))
ts.plot(z)
ts.plot(fz)
return(list(z=z, fz=fz, begin=begin, ind=ind))

}


 ergroup2 <-  function(rest, n, q)
{
k <- length(beta)
   ind <- 0
   r0 <- rep(0,rest+1)
```

```
    r1 <- rep(0:1,c(rest,1))
    p <- runif(1)
    if (p < q) {x1 <- r0; begin <- 1} else {x1 <- r1; begin <- 0}
    for (i in 2:n) {
p <- runif(1)
if ( p<q) {
x1 <- c(x1, r0)
ind <- ind+1
} else {
x1 <- c(x1, r1)
}
}
return(x1)
}


eres <- function(stim, beta, sig, pl=TRUE)
{
k <- length(beta)
z <- c(rep(0, k), stim)
fz <- filter(z, beta, sides=1)
l <- length(fz)
fz <- fz[(k+1):l]
fz <- fz+sig*rnorm(fz)
z <-  z[(k+1):l]
if(pl==TRUE)
{ts.plot(fz)
lines(z, col=2, lty=2)}
return(list(z=z, fz=fz))
}


############### Estimation #################
##cohtest without drawing
##return freq
fcohtest <- function(z, fz, ioptw, M)
{ cohtest.no(z, fz, ioptw, M) -> dats
length(dats$f11) -> q
c(0:(q-1)/(2*(q-1))) -> fre
for(i in 1:(q-1))
{
if(dats$f11[i]>=dats$f11[i+1]) break
}
ind2 <- dats$f22==max(dats$f22)
if(i==q-1) fre1=fre[dats$f11==max(dats$f11)]
```

```
else fre1=fre[i]
return(list(fre1=fre1, fre2=fre[ind2], coh=dats$coh[i],
        fstat=dats$fstat[i], ft=dats$ft))
}


fcohtest.p <- function(z, fz, ioptw, M)
{ cohtest.no(z, fz, ioptw, M) -> dats
length(dats$f11) -> q
c(0:(q-1)/(2*(q-1))) -> fre
ind <- dats$f11==max(dats$f11)
return(list(fre=fre[ind], coh=dats$coh[ind],
        fstat=dats$fstat[ind], ft=dats$ft))
}




Ffre <- function(z, fz, ioptw, M)
{
cohtest.no(z, fz, ioptw, M) -> dats
length(dats$f11) -> q
c(0:(q-1)/(2*(q-1))) -> fre
return(fre)
}


coher.label <- function(x1,x2,ioptw,M)
#-------------------------------------------------------------
#
#   Function to estimate the spectra, coherence and phase
#   of bivariate series x1 and x2 using window ioptw and M
#
#   Arguments:
#     x1, x2:    time series
#     ioptw:     window option 1 ... 8
#     M:         scale parameter for window
#
#   Value: None
#   Need library(calibrate)
#
#-------------------------------------------------------------
{
  f11 <- spe(x1,ioptw,M)$f
  f22 <- spe(x2,ioptw,M)$f
```

```
 f12 <- xspe(x1,x2,ioptw,M)
 c12 <- f12$re
 q12 <- f12$im

 amp <- polar(c12,q12)$amp
 phi <- polar(c12,q12)$phase
 coh <- amp^2/(f11*f22)

 par(mfrow=c(2,2))
 label <- fcohtest(x1, x2, ioptw, M)

 sp.plot(f11, y.lab = "Spectrum f11")
  axis(1, label$fre1, round(label$fre1, 3))
 abline(v=label$fre1, col='grey70')
 textxy(label$fre1, max(f11)/2, label$fre1)
 sp.plot(f22, y.lab = "Spectrum f22")
  axis(1, label$fre2, round(label$fre2, 3))
 abline(v=label$fre2, col='grey70')

 ilam2 <- c(2,2./3.,.795,.539,.586,1,1.2,1.66)
 n <- length(x1)

 w12 <- sqrt(coh)
 se <- sqrt(abs(1-coh)*M*ilam2[ioptw]/(2*n))
 lo <- w12-2*se
 up <- w12+2*se
 ci.plot(w12, lo, up, y.lab="Coherence")
 axis(1, label$fre1, round(label$fre1, 3))
 abline(v=label$fre1, col='grey70')
 textxy(label$fre1, label$coh, label$coh)

 se <- amp*sqrt(abs((1/coh)-1)*M*ilam2[ioptw]/(2*n))
 lo <- phi-2*se
 up <- phi+2*se
 ci.plot(phi, lo, up, y.lab="Phase")

 par(mfrow=c(1,1))
 invisible()
}

cohtest.label <- function(x1,x2,ioptw,M)
#-----------------------------------------------------------
#
# Function to estimate the spectra, coherence and phase
```

```
#   of bivariate series x1 and x2 using window ioptw and M
#
#   Arguments:
#     x1, x2:    time series
#     ioptw:     window option 1 ... 8
#     M:         scale parameter for window
#
#   Value: None
#   Need library(calibrate)
#
#-------------------------------------------------------
{
  f11 <- spe(x1,ioptw,M)$f
  f22 <- spe(x2,ioptw,M)$f
  f12 <- xspe(x1,x2,ioptw,M)
  c12 <- f12$re
  q12 <- f12$im

  amp <- polar(c12,q12)$amp
  phi <- polar(c12,q12)$phase
  coh <- amp^2/(f11*f22)

#  n <- length(f11)
  n <- length(x1)
  par(mfrow=c(2,2))
  label <- fcohtest(x1, x2, ioptw, M)

  sp.plot(f11, y.lab = "Spectrum f11")`
   axis(1, label$fre1, round(label$fre1, 3))
 abline(v=label$fre1, col='grey70')
 textxy(label$fre1, max(f11)/2, label$fre1)
  sp.plot(f22, y.lab = "Spectrum f22")
   axis(1, label$fre2, round(label$fre2, 3))
 abline(v=label$fre2, col='grey70')

  ilam2 <- c(2,2./3.,.795,.539,.586,1,1.2,1.66)

  const <- (n - M*ilam2[ioptw])/(M*ilam2[ioptw])
  fstat <- const*coh/(1-coh)
  nc <- 2*const
  ft <- qf(.95,2,nc)
  sp.plot(fstat, y.lab = "F-stat")
  lines(c(0:(n - 1))/(2 * (n - 1)), rep(ft,n),lty=2)
  axis(1, label$fre1, round(label$fre1, 3))
```

```
abline(v=label$fre1, col='grey70')
textxy(label$fre1, label$fstat, label$fstat)


  sp.plot(phi, y.lab = "Phase")
  par(mfrow=c(1,1))
  invisible()
return(list(fstat=fstat, cons=const, f11=f11, f22=f22, coh=coh));
}


co.label <- function(x1,x2,ioptw,M)
#-----------------------------------------------------------
#
#  Function to estimate the spectra, coherence and phase
#  of bivariate series x1 and x2 using window ioptw and M
#
#  Arguments:
#    x1, x2:    time series
#    ioptw:     window option 1 ... 8
#    M:         scale parameter for window
#
#  Value: None
#  Need library(calibrate)
#
#-----------------------------------------------------------
{
 f11 <- spe(x1,ioptw,M)$f
 f22 <- spe(x2,ioptw,M)$f
 f12 <- xspe(x1,x2,ioptw,M)
 c12 <- f12$re
 q12 <- f12$im

 amp <- polar(c12,q12)$amp
 phi <- polar(c12,q12)$phase
 coh <- amp^2/(f11*f22)

 par(mfrow=c(2,2))
 label <- fcohtest(x1, x2, ioptw, M)

 sp.plot(f11, y.lab = "Spectrum f11")
  axis(1, label$fre1, round(label$fre1, 3))
 abline(v=label$fre1, col='grey70', lty=2)
 textxy(label$fre1, max(f11)/2, label$fre1)
```

```
sp.plot(f22, y.lab = "Spectrum f22")
 axis(1, label$fre2, round(label$fre2, 3))
abline(v=label$fre2, col='grey70', lty=2)

ilam2 <- c(2,2./3.,.795,.539,.586,1,1.2,1.66)
n <- length(x1)

w12 <- sqrt(coh)
se <- sqrt(abs(1-coh)*M*ilam2[ioptw]/(2*n))
lo <- w12-2*se
up <- w12+2*se
ci.plot(w12, lo, up, y.lab="Coherence")
axis(1, label$fre1, round(label$fre1, 3))
abline(v=label$fre1, col='grey70', lty=2)
textxy(label$fre1, label$coh, label$coh)

 const <- (n - M*ilam2[ioptw])/(M*ilam2[ioptw])
 fstat <- const*coh/(1-coh)
 nc <- 2*const
 ft <- qf(.95,2,nc)
 sp.plot(fstat, y.lab = "F-stat")
 lines(c(0:(n - 1))/(2 * (n - 1)), rep(ft,n),lty=2)
 axis(1, label$fre1, round(label$fre1, 3))
abline(v=label$fre1, col='grey70', lty=2)
textxy(label$fre1, label$fstat, label$fstat)


par(mfrow=c(1,1))
invisible()
}




cohtest.no <- function(x1,x2,ioptw,M)
#---------------------------------------------------------
#
# Function to estimate the spectra, coherence and phase
# of bivariate series x1 and x2 using window ioptw and M
#
# Arguments:
#   x1, x2:    time series
#   ioptw:     window option 1 ... 8
#   M:         scale parameter for window
#
```

```
#   Value: None
#
#------------------------------------------------------------
{
   f11 <- spe(x1,ioptw,M)$f
   f22 <- spe(x2,ioptw,M)$f
   f12 <- xspe(x1,x2,ioptw,M)
   c12 <- f12$re
   q12 <- f12$im

   amp <- polar(c12,q12)$amp
   phi <- polar(c12,q12)$phase
   coh <- amp^2/(f11*f22)

#   q <- length(f11)
   n <- length(x1)

   ilam2 <- c(2,2./3.,,.795,.539,.586,1,1.2,1.66)

   const <- (n - M*ilam2[ioptw])/(M*ilam2[ioptw])
   fstat <- const*coh/(1-coh)
   nc <- 2*const
   ft <- qf(.95,2,nc)

return(list(fstat=fstat, cons=const, f11=f11, f22=f22,
            coh=coh, ft=ft));
}




cohtest.fre <- function(x1,x2,ioptw,M, fre)
#------------------------------------------------------------
#
#   Function to estimate the spectra, coherence and phase
#   of bivariate series x1 and x2 using window ioptw and M
#
#   Arguments:
#     x1, x2:     time series
#     ioptw:      window option 1 ... 8
#     M:          scale parameter for window
#
#   Value: None
```

```
#
#-------------------------------------------------------------
{
  f11 <- spe(x1,ioptw,M)$f
  f22 <- spe(x2,ioptw,M)$f
  f12 <- xspe(x1,x2,ioptw,M)
  c12 <- f12$re
  q12 <- f12$im

  amp <- polar(c12,q12)$amp
  phi <- polar(c12,q12)$phase
  coh <- amp^2/(f11*f22)

  q <- length(f11)
  n <- length(x1)

  ilam2 <- c(2,2./3.,.795,.539,.586,1,1.2,1.66)

  const <- (n - M*ilam2[ioptw])/(M*ilam2[ioptw])
  fstat <- const*coh/(1-coh)
  nc <- 2*const
  ft <- qf(.95,2,nc)

  c(0:(q-1)/(2*(q-1))) -> allfre
  ind <- which(allfre==fre)

  return(list(coh=coh[ind], fstat=fstat[ind], ft=ft, nc=nc))
}



###### Estimate the optimal M
Mest <- c(seq(1, 50, 2), seq(51, 248, 10))
Mopt <- function(x, y, Mest, ioptw, kn)
{ mse <- NA
  for(i in Mest)
  {
    g <- filtest(x,y,ioptw,M=i,kn)
    bhat <- c(g$b0,g$coef)
    ms <- mean((beta-bhat)^2) ##unbiased
    mse <- c(mse, ms)
  }
  mse <- mse[-1]
```

```
   plot(Mest, mse, typ='l')
   return(Mest[mse==min(mse)])
}


optM <- function(x, y, Mest, ioptw, kn)
{ mse <- NA
  for(i in Mest)
  {
   g <- estfilt(x,y,ioptw,M=i,kn)
   bhat <- g
   ms <- mean((beta-bhat)^2) ##unbiased
   mse <- c(mse, ms)
  }
  mse <- mse[-1]
  plot(Mest, mse, typ='l')
  return(Mest[mse==min(mse)])
}




#########without noise
######## the effect of M
Meffect <- function(z, fz, Mvec, ioptw, kn)
{
 bhat <- beta
 for(i in Mvec)
  {
   g <- filtest(z, fz, ioptw, M=i, kn)
   bh <- c(g$b0, g$coef)
   bhat <- cbind(bhat, bh)
  }
 return(bhat[,-1])
}




effectM <- function(z, fz, Mvec, ioptw, kn)
{
 bhat <- beta
 for(i in Mvec)
  {
   g <- estfilt(z, fz, ioptw, M=i, kn)
   bh <- g
   bhat <- cbind(bhat, bh)
  }
 return(bhat[,-1])
```

```
}



################ When we add noise, then estimate HRF
HRFest <- function(x, Y, M, ioptw, kn) ##Y is matrix
{
 sn <- dim(Y)[2]
 bhat <- NULL
   for(i in 1:sn)
     {
      g <- filtest(x, Y[, i], ioptw, M, kn)
      bh <- c(g$b0, g$coef)
      bhat <- cbind(bhat, bh)
     }
 return(bhat)
}

estHRF <- function(x, Y, M, ioptw, kn) ##Y is matrix
{
 sn <- dim(Y)[2]
 bhat <- NULL
   for(i in 1:sn)
     {
      g <- estfilt(x, Y[, i], ioptw, M, kn)
      bh <- g
      bhat <- cbind(bhat, bh)
     }
 return(bhat)
}



estHRF.norm <- function(x, Y, M, ioptw, kn) ##Y is matrix
{
 sn <- dim(Y)[2]
 bhat <- NULL
   for(i in 1:sn)
     {
      g <- estfilt(x, Y[, i], ioptw, M, kn)
      bh <- g/sqrt(sum(g^2))
      bhat <- cbind(bhat, bh)
     }
 return(bhat)
}
```

```
## If Y is matrix
## return M
MoptY <- function(x, Y, Mest, ioptw, kn)
{
 sn <- dim(Y)[2]
 bm <- matrix(rep(beta, sn), ,sn)
 mse <- NA
 for(i in Mest)
  {
   bhat <- HRFest(x, Y, i, ioptw, kn)
   ms <- mean((bhat-bm)^2)
   mse <- c(mse, ms)
  }
 mse <- mse[-1]
 return(Mest[mse==min(mse)])
}


# Smallest MSE estimation

MSEsmall <- function(x, Y, M, ioptw, kn)
{
 sn <- dim(Y)[2]
 bm <- matrix(rep(beta, sn), ,sn)
 mse <- NULL
 bhat <- HRFest(x, Y, M, ioptw, kn)
 ms <- apply((bhat-bm)^2, 2, mean)
 return(bhat[, ms==min(ms)])
}


optMY <- function(x, Y, Mest, ioptw, kn)
## Y is matrix
## return M
{
 sn <- dim(Y)[2]
 bm <- matrix(rep(beta, sn), ,sn)
 mse <- NULL
 for(i in Mest)
  {
   bhat <- estHRF(x, Y, i, ioptw, kn)
   ms <- mean((bhat-bm)^2)
```

```
   mse <- c(mse, ms)
  }
 return(Mest[mse==min(mse)])
}

smallMSE <- function(x, Y, M, ioptw, kn)
{
 sn <- dim(Y)[2]
 bm <- matrix(rep(beta, sn), ,sn)
 mse <- NULL
 bhat <- estHRF(x, Y, M, ioptw, kn)
 ms <- apply((bhat-bm)^2, 2, mean)
 return(bhat[, ms==min(ms)])
}




################ Plot ###############
###For every sigma, we have optimal M
###Then, draw the plot, and return the values

quan025 <- function(x) quantile(x, probs=c(.025, .975), type=4)


HRFplot <- function(x, Y, Mstar, ioptw, kn,
                    ylim=FALSE, beta=beta)
{
 sn <- dim(Y)[2]
 bhat <- HRFest(x, Y, Mstar, ioptw, kn)
 if(ylim[1]==FALSE) plot(beta, ylim=c(min(c(-1,bhat)),
        max(c(1,bhat))), typ='n')
 else plot(beta, ylim=ylim, typ='n')
 for(i in 1:sn) lines(bhat[, i], col="grey")
 lines(beta, lwd=.5)
 lines(apply(bhat, 1, mean), col='red', lwd=2)
 quan <- apply(bhat, 1, quan025)
 lines(quan[1,], col='red')
 lines(quan[2,], col='red')
 title(paste('Estimated HRF',': M=', Mstar, '; Sig=', sig))
}

plotHRF <- function(x, Y, Mstar, ioptw, kn, ylim=FALSE,
                    beta=beta, tit=TRUE)
{
```

```
 sn <- dim(Y)[2]
 bhat <- estHRF(x, Y, Mstar, ioptw, kn)
 if(ylim[1]==FALSE) plot(beta, ylim=c(min(c(-1,bhat)),
        max(c(1,bhat))), typ='n', ylab="HRF", xlab="Time")
 else plot(beta, ylim=ylim, typ='n', ylab="HRF", xlab="Time")
 for(i in 1:sn) lines(bhat[, i], col="grey")
# lines(smallMSE(x, Y, Mstar, ioptw, kn), col='orange')
 lines(beta, lwd=.5)
 lines(apply(bhat, 1, mean), col='red', lwd=2)
 quan <- apply(bhat, 1, quan025)
 lines(quan[1,], col='red')
 lines(quan[2,], col='red')
 if(tit==TRUE) title(paste('Estimated HRF',': M=',
                           Mstar, '; Sig=', sig))
}


plotHRF.norm <- function(x, Y, Mstar, ioptw, kn, ylim=FALSE)
{
 sn <- dim(Y)[2]
 bhat <- estHRF.norm(x, Y, Mstar, ioptw, kn)
 bet <- beta/sqrt(sum(beta^2))
 if(ylim[1]==FALSE) plot(bet, ylim=c(min(c(bet,bhat)),
        max(c(bet,bhat))), typ='n')
   else plot(bet, ylim=ylim, typ='n')
 for(i in 1:sn) lines(bhat[, i], col="grey")
 lines(beta, lwd=2)
 lines(apply(bhat, 1, mean), col='red', lwd=2)
 quan <- apply(bhat, 1, quan025)
 lines(quan[1,], col='blue')
 lines(quan[2,], col='blue')
 title(paste('Estimated HRF',': M=', Mstar, '; Sig=', sig))
}


hrfdraw <- function(esthat, yrange, beta=beta, color, lty)
 {
 ndat <- dim(esthat)[2]
 plot(beta, ylim=yrange, typ='n', ylab="HRF", xlab="Time")
 for(i in 1:ndat) lines(esthat[, i], col="grey")
 lines(beta, lwd=2, col=color, lty=lty)
 lines(apply(esthat, 1, mean), col='purple', lwd=2)
 quan <- apply(esthat, 1, quan025)
 lines(quan[1,], col='orange')
```

```
lines(quan[2,], col='orange')
}


drawpoly <- function(hhat, yrange, beta, color, lty)
{plot(beta, ylim=yrange, typ='n')
quan <- apply(hhat, 1, quan025)
x <- 1:dim(hhat)[1]
xx <- c(x, rev(x))
yy <- c(quan[1,], rev(quan[2,]))
polygon(xx, yy, col="gray", border=NA, ylab="HRF", xlab="Time")
lines(beta, lwd=2, col=color, lty=lty)
lines(apply(hhat, 1, mean), col='red', lwd=2)}


############ without smooth
estfilt.nonsmooth <- function(x1,x2,ioptw,M,kn, plot=FALSE)
#-----------------------------------------------------------
#
#  Function to estimate the best filter coefficients
#  of bivariate series x1 and x2 using window ioptw and M.
#  It uses the inverse Fourier transform on A=f21/f11.
#
#  Arguments:
#    x1, x2:     time series
#    ioptw:      window option 1 ... 8
#    M:          scale parameter for window
#    kn+1:        number of beta's
#
#  Value: beta[1:(kn+1)] (coefficients of the transfer function)
#
#  See Theorem 4.1.6 of TIMESLAB (p.321)
#
#-----------------------------------------------------------
{
  par(mfrow=c(2, 2))
  sp <- spectrum(cbind(x1, x2), plot=plot)
  if(plot)
  { plot(sp, plot.type="coherency")
    plot(sp, plot.type="phase")}
  f11 <- sp$spec[, 1]
  f22 <- sp$spec[, 2]
  f21 <- complex(modulus=sqrt(sp$coh*f11*f22), argument=sp$phase)
  c21 <- Re(f21)
```

```
   q21 <- Im(f21)

   c211 <- c21/f11
   q211 <- q21/f11

   q <- length(c21)
   c211 <- c(c211,c211[(q-1):2])        # extend to [0,1] via sym
   q211 <- c(q211,-q211[(q-1):2])       # extend to [0,1] via antisym

   z <- c211 + q211*1i                  # Form complex array for FFT

   beta <- Re(fft(z,inv=T))/(2*q-2)
   if(plot) plot(beta, typ='l')
   par(mfrow=c(1, 1))
   beta[1:(kn+1)]
}



################# bandwidth Selection ##########
mcv <- function(x, y, m, Q, ioptw=ioptw, kn=kn)
## m is the fraction of the segment in the time series
## Q is the number of segments
{
n <- length(y)
if(length(x) != n) stop("x and y should have equal length!")
m <- m*n
Mest <- 1:(n-m*Q-1)
mse <- matrix(NA, length(Mest), Q)
for(i in 1:Q)
{
P <- n-(Q+1-i)*m
fz <- y[1:P]
z <- x[1:P]
for(M in Mest)
{
h <- estfilt(z, fz, ioptw, M, kn)
pts <- filter(x[(P-kn):(P+m)], h, sides=1)
pts <- pts[-(1:(kn+1))]
mse[M,i] <- mean((pts-y[(P+1):(P+m)])^2)
}
}
total.mse <- apply(mse, 1, sum)
Mcv <- Mest[total.mse==min(total.mse)]
return(Mcv)
```

```
}


mcvss <- function(X, y, m, Q, ioptw=ioptw, kn=kn)
## m is the fraction of the segment in the time series
## Q is the number of segments
{
n <- length(y)
if(dim(X)[1] != n) stop("x and y should have equal length!")
r <- dim(X)[2]
m <- m*n
Mest <- 1:(n-m*Q-1)
mse <- array(NA, c(length(Mest), length(Mest), Q))
for(i in 1:Q)
{
P <- n-(Q+1-i)*m
fz <- y[1:P]
Z <- X[1:P,]
for(M1 in Mest)
for(M2 in Mest)
{
h <- estfilts(fz, Z, ioptw, c(M1, M2), kn)
pts <- NULL
for(j in 1:r)
pts <- cbind(pts, filter(X[(P-kn):(P+m), j],
                    h[,j], sides=1))
pts <- apply(pts, 1, sum)
pts <- pts[-(1:(kn+1))]
mse[M1, M2, i] <- mean((pts-y[(P+1):(P+m)])^2)
}
}
total.mse <- apply(mse, c(1, 2), sum)
Mcv <- which(total.mse==min(total.mse), arr.ind=T)
return(Mcv)
}


optMs <- function(X, y, Mest, beta, ioptw, kn)
{ mse <- NA
  for(i in Mest)
  {
    g <- estfilts(y,X,ioptw,M=i,kn)
    bhat <- g
```

```
    ms <- mean((beta-bhat)^2) ##unbiased
    mse <- c(mse, ms)
   }
  mse <- mse[-1]
  return(Mest[mse==min(mse)])
}


mcvs <- function(X, y, m, Q, ioptw=ioptw, kn=kn)
## m is the fraction of the segment in the time series
## Q is the number of segments
{
n <- length(y)
if(dim(X)[1] != n) stop("x and y should have equal length!")
r <- dim(X)[2]
m <- m*n
Mest <- 1:(n-m*Q-1)
mse <- matrix(NA, length(Mest), Q)
for(i in 1:Q)
{
P <- n-(Q+1-i)*m
fz <- y[1:P]
Z <- X[1:P,]
for(k in 1:length(Mest))
{
h <- estfilts(fz, Z, ioptw, Mest[k], kn)
pts <- NULL
for(j in 1:r)
pts <- cbind(pts, filter(X[(P-kn):(P+m), j],
                         h[,j], sides=1))
pts <- apply(pts, 1, sum)
pts <- pts[-(1:(kn+1))]
mse[k, i] <- mean((pts-y[(P+1):(P+m)])^2)
}
}
total.mse <- apply(mse, 1, sum)
Mcv <- Mest[total.mse==min(total.mse)]
return(Mcv)
}


require(splines)
interp  <- function(ts, TR)
{
n <- length(ts)
```

```
time <- seq(0, , by=TR, length=n)
ns <- n*TR-1
x0.ispl <- interpSpline(time, ts)
y <- predict(x0.ispl, 0:ns)$y
return(y)
}


# MultiVM.R

######### HRF matrix form ################  .
####  Input: y -- the BOLD response
####         X -- the matrix for multiple stimuli
####         M -- the vector of length r,
####              specify the smoothing parameter for each HRF
#### Output: hs -- the matrix for the HRFs estimates
###########################################

## estfilt() for 1 dimensional case.

estfilts <- function(y, X, ioptw, M, kn)
{
 fXX <- spes(X, ioptw, M)
 fyX <- xspes(y, X, ioptw, M)
 H <- divarray(fyX, fXX)
 return(Htoh(H, kn))
}


spes <- function(X, ioptw, M)
### spe matrix for X ###
{
 n <- dim(X)[1]
 r <- dim(X)[2]
 fXX <- array(NA, c(r, r, ((n / 2) + 1)))

 for(i in 1:r)
for(j in i:r)
 {
  if(i==j) fXX[i, i,] <- spe(X[, i], ioptw, M[i])$f
  else
{
 f21 <- xspe(X[, i], X[, j], ioptw, M[j])
 fXX[i, j, ] <- f21$re + 1i * f21$im
 f21 <- xspe(X[, i], X[, j], ioptw, M[i])
```

```
 fXX[j, i, ] <- f21$re - 1i * f21$im
}
 }


 return(fXX)
}


xspes <- function(y, X, ioptw, M)
### xspe matrix for y and X ###
{
 n <- dim(X)[1]
 r <- dim(X)[2]
 fyX <- array(NA, c(1, r, ((n / 2) + 1)))

 for(i in 1:r)
  {
   f21 <- xspe(y, X[, i], ioptw, M[i])
   fyX[1, i, ] <- f21$re + 1i * f21$im
  }
 return(fyX)
}


divarray <- function(fyX, fXX)
### fyX/fXX #####
{
 q <- dim(fXX)[3]
 H <- NULL
 for(i in 1:q)
  {
   H <- rbind(H, fyX[,,i]%*% solve(fXX[,,i]))
  }
 return(H)
}


Htoh <- function(H, kn)
### estimate h matrix
{
 q <- dim(H)[1]
 H1 <- rbind(H, Conj(as.matrix(H[(q-1):2, ])))
 h <- apply(H1, 2, ifft)
 h <- Re(h)/(2*q-2)
 return(h[1:(kn+1), ])
}
```

```
ifft <- function(x) fft(x, inv=T)



############## WLS method #########################
err <- function(fyX, fXX, fyy)
## estimate the error spectrum ##
{
 q <- dim(fXX)[3]
 fee <- NULL
 for(i in 1:q)
   {
    fee <- c(fee, fyy[i]-fyX[,,i]%*% solve(fXX[,,i])
               %*% Conj(as.matrix(fyX[,,i])))
   }
 return(Re(fee))
}

esth.wls <- function(y, X, ioptw, Merr, M, kn)
{
 fXX <- spes(X, ioptw, Merr)
 fyX <- xspes(y, X, ioptw, Merr)
 fyy <- spe(y, ioptw, floor(mean(Merr)))$f
 fee <- err(fyX, fXX, fyy)
 fyX <- array.smooth(fyX, fee, ioptw, M)
 fXX <- array.smooth(fXX, fee, ioptw, M)
 H <- divarray(fyX, fXX)
 return(Htoh.w(H, kn))
}

esterr <- function(y, X, ioptw, Merr)
{
 fXX <- spes(X, ioptw, Merr)
 fyX <- xspes(y, X, ioptw, Merr)
 fyy <- spe(y, ioptw, floor(mean(Merr)))$f
 fee <- err(fyX, fXX, fyy)
 return(fee)
}

array.smooth <- function(fm, fee, ioptw, M)
{
 c <- dim(fm)
 fm1 <- array(NA, c(c[1], c[2], 2*c[3]-2))
 for(i in 1:c[1])
 for(j in 1:c[2])
```

```
{
 f1 <- fm[i, j, ]/fee
 f1 <- c(f1, Conj(f1[(c[3]-1):2]))
 fm1[i, j, ] <- smooth.fre(f1, ioptw, M[i])
}
 return(fm1)
}


Htoh.w <- function(H, kn)
### estimate h matrix
{
 n <- dim(H)[1]
 h <- apply(H, 2, ifft)
 h <- Re(h)/n
 return(h[1:(kn+1), ])
}



#### needs further modification
Fcoh <- function(y, X, ioptw, M, kn, fre, phase=F)
{
 dim(X)[1] -> n
 dim(X)[2] -> r
 fXX <- spes(X, ioptw, M)
 fyX <- xspes(y, X, ioptw, M)
 fyy <- spe(y, ioptw, floor(mean(M)))$f

 q <- length(fyy)

 par(mfrow=c(2,2))

 spX.plot(fXX, fre, y.lab = "Spectrum f11")
 sp.plot(fyy, y.lab = "Spectrum f22")

 fee <- err(fyX, fXX, fyy)
 coh <- 1-fee/fyy
 sp.plot(coh, y.lab="Coherence")

 ilam2 <- c(2,2./3.,.795,.539,.586,1,1.2,1.66)
 const <- (n - M*ilam2[ioptw])/(M*ilam2[ioptw])
 fstat <- const*coh/(1-coh)
 nc <- 2*(const-r)
 ft <- qf(.95,2,nc)
 sp.plot(fstat, y.lab = "F-stat")
```

```
 lines(c(0:(n - 1))/(2 * (n - 1)), rep(ft,n),lty=2)

 par(mfrow=c(1,1))
 invisible()

 if(phase==T)
  {
   phi <- NULL
   for(i in 1:r)
   phi <- cbind(phi, atan2(Im(fyX[1, i,]),Re(fyX[1, i,])))
   plot(c(0:(q - 1))/(2 * (q - 1)), phi[, 1], ylim=c(min(phi),
           max(phi)), ylab="phase",
xlab="frequency", type='n')
   for(i in 1:r)
   lines(c(0:(q-1))/(2*(q-1)),phi[, i], col=i)
  }
}


spX.plot <- function(fXX, fre, y.lab="Spectrum", p.type="l")
# ------------------------------------------------------
#
#    Function for plotting spectral quantities
#
# ------------------------------------------------------
{
   q <- dim(fXX)[3]
   r <- dim(fXX)[2]
   plot(c(0:(q-1))/(2*(q-1)),Re(fXX[1, 1, ]),
           ylim=c(0, max(Re(fXX))), xlab="frequency",
        ylab=y.lab,type='n')
   for(i in 1:r)
   lines(c(0:(q-1))/(2*(q-1)),Re(fXX[i, i, ]), col=i)
   axis(3, fre[i], round(fre[i], 3))
   abline(v=fre[i], col=i)
}


# MultiV.R

######### HRF matrix form ################
####  Input: y -- the BOLD response
####          X -- the matrix for multiple stimuli
#### Output: hs -- the matrix for the HRFs estimates
```

```
#############################################

## estfilt() for 1 dimensional case.

estfilts <- function(y, X, ioptw, M, kn)
{
 fXX <- spes(X, ioptw, M)
 fyX <- xspes(y, X, ioptw, M)
 H <- divarray(fyX, fXX)
 return(Htoh(H, kn))
}


spes <- function(X, ioptw, M)
### spe matrix for X ###
{
 n <- dim(X)[1]
 r <- dim(X)[2]
 fXX <- array(NA, c(r, r, ((n / 2) + 1)))

 for(i in 1:r)
for(j in i:r)
 {
   if(i==j) fXX[i, i,] <- spe(X[, i], ioptw, M)$f
   else
{
 f21 <- xspe(X[, i], X[, j], ioptw, M)
 fXX[i, j, ] <- f21$re + 1i * f21$im
 fXX[j, i, ] <- f21$re - 1i * f21$im
}
 }

 return(fXX)
}

xspes <- function(y, X, ioptw, M)
### xspe matrix for y and X ###
{
 n <- dim(X)[1]
 r <- dim(X)[2]
 fyX <- array(NA, c(1, r, ((n / 2) + 1)))

 for(i in 1:r)
  {
```

```
    f21 <- xspe(y, X[, i], ioptw, M)
    fyX[1, i, ] <- f21$re + 1i * f21$im
   }
 return(fyX)
}


divarray <- function(fyX, fXX)
### fyX/fXX #####
{
 q <- dim(fXX)[3]
 H <- NULL
 for(i in 1:q)
   {
    H <- rbind(H, fyX[,,i]%*% solve(fXX[,,i]))
   }
 return(H)
}


Htoh <- function(H, kn)
### estimate h matrix
{
 q <- dim(H)[1]
 H1 <- rbind(H, Conj(as.matrix(H[(q-1):2, ])))
 h <- apply(H1, 2, ifft)
 h <- Re(h)/(2*q-2)
 return(h[1:(kn+1), ])
}


ifft <- function(x) fft(x, inv=T)



############## WLS method ########################
err <- function(fyX, fXX, fyy)
## estimate the error spectrum ##
{
 q <- dim(fXX)[3]
 fee <- NULL
 for(i in 1:q)
  {
   fee <- c(fee, fyy[i]-fyX[,,i]%*% solve(fXX[,,i])
            %*% Conj(as.matrix(fyX[,,i])))
  }
 return(Re(fee))
}
```

```
esth.wls <- function(y, X, ioptw, Merr, M, kn)
{
 fXX <- spes(X, ioptw, Merr)
 fyX <- xspes(y, X, ioptw, Merr)
 fyy <- spe(y, ioptw, Merr)$f
 fee <- err(fyX, fXX, fyy)
 fyX <- array.smooth(fyX, fee, ioptw, M)
 fXX <- array.smooth(fXX, fee, ioptw, M)
 H <- divarray(fyX, fXX)
 return(Htoh.w(H, kn))
}


esterr <- function(y, X, ioptw, Merr)
{
 fXX <- spes(X, ioptw, Merr)
 fyX <- xspes(y, X, ioptw, Merr)
 fyy <- spe(y, ioptw, Merr)$f
 fee <- err(fyX, fXX, fyy)
 return(fee)
}


array.smooth <- function(fm, fee, ioptw, M)
{
 c <- dim(fm)
 fm1 <- array(NA, c(c[1], c[2], 2*c[3]-2))
 for(i in 1:c[1])
for(j in 1:c[2])
{
 f1 <- fm[i, j, ]/fee
 f1 <- c(f1, Conj(f1[(c[3]-1):2]))
 fm1[i, j, ] <- smooth.fre(f1, ioptw, M)
}
 return(fm1)
}


Htoh.w <- function(H, kn)
### estimate h matrix
{
 n <- dim(H)[1]
 h <- apply(H, 2, ifft)
 h <- Re(h)/n
 return(h[1:(kn+1), ])
}
```

```
Fcoh <- function(y, X, ioptw, M, fre, phase=F)
{
 dim(X)[1] -> n
 dim(X)[2] -> r
 fXX <- spes(X, ioptw, M)
 fyX <- xspes(y, X, ioptw, M)
 fyy <- spe(y, ioptw, M)$f

 q <- length(fyy)

 par(mfrow=c(2,2))

 spX.plot(fXX, fre, y.lab = "Spectrum f11")
 sp.plot(fyy, y.lab = "Spectrum f22")

 fee <- err(fyX, fXX, fyy)
 coh <- 1-fee/fyy
 sp.plot(coh, y.lab="Coherence")

 ilam2 <- c(2,2./3.,.795,.539,.586,1,1.2,1.66)
 const <- n/(M*ilam2[ioptw])
 fstat <- const*coh/(1-coh)/r
 nc <- 2*(const-r)
 ft <- qf(.95,2,nc)
 sp.plot(fstat, y.lab = "F-stat")
 lines(c(0:(q-1))/(2*(q-1)), rep(ft,q),lty=2)

 par(mfrow=c(1,1))
 invisible()

 if(phase==T)
  {
   phi <- NULL
   for(i in 1:r)
   phi <- cbind(phi, atan2(Im(fyX[1, i,]),Re(fyX[1, i,])))
   plot(c(0:(q - 1))/(2 * (q - 1)), phi[, 1], ylim=c(min(phi),
            max(phi)), ylab="phase",
xlab="frequency", type='n')
   for(i in 1:r)
   lines(c(0:(q-1))/(2*(q-1)),phi[, i], col=i)
  }
}
```

```
spX.plot <- function(fXX, fre, y.lab="Spectrum", p.type="l")
# ----------------------------------------------------
#
#   Function for plotting spectral quantities
#
# ----------------------------------------------------
{
    q <- dim(fXX)[3]
    r <- dim(fXX)[2]
    plot(c(0:(q-1))/(2*(q-1)),Re(fXX[1, 1, ]),
                ylim=c(0, max(Re(fXX))), xlab="frequency",
            ylab=y.lab,type='n')
    for(i in 1:r)
    {lines(c(0:(q-1))/(2*(q-1)),Re(fXX[i, i, ]), col=i)
    axis(3, fre[i], round(fre[i], 3))
    abline(v=fre[i], col=i)}
}



################################################################
############## Partial Coherence ###############################
################################################################

pcoh <- function(y, X, ioptw, M, fre)
{
 n <- dim(X)[1]
 r <- dim(X)[2]
 spX <- spes(X, ioptw, M)
 spYX <- xspes(y, X, ioptw, M)
 spyy <- spe(y, ioptw, M)$f

 par(mfrow=c(2,2))

 spX.plot(spX, fre, y.lab = "Spectrum f11")
 sp.plot(spyy, y.lab = "Spectrum f22")

 q <- n/2+1
 coh <- fstat <- array(NA, c(r, q))
 ilam2 <- c(2,2./3.,.795,.539,.586,1,1.2,1.66)
 const <- (n - (r-1)*M*ilam2[ioptw])/(M*ilam2[ioptw])
 nc <- 2*(const-1)
 for(i in 1:r)
```

```
{
 ryi <- sqrt(ocoh(y, as.matrix(X[, i]), ioptw, M))
 ryj <- sqrt(ocoh(y, as.matrix(X[,-i]), ioptw, M))
 rij <- sqrt(ocoh(X[,i], as.matrix(X[, -i]), ioptw, M))
 coh[i,] <- Re((ryi-ryj*rij)^2/(1-ryj^2)/(1-rij^2))
 fstat[i,] <- const*coh[i,]/(1-coh[i,])
}
 plot(c(0:(q-1))/(2*(q-1)), rep(1, q) ,xlab="frequency",
        ylab="Partial Coherence", ylim=c(0, 1), type='n')
 for(i in 1:r)
{
 lines(c(0:(q-1))/(2*(q-1)), coh[i,], col=i)
 axis(3, fre[i], round(fre[i], 3))
     abline(v=fre[i], col=i)
   }
 ft <- qf(.95,2,nc)
 plot(c(0:(q-1))/(2*(q-1)), 1:q ,xlab="frequency",
        ylab="F(Partial)", ylim=c(min(fstat),
             max(fstat)), type='n')
 for(i in 1:r)
{
 lines(c(0:(q-1))/(2*(q-1)), fstat[i,], col=i)
 axis(3, fre[i], round(fre[i], 3))
     abline(v=fre[i], col=i)
   }
 lines(c(0:(q-1))/(2*(q-1)), rep(ft,q),lty=2)

 par(mfrow=c(1,1))
 invisible()
}



ocoh <- function(y, X, ioptw, M)
{
 dim(X)[1] -> n
 fXX <- spes(X, ioptw, M)
 fyX <- xspes(y, X, ioptw, M)
 fyy <- spe(y, ioptw, M)$f
 fee <- err(fyX, fXX, fyy)
 coh <- 1-fee/fyy
 return(coh)
}
```

```
## Only for two stimuli
pcoh.fre <- function(y, X, ioptw, M, fre)
{
 n <- dim(X)[1]
 r <- dim(X)[2]
 spX <- spes(X, ioptw, M)
 spYX <- xspes(y, X, ioptw, M)
 spyy <- spe(y, ioptw, M)$f

 q <- n/2+1
 coh <- fstat <- array(NA, c(r, q))
 cohx <- fstatx <- rep(NA, r)
 ilam2 <- c(2,2./3.,.795,.539,.586,1,1.2,1.66)
 const <- (n - (r-1)*M*ilam2[ioptw])/(M*ilam2[ioptw])
 nc <- 2*(const-1)
 for(i in 1:r)
 {
 ryi <- sqrt(ocoh(y, as.matrix(X[, i]), ioptw, M))
 ryj <- sqrt(ocoh(y, as.matrix(X[,-i]), ioptw, M))
 rij <- sqrt(ocoh(X[,i], as.matrix(X[, -i]), ioptw, M))
 coh[i,] <- Re((ryi-ryj*rij)^2/(1-ryj^2)/(1-rij^2))
 fstat[i,] <- const*coh[i,]/(1-coh[i,])
 ind <- abs(c(0:(q-1))/(2*(q-1))-fre[i])
 ind <- which(ind==min(ind), arr.ind=T)
 cohx[i] <- coh[i, ind]
 fstatx[i] <- fstat[i, ind]
 }
 ft <- qf(.95,2,nc)
 return(list(coh=cohx, fstat=fstatx, ft=ft))
}

ocoh <- function(y, X, ioptw, M)
{
 dim(X)[1] -> n
 fXX <- spes(X, ioptw, M)
 fyX <- xspes(y, X, ioptw, M)
 fyy <- spe(y, ioptw, M)$f
 fee <- err(fyX, fXX, fyy)
 coh <- 1-fee/fyy
 return(coh)
}

## For more than 2 types of stimuli
```

```
pcoh3.fre <- function(y, X, ioptw, M, fre)
{
 n <- dim(X)[1]
 r <- dim(X)[2]
 spX <- spes(X, ioptw, M)
 spYX <- xspes(y, X, ioptw, M)
 spyy <- spe(y, ioptw, M)$f

 q <- n/2+1
 coh <- fstat <- array(NA, c(r, q))
 cohx <- fstatx <- rep(NA, r)
 ilam2 <- c(2,2./3.,.795,.539,.586,1,1.2,1.66)
 const <- (n - (r-1)*M*ilam2[ioptw])/(M*ilam2[ioptw])
 nc <- 2*(const-1)
 for(i in 1:r)
{
 syi <- xspes(y, as.matrix(X[, i]), ioptw, M)
 syj <- xspes(y, as.matrix(X[,-i]), ioptw, M)
 sjj <- spes(as.matrix(X[,-i]), ioptw, M)
 sij <- xspes(X[, i], as.matrix(X[,-i]), ioptw, M)
 psyij <- NULL
 for(k in 1:q)
  {
   psyij <- c(psyij, syi[k]-syj[,,k]%*% solve(sjj[,,k])
                %*% Conj(as.matrix(sij[,,k])))
  }
 syy <- spe(y, ioptw, M)$f
 psyyj <- NULL
 for(k in 1:q)
  {
   psyyj <- c(psyyj, syy[k]-syj[,,k]%*% solve(sjj[,,k])
                %*% Conj(as.matrix(syj[,,k])))
  }
 sii <- spe(X[, i], ioptw, M)$f
 psiij <- NULL
 for(k in 1:q)
  {
   psiij <- c(psiij, sii[k]-sij[,,k]%*% solve(sjj[,,k])
                %*% Conj(as.matrix(sij[,,k])))
  }
 coh[i,] <- Re(psyij*Conj(psyij))/Re(psyyj)/Re(psiij)
 fstat[i,] <- const*coh[i,]/(1-coh[i,])
 ind <- abs(c(0:(q-1))/(2*(q-1))-fre[i])
 ind <- which(ind==min(ind), arr.ind=T)
```

```
  cohx[i] <- coh[i, ind]
  fstatx[i] <- fstat[i, ind]
}
 ft <- qf(.95,2,nc)
 return(list(coh=cohx, fstat=fstatx, ft=ft))
}


pcoh3.fre.faster <- function(y, X, ioptw, M, fre)
{
 n <- dim(X)[1]
 r <- dim(X)[2]
 spX <- spes(X, ioptw, M)
 spYX <- xspes(y, X, ioptw, M)
 spyy <- spe(y, ioptw, M)$f

 q <- n/2+1
 coh <- fstat <- array(NA, c(r, q))
 cohx <- fstatx <- rep(NA, r)
 ilam2 <- c(2,2./3.,.795,.539,.586,1,1.2,1.66)
 const <- (n - (r-1)*M*ilam2[ioptw])/(M*ilam2[ioptw])
 nc <- 2*(const-1)
 for(i in 1:r)
{
 syi <- xspes(y, X[, i], ioptw, M)
 syj <- xspes(y, X[,-i], ioptw, M)
 sjj <- spes(X[,-i], ioptw, M)
 sij <- xspes(X[, i], X[,-i], ioptw, M)
 psyij <- NULL
 for(k in 1:q)
  {
   psyij <- c(psyij, syi[k]-syj[,,k]%*% solve(sjj[,,k])
                     %*% Conj(as.matrix(sij[,,k])))
  }
 syy <- spe(y, ioptw, M)$f
 psyyj <- NULL
 for(k in 1:q)
  {
   psyyj <- c(psyyj, syy[k]-syj[,,k]%*% solve(sjj[,,k])
                     %*% Conj(as.matrix(syj[,,k])))
  }
 sii <- spe(X[, i], ioptw, M)$f
 psiij <- NULL
 for(k in 1:q)
  {
```

```
    psiij <- c(psiij, sii[k]-sij[,,k]%*% solve(sjj[,,k])
                    %*% Conj(as.matrix(sij[,,k])))
  }
  coh[i,] <- Re(psyij*Conj(psyij))/Re(psyyj)/Re(psiij)
  fstat[i,] <- const*coh[i,]/(1-coh[i,])
  ind <- abs(c(0:(q-1))/(2*(q-1))-fre[i])
  ind <- which(ind==min(ind), arr.ind=T)
  cohx[i] <- coh[i, ind]
  fstatx[i] <- fstat[i, ind]
}
ft <- qf(.95,2,nc)
return(list(coh=cohx, fstat=fstatx, ft=ft))
}


pcoh3 <- function(y, X, ioptw, M, fre)
{
n <- dim(X)[1]
r <- dim(X)[2]
spX <- spes(X, ioptw, M)
spYX <- xspes(y, X, ioptw, M)
spyy <- spe(y, ioptw, M)$f

par(mfrow=c(2,2))

spX.plot(spX, fre, y.lab = "Spectrum f11")
sp.plot(spyy, y.lab = "Spectrum f22")

q <- n/2+1
coh <- fstat <- array(NA, c(r, q))
ilam2 <- c(2,2./3.,.795,.539,.586,1,1.2,1.66)
const <- (n - (r-1)*M*ilam2[ioptw])/(M*ilam2[ioptw])
nc <- 2*(const-1)
for(i in 1:r)
{
syi <- xspes(y, as.matrix(X[, i]), ioptw, M)
syj <- xspes(y, as.matrix(X[,-i]), ioptw, M)
sjj <- spes(as.matrix(X[,-i]), ioptw, M)
sij <- xspes(X[, i], as.matrix(X[,-i]), ioptw, M)
psyij <- NULL
for(k in 1:q)
  {
    psyij <- c(psyij, syi[k]-syj[,,k]%*% solve(sjj[,,k])
                    %*% Conj(as.matrix(sij[,,k])))
```

```
   }
 syy <- spe(y, ioptw, M)$f
 psyyj <- NULL
 for(k in 1:q)
  {
    psyyj <- c(psyyj, syy[k]-syj[,,k]%*% solve(sjj[,,k])
                    %*% Conj(as.matrix(syj[,,k])))
  }
 sii <- spe(X[, i], ioptw, M)$f
 psiij <- NULL
 for(k in 1:q)
  {
    psiij <- c(psiij, sii[k]-sij[,,k]%*% solve(sjj[,,k])
                    %*% Conj(as.matrix(sij[,,k])))
  }
 coh[i,] <- Re(psyij*Conj(psyij))/Re(psyyj)/Re(psiij)
 fstat[i,] <- const*coh[i,]/(1-coh[i,])
}
 plot(c(0:(q-1))/(2*(q-1)), rep(1, q) ,xlab="frequency",
       ylab="Partial Coherence", ylim=c(0, 1), type='n')
 for(i in 1:r)
{
 lines(c(0:(q-1))/(2*(q-1)), coh[i,], col=i)
 axis(3, fre[i], round(fre[i], 3))
    abline(v=fre[i], col=i)
    }
 ft <- qf(.95,2,nc)
 plot(c(0:(q-1))/(2*(q-1)), 1:q ,xlab="frequency",
       ylab="F(Partial)", ylim=c(min(fstat),
       max(fstat)), type='n')
 for(i in 1:r)
{
 lines(c(0:(q-1))/(2*(q-1)), fstat[i,], col=i)
 axis(3, fre[i], round(fre[i], 3))
    abline(v=fre[i], col=i)
    }
 lines(c(0:(q-1))/(2*(q-1)), rep(ft,q),lty=2)

 par(mfrow=c(1,1))
 invisible()
}
```

```
######### Activation map ######################
act.fre <- function(y, X, ioptw, M, fre)
{
 dim(X)[1] -> n
 dim(X)[2] -> r
 fXX <- spes(X, ioptw, M)
 fyX <- xspes(y, X, ioptw, M)
 fyy <- spe(y, ioptw, M)$f
 fee <- err(fyX, fXX, fyy)
 H <- divarray(fyX, fXX)

 q <- length(fyy)
 c(0:(q-1)/(2*(q-1))) -> allfre
 ind <- which(abs(allfre-sum(fre))<10^{-4})
 fXX <- fXX[,,ind]
 H <- H[ind,]
 fee <- fee[ind]
 ilam2 <- c(2,2./3.,.795,.539,.586,1,1.2,1.66)
 const <- n/(M*ilam2[ioptw])
 F <- const*t(H)%*%fXX%*%t(t(Conj(H)))/r/fee
 Falpha <- qf(.95, 2, 2*(const-r))
 Falpha2 <- qf(.99, 2, 2*(const-r))

 return(list(F=Re(F), Falpha=Falpha, Falpha2=Falpha2))
}

################# Testing the difference ################
t.fre <- function(y, X, ioptw, M, fre, vc)
{
 n <- dim(X)[1]
       r <- dim(X)[2]
 c2 <- as.matrix(rep(vc, 2))
       spX <- spes(X, ioptw, M)
       spYX <- xspes(y, X, ioptw, M)
 spyy <- spe(y, ioptw, M)$f

 q <- dim(spX)[3]
 c(0:(q-1)/(2*(q-1))) -> allfre
 ind <- which(abs(allfre-fre)<10^{-4})
 fxx0 <- spX[,,ind]
 fxx1 <- cbind(Re(fxx0), -Im(fxx0))
 fxx2 <- cbind(Im(fxx0), Re(fxx0))
 fxx <- rbind(fxx1, fxx2)
 H <- divarray(spYX, spX)[ind, ]
```

```
H2 <- c(Re(H), Im(H))
see <- err(spYX, spX, spyy)[ind]
ilam2 <- c(2,2./3.,.795,.539,.586,1,1.2,1.66)
      const <- n/(M*ilam2[ioptw])
t <- t(c2)%*%t(t(H2))/sqrt(2*(const-r)/const*see*t(c2)
              %*%solve(fxx)%*%c2)
thres.l <- qt(.05, 2*(const-r))
thres.r <- qt(.95, 2*(const-r))
thres.b <- qt(.975, 2*(const-r))
return(list(t=t, l=thres.l, r=thres.r, b=thres.b))
}



#    a.R

actv.fre <- function(y, X, ioptw, M, fre)
#
{
dim(X)[1] -> n
dim(X)[2] -> r
fXX <- spes(X, ioptw, M)
fyX <- xspes(y, X, ioptw, M)
fyy <- spe(y, ioptw, M)$f
fee <- err(fyX, fXX, fyy)
H <- divarray(fyX, fXX)

q <- length(fyy)
c(0:(q-1)/(2*(q-1))) -> allfre
ind <- which(abs(allfre-sum(fre))<10^{-4})
fXX <- fXX[,,ind]
H <- H[ind,]
fee <- fee[ind]
coh <- 1 - fee/fyy[ind]
ilam2 <- c(2,2./3.,.795,.539,.586,1,1.2,1.66)
const <- n/(M*ilam2[ioptw])
F <- const*t(H)%*%fXX%*%t(t(Conj(H)))/r/fee
Falpha <- qf(.95, 2, 2*(const-r))
Falpha2 <- qf(.99, 2, 2*(const-r))

return(list(coh=coh, F=Re(F), Falpha=Falpha, Falpha2=Falpha2))
}
```

Bibliography

1. P. Bai, X. Huang, and Y.K. Truong. Nonparametric estimation of hemodynamic response function: A frequency-domain approach. In J. Rojo, editor, *Optimality: The Third Erich L. Lehmann Symposium*, volume 57 of *IMS Lecture Notes Monograph Series*, pages 190–215. Institute of Mathematical Statistics, 2009.

2. G.M. Boynton, S.A. Engel, G.H. Glover, and D.J. Heeger. Linear systems analysis of functional magnetic resonance imaging in Human V1. *Journal of Neuroscience*, 16(13):4207–4221, 1996.

3. R.L. Buckner, P.A. Bandettini, K.M. O'Craven, R.L. Savoy, S.E. Petersen, M.E. Raichle, and B.R. Rosen. Detection of cortical activation during averaged single trials of a cognitive task using functional magnetic resonance imaging. *Proceedings of the National Academy of Sciences of the United States of America*, 93(25):14878–14883, 1996.

4. E. Bullmore, M. Brammer, S.C.R. Williams, S. Rabe-Hesketh, N. Janot, A. David, J. Mellers, R. Howard, and P. Sham. Statistical methods of estimation and inference for functional MR image analysis. *Magnetic Resonance in Medicine*, 35(2):261–276, 1996.

5. R. Casanova, S. Ryali, J. Serences, L. Yang, R. Kraft, P.J. Laurienti, and J.A. Maldjian. The impact of temporal regularization on estimates of the BOLD hemodynamic response function: a comparative analysis. *NeuroImage*, 40(4): 1606–1618, 2008.

6. W. Chen. On estimating hemodynamic response functions. PhD thesis, The University of North Carolina at Chapel Hill, NC, 2012.

7. M.S. Cohen. Parametric analysis of fMRI data using linear systems methods. *NeuroImage*, 6(2):93–103, 1997.

8. R.W. Cox. AFNI: Software for analysis and visualization of functional magnetic resonance neuroimages. *Computers and Biomedical Research*, 29(3): 162–173, 1996.

9. A.M. Dale and R.L. Buckner. Selective averaging of rapidly presented individual trials using fMRI. *Human Brain Mapping*, 5(5):329–340, 1997.

10. K.J. Friston, K.J. Worsley, R.S.J. Frackowiak, J.C. Mazziotta, and A.C. Evans. Assessing the significance of focal activations using their spatial extent. *Human Brain Mapping*, 1(3):210–220, 1993.

11. K.J. Friston, P. Jezzard, and R. Turner. Analysis of functional MRI time-series. *Human Brain Mapping*, 1(2):153–171, 1994.

12. K.J. Friston, C.D. Frith, R. Turner, and R.S.J. Frackowiak. Characterizing evoked hemodynamics with fMRI. *NeuroImage*, 2(2PA):157–165, 1995.

13. K.J. Friston, A.P. Holmes, K.J. Worsley, J.B. Poline, C.D. Frith, R.S.J. Frackowiak, et al. Statistical parametric maps in functional imaging: A general linear approach. *Hum Brain Mapp*, 2(4):189–210, 1995.

14. K.J. Friston, P. Fletcher, O. Josephs, A. Holmes, M.D. Rugg, and R. Turner. Event-related fMRI: Characterizing differential responses. *NeuroImage*, 7(1): 30–40, 1998.

15. K.J. Friston, O. Josephs, G. Rees, and R. Turner. Nonlinear event-related re-

sponses in fMRI. *Magnetic Resonance in Medicine*, 39:41–52, 1998.

16. K.J. Friston, W. Penny, C. Phillips, S. Kiebel, G. Hinton, and J. Ashburner. Classical and Bayesian inference in neuroimaging: Theory. *NeuroImage*, 16 (2):465–483, 2002.

17. K.J. Friston, J. Ashburner, S.J. Kiebel, T.E. Nichols, and W.D. Penny, editors. *Statistical parametric mapping: The analysis of functional brain images*. Academic Press, 2007. URL http://www.fil.ion.ucl.ac.uk/spm/.

18. C.R. Genovese. A Bayesian time-course model for functional magnetic resonance imaging data. *Journal of the American Statistical Association*, 95(451): 691–719, 2000.

19. G.H. Glover. Deconvolution of impulse response in event-related BOLD fMRI. *NeuroImage*, 9(4):416–429, 1999.

20. C. Gössl, L. Fahrmeir, and D.P. Auer. Bayesian modeling of the hemodynamic response function in BOLD fMRI. *NeuroImage*, 14(1):140–148, 2001.

21. R.N.A. Henson, T. Shallice, M.L. Gorno-Tempini, and R.J. Dolan. Face repetition effects in implicit and explicit memory tests as measured by fMRI. *Cerebral Cortex*, 12(2):178–186, 2002. ISSN 1047-3211.

22. J. Hykin, R. Bowtell, P. Glover, R. Coxon, LD Blumhardt, and P. Mansfield. Investigation of the linearity of functional activation signal changes in the brain using echo planar imaging (EPI) at 3.0 T. In *Proceedings of the SMR and ESMRB, Joint Meeting*, page 795, 1995.

23. J. Jacobs, C. Hawco, E. Kobayashi, R. Boor, P. LeVan, U. Stephani, M. Siniatchkin, and J. Gotman. Variability of the hemodynamic response as a function of age and frequency of epileptic discharge in children with epilepsy. *NeuroImage*, 40(2):601–614, 2008.

24. M. Jenkinson, C.F. Beckmann, T.E. Behrens, M.W. Woolrich, and S.M. Smith. Fsl. *NeuroImage*, 62:782–790, 2012. URL http://www.fmrib.ox.ac.uk/fsl/.

25. O. Josephs, R. Turner, and K. Friston. Event-related fMRI. *Human Brain Mapping*, 5(4):243–248, 1997.

26. F. Kruggel and D.Y. von Cramon. Modeling the hemodynamic response in single-trial functional MRI experiments. *Magnetic Resonance in Medicine*, 42 (4):787–797, 1999.

27. N. Lange and S.L. Zeger. Non-linear fourier time series analysis for human brain mapping by functional magnetic resonance imaging. *Journal of the Royal Statistical Society: Series C (Applied Statistics)*, 46(1):1–29, 1997.

28. M.A. Lindquist. The statistical analysis of fMRI data. *Statistical Science*, 23 (4):439–464, 2008.

29. M.A. Lindquist and T.D. Wager. Validity and power in hemodynamic response modeling: A comparison study and a new approach. *Human Brain Mapping*, 28(8):764–784, 2007.

30. M.A. Lindquist, J. Meng Loh, L.Y. Atlas, and T.D. Wager. Modeling the hemodynamic response function in fMRI: Efficiency, bias and mis-modeling. *NeuroImage*, 45(1S1):187–198, 2009.

31. Y. Lu, A.P. Bagshaw, C. Grova, E. Kobayashi, F. Dubeau, and J. Gotman. Using voxel-specific hemodynamic response function in EEG-fMRI data analysis. *NeuroImage*, 32(1):238–247, 2006.

32. S. Makni, C. Beckmann, S. Smith, and M. Woolrich. Bayesian deconvolution fMRI data using bilinear dynamical systems. *NeuroImage*, 42(4):1381–1396, 2008.

33. J.L. Marchini and B.D. Ripley. A new statistical approach to detecting significant activation in functional MRI. *NeuroImage*, 12(4):366–380, 2000.

34. P.P. Mitra, S. Ogawa, X. Hu, and K. Ugurbil. The nature of spatiotemporal changes in cerebral hemodynamics as manifested in functional magnetic resonance imaging. *Magnetic Resonance in Medicine*, 37(4):511–518, 1997.

35. J.A. Mumford and T. Nichols. Modeling and inference of multisubject fMRI data. *IEEE Engineering in Medicine and Biology Magazine*, 25(2):42–51, 2006.

36. J.C. Rajapakse, F. Kruggel, J.M. Maisog, and D.Y. Von Cramon. Modeling hemodynamic response for analysis of functional MRI time-series. *Human Brain Mapping*, 6(4):283–300, 1998.

37. W. Richter, K. Ugurbil, A. Georgopoulos, and S.G. Kim. Time-resolved fMRI of mental rotation. *NeuroReport*, 8(17):3697–3702, 1997.

38. J. Steffener, MH Tabert, and Y. Stern. Investigating hemodynamic response variability at the group level using basis functions. *NeuroImage*, 47:56–56, 2009.

39. B.D. Ward and L. Revision. Deconvolution analysis of fMRI time series data. *AFNI 3dDeconvolve Documentation,* Medical College of Wisconsin, 2000.

40. A.M. Wink, H. Hoogduin, and J.B.T.M. Roerdink. Data-driven haemodynamic response function extraction using Fourier-wavelet regularised deconvolution. *BMC Medical Imaging*, 8(1):7, 2008. ISSN 1471-2342. doi: 10.1186/1471-2342-8-7. URL http://www.biomedcentral.com/1471-2342/8/7.

41. M.W. Woolrich, M. Jenkinson, J.M. Brady, and S.M. Smith. Fully Bayesian spatio-temporal modeling of fMRI data. *IEEE Transactions on Medical Imaging*, 23(2):213–231, 2004.

42. E. Zarahn. Testing for neural responses during temporal components of trials with BOLD fMRI. *NeuroImage*, 11(6):783–796, 2000.

43. C. Zhang, Y. Jiang, and T. Yu. A comparative study of one-level and two-level semiparametric estimation of hemodynamic response function for fMRI data. *Statistics in Medicine*, 26(21):3845–3861, 2007.

44. C. Zhang, Y. Lu, T. Johnstone, T. Oakes, and R.J. Davidson. Efficient modeling and inference for event-related fMRI data. *Computational Statistics & Data Analysis*, 52(10):4859–4871, 2008.

5 Independent Component Analysis: An Overview

Dong Wang
University of North Carolina at Chapel Hill

Seonjoo Lee
Columbia University, New York

Haipeng Shen
University of Hong Kong, China

Young K. Truong
University of North Carolina at Chapel Hill

CONTENTS

5.1 INTRODUCTION

Independent component analysis (ICA) plays an important role in *Blind Source Separation* (BSS) problems. In many applications, multichannel data or sensor biological observations are often acquired through some recording devices, with each sensor receiving different mixed source signals. Both the sources and the mixing mechanism are usually not available, but are related to the sensor observations given by the model:

$$x = As, \tag{5.1}$$

where

$\mathbf{x}$: k-dimensional random vector of sensor observations,

$\mathbf{A}$: $k \times m$ matrix to be viewed as the mixing mechanism,

$\mathbf{s}$: m-dimensional random vector of latent sources.

Given T copies (or samples) of $\mathbf{x}$, the primary goal of BSS is to identify or separate the sources $\mathbf{s}$. It is equally important to determine the mixing matrix $\mathbf{A}$. But determining the solution from an array of sensors is a classical but difficult problem in signal processing. It is an ill-posted problem with multiple ways to unmix or identify the latent (blind) sources. In a seminal paper, Herault and Jutten (1986) proposed a solution by separating the latent source vector into independent components (ICs) [21]. Another important contribution to the new era of BSS is the introduction of their adaptive algorithm based on an independence test using non-linear functions [26]. Since then, BSS has become a very active research area in signal processing. We will highlight some of the popular algorithms here and refer the reader to several books on this topic [25, 9, 11, 33].

The first algorithm to be mentioned is a popular *minimum mutual information*-based ICA method [10], followed with the *Infomax* method using the maximum entropy principle [5]. This method was further developed using the *natural gradient* [2, 1] which is also connected to maximum likelihood estimation (MLE). Then came computationally more efficient methods based on the *fixed-point* or *FastICA algorithm* [23, 22, 24]. Most of the recent contributions have focused mainly on practical applications by extending the above methods [25, 33].

It is important to know that ICA is an *unsupervised learning* problem and has many interesting useful applications ranging from medical signal analysis to text mining. The method has been used as *dimension reduction* and *feature transformation* techniques. Let $\mathbf{X}$ and $\mathbf{S}$ denote the matrices constructed from the observation and source vectors, respectively. Then ICA can be viewed as a matrix factorization method given by

$$\mathbf{X} = \mathbf{AS}.$$

Many important features can be extracted by interpreting the matrix multiplication in the form of an *outer product*. That is, the matrix $\mathbf{X}$ is viewed as the sum of columns of $\mathbf{A}$ times rows of $\mathbf{S}$. This viewpoint is very useful for temporal and spatial feature extraction in neuroimaging data analyses. Based on some biological observations or medical image data, there have been several books discussing algorithms for the separation of independent sources [25, 30, 9, 11, 33].

Note that most of these algorithms have been developed by ignoring the auto-correlated or temporal structures of the latent sources. Some have referred to them as *instantaneous ICA* [11]. From the viewpoint of statistical theory, these methods can be synthesized as an application of *maximum likelihood estimation* (MLE) to the problem of estimating the *probability density function* (pdf) of each latent source. The aforementioned algorithms all employed various parametric functional forms of the source pdf. The desire to estimate these densities with greater flexibility has led to the problem of *nonparametric density estimation*, which will be addressed in Chapter 6. Using pdf, however, is not the only way to characterize the sources. In fact, some sources may be better described using their physiological or biological temporal or spatial features.

For example, in *electroencephalogram (EEG)* data, various rhythms have distinct spectral characteristics that are easier to identify than their counterparts using pdf. These types of sources will be referred to as colored sources and the process of separating them is called the *independent colored source analysis* (ICSA). Chapter 7 is devoted to this topic. The discussion so far has been centered around the data (neuroimaging) from one subject. The problem of handling multiple subjects or a group of individuals will be given in the rest of this chapter followed by Chapter 8, which presents an approach to *group blind source separation* (GBSS) using the idea of ICSA.

5.2 NEUROIMAGING DATA ANALYSIS

Independent component analysis (ICA) is an important data-driven approach in recovering hidden source signals [25, 30, 9, 11, 33]. ICA assumes that the observed vector-valued signal is a linear mixture of the hidden source signals and that these source signals are statistically independent. ICA has been used widely in neuroimaging data analysis because of its power in extracting hidden features.

In order to apply ICA to neuroimaging data analysis, each observation is expressed by a two-dimensional matrix, with one dimension usually corresponding to the spatial information and the other representing the temporal process. In what follows, we will refer to the spatial information as spatial maps and the temporal processes as temporal courses since they commonly are used in the ICA literature, especially in fMRI studies. This type of ICA analysis is called *single-subject ICA* because ICA is applied to one single data sample. The single sample can come from one subject, one session, or one run.

The goal of single-subject ICA is to decompose each matrix into a set of spatial maps and the associated temporal courses. These extracted signals can provide important information in understanding the intrinsic functional and structural organization of the brain. Either the spatial maps or the temporal course can be viewed as hidden sources in the single-subject ICA. In other words, the statistical independence assumption can be imposed either on the spatial maps or the temporal courses. The corresponding ICA analysis is named *spatial ICA* (sICA) or *temporal ICA* (tICA), respectively.

Recently, it has become much more common that multiple data observations are collected during experiments, e.g., neuroimaging data from multiple subjects. When the subjects are from the same group and the data collected under similar experimental conditions, common features usually exist among subjects, either spatial maps or temporal courses. One interesting question is how we can identify or estimate this common information from multiple subjects or groups of subjects.

To deal with a multi-subject analysis while keeping the power of single-subject ICA, *group ICA* (GICA) has been developed and studied by various researchers. It is now a standard tool for analyzing multiple or groups of neuroimaging data, including fMRI, EEG, MEG, and PET, among others. GICA adopts the main assumptions of the single-subject ICA, namely that the sources are independent and the mixing process is linear. Similar to the single-subject ICA, the independence assumption can be imposed either on spatial maps or temporal courses. In the remainder of this review, we will illustrate each GICA method using the sICA analysis since tICA GICA approaches can be conducted in a similar way as the sICA type analysis on the transpose of the data matrices.

One goal of GICA is to identify or estimate common features from multiple subjects or groups of subjects. In what follows, we will refer to the common features as group-level components and the features of each subject as subject-level components. One difference between GICA and single-subject ICA is that we need to specify the relationship between the group- and subject-level components, which is usually treated as a prior assumption for GICA approaches. The *subject-level features* can be assumed to be the same for all subjects, and hence the common *group-level features* are also the same as the subject-level features. Moreover, we could allow variabilities among subjects, i.e., different subjects have distinct features. The common group-level features can be modeled as a function, usually as the group mean, of the subject-level features. Furthermore, we can assume that all subjects have their own features and no group-level components exist. In summary, the subject-level features can be modeled as the same as, as a function of, or different from their group-level components.

Recall that single-subject ICA decomposes each observation into two parts, the spatial maps and the temporal courses. The aforementioned three types of relationships between group- and subject-level features can all be imposed on these two parts, which leads to various combinations of group structure assumptions on the spatial maps and temporal courses. The group structure assumption is important prior information for most GICA approaches. We can classify GICA approaches according to their group structure assumptions. We will review various group structure assumptions in the next section.

To summarize, each GICA approach assumes that the hidden source signals, either spatial maps in the sICA-type analysis or temporal courses in the tICA-type, are mutually independent and the mixing process is linear. Moreover, the GICA approach needs to specify one or more group structure assumptions that expresses our prior knowledge on the data. In the remaining sections of this chapter, we review various GICA approaches based on the type of group structure assumptions. All approaches are illustrated using the sICA-type analysis with applications in fMRI studies. GICA approaches for tICA-type analyses and other neuroimaging modalities can be conducted in a similar way.

5.3 SINGLE-SUBJECT AND THE GROUP STRUCTURE ASSUMPTIONS

Let X_i, $i = 1, \ldots, n$, of size $T \times V$ be the ith data sample. We use T to denote the number of time points and hence each column of X_i is a temporal course at a particular location. Similarly, we use V to represent the number of spatial locations and thus each row of X_i contains all the spatial information, or a vectorized spatial map, at a specific time point.

The single-subject ICA that is applied on a single data matrix X_i decomposes X_i into a set of spatial maps, represented by the rows of S_i, and the temporal courses, represented by the columns of A_i. For the noise-free ICA model, it can be written as

$$X_i = A_i S_i.$$

Let r denote the number of independent components. Then the matrix A_i is of size $T \times r$ and S_i is of size $r \times V$.

As mentioned in the introduction, group structure is the way we model the relationship between the subject- and group-level spatial maps and temporal courses. Group structure plays an important role in developing GICA approaches. Since we review sICA-type GICA approaches in this chapter, in the following sections we summarize several group structure assumptions based on spatial maps.

Let A and S, which are of the same sizes as A_i and S_i, respectively, represent the group-level temporal courses and spatial maps. The following types of group structure assumptions are commonly used in GICA approaches.

1. Homogeneous in both space and time: $S_i = S$ and $A_i = A$ for all i. In this case, all subject-level components are exactly the same as the group-level components. In other words, all subjects share the same set of spatial maps and temporal courses.

2. Homogeneous in both space and time but with subject-specific weights: $A_i = A\Lambda_i$ and $S_i = S$. Here, Λ_i is a diagonal matrix with the jth diagonal entry being the weight of the jth independent component for subject i. In this case, all subjects share the same set of spatial maps and temporal courses but the weight for each pair of spatial maps and temporal courses is assumed to be different across subjects.

3. Homogeneous in space: $S_i = S$. In this case, all subjects share the same set of spatial maps but the temporal courses are assumed to be different.

4. Inhomogeneous in space. In this case, subject-level spatial maps are assumed to be different across subjects and are modeled as a function of the group-level features, i.e., $S_i = f(S)$. Here $f(\cdot) : \mathbb{R}^{r \times V} \to \mathbb{R}^{r \times V}$ is a function indicating that the group-level components can be extracted from the subject-level features.

In this chapter, we only illustrate the group structures for sICA-type analysis. For tICA-type analysis, we can model the group structure in a similar way with S_i denoting the temporal courses and A_i being the spatial maps.

5.4   HOMOGENEOUS IN SPACE

In this setting, all subject-level spatial maps are assumed to be the same as the group-level components and the temporal courses are different across subjects. This type of group structure and GICA approach was proposed by Calhoun et al. [8] and is now one of the most widely used group structures in GICA approaches.

The model assumes that subjects share the same set of spatial maps, i.e., the subject-level spatial maps are the same as the group-level components or $S_i = S$. Under this particular group structure, the single-subject ICA decomposition of each matrix can be written as

$$X_i = A_i S$$

by the noise-free ICA model. If we concatenate the data matrices X_i and the mixing matrices A_i along the temporal direction and denote them by

$$X = [X_1^T, X_2^T, \ldots, X_n^T]^T$$

and

$$A = [A_1^T, A_2^T, \ldots, A_n^T]$$

respectively, mathematically we can combine all single-subject ICA decompositions into one equation as

$$X = AS.$$

The above equation has the same form as the single-subject ICA. This type of group structure naturally leads to a class of GICA approaches that concatenate the data matrices first followed by a single-subject ICA approach [8].

In order to reduce noise in the data, GICA approaches based on this group structure usually adopt a two-step dimension reduction prior to the single-subject ICA. At the first step, a subject-level dimension reduction is applied to each data matrix. Erhardt et al. [14] summarized several subject-level dimension reduction techniques, including subject-specific *principal components analysis* (PCA) or probabilistic PCA (PPCA), spatial concatenation PCA or PPCA, and group data mean PCA or PPCA. These dimension-reduced data matrices are then concatenated along the temporal direction to form one matrix containing all dimension-reduced matrices from the subject-level dimension reduction step. At the second step, a group-level dimension reduction is applied to the concatenated matrix by PCA [8] or *canonical correlation analysis* (CCA) [32]. The number of components retained after the second step dimension reduction is usually the same as the number of independent components and hence the mixing matrix is a square matrix.

After the two-step dimension reduction, we obtain a dimension-reduced matrix. Any standard ICA algorithm, e.g., FastICA, then can be applied to this matrix. The group-level spatial maps S are obtained by this single-subject ICA step.

We only illustrate this approach through temporal *concatenation* assuming common spatial maps. We also can concatenate the spatial maps by assuming common temporal courses via a tICA-type analysis [31].

So far we described how to obtain the group-level spatial maps under this group structure assumption. We also can obtain the subject-level spatial maps through another step called the *back-reconstruction* step [8]. For a more detailed comparison of various dimension reduction and back-reconstruction approaches, please see Erhardt et al. [14].

5.5   HOMOGENEOUS IN BOTH SPACE AND TIME

For this type of group structure, all subject-level components, both spatial maps and temporal courses, are assumed to be the same as the group-level components. Mathematically, we have $A_i = A$, $S_i = S$ and the ICA decomposition for each subject can be written as

$$X_i = AS.$$

Denote $\bar{X}$ as the average matrix of the data samples, i.e.,

$$\bar{X} = \frac{1}{n} \sum_{i=1}^{n} X_i,$$

where n is the number of subjects. Since all X_i are assumed to have the same ICA decomposition, mathematically the averaged data matrix $\bar{X}$ still would have the same ICA decomposition as each individual subject, i.e.,

$$\bar{X} = AS.$$

The above observation means that we can average the data matrices first and then apply any standard single-subject ICA approach to the averaged data matrix $\bar{X}$ to obtain the group-level components A and S [27, 28].

The computational cost of this approach probably is the lowest among all GICA approaches. This approach, however, makes strong assumptions on group spatial maps and temporal courses. We can relax the assumptions by allowing subject-specific weights on each individual observation. This approach will be reviewed in the next section.

5.6   HOMOGENEOUS IN BOTH SPACE AND TIME WITH SUBJECT-SPECIFIC WEIGHTS

For this type of group structure, the subject-level temporal courses and spatial maps are both assumed to be the same as the group-level components but each subject has its own weights on the components. This group structure is similar to the one used in the parallel factor analysis or PARAFAC model [20], when organizing the data matrices as a three-way tensor. Mathematically, the model with this group structure can be expressed as

$$X_i = A\Lambda_i S + E_i,$$

where E_i is the subject-level additive noise and Λ_i is a diagonal matrix representing the subject-specific weights. To be more specific, the jth diagonal entry of Λ_i is the weight of the jth spatial map and the associated temporal course for the ith subject.

The *tensor probabilistic independent component analysis* [4] is one of the most used GICA approaches with this type of group structure assumption. It adopts a similar idea to the *alternating least squares* approach for solving the PARAFAC model, using a combination of the single-subject probabilistic independent component analysis [PICA; 3].

Let X denote the temporal concatenation of matrices X_i as in the homogeneous in space case, i.e.,

$$X = [X_1^T, X_2^T, \ldots, X_n^T]^T$$

and B denote the temporal concatenation of $A\Lambda_i$, i.e.,

$$B = [(A\Lambda_1)^T, (A\Lambda_2)^T, \ldots, (A\Lambda_n)^T]^T.$$

Then similar to the homogeneous in space case, we can express these concatenated matrices as

$$X = BS + E,$$

where E is the noise matrix. Note that the above expression is the same as the single-subject probabilistic ICA model. Moreover, the matrix B has a special structure. Namely, the jth column of the matrix B consists of n scaled copies of the jth column of the matrix A, where the scaling factors for each copy are the jth diagonal entries of the matrices Λ_i. Due to this particular structure, once we have an estimate of B, we can estimate the group-level temporal courses A and the subject-specific weights Λ_i by *singular value decomposition* (SVD).

Based on these observations, the tensor PICA approach iteratively decomposes the matrices X and B to update estimates of A, S and Λ_i. The algorithm terminates when it reaches the convergence criterion. For more details about this approach, please see Beckmann and Smith [4].

5.7 INHOMOGENEOUS IN SPACE

In this setting, the group structure assumption is hierarchical. Each subject is assumed to have its own temporal courses and spatial maps, i.e., $X_i = A_i S_i$. The subject-level spatial maps are viewed as a function, $f(\cdot)$, of the group-level components, i.e.,

$$S_i = f(S).$$

Since this type of group structure assumes distinct subject-level spatial maps across subjects, it naturally leads to a two-step GICA approach. At the first stage, standard single-subject ICA is applied to each subject to obtain subject-level components. At the second stage, the group-level spatial maps are extracted from the subject-level components, usually by taking the average.

For single-subject ICA, recall that there does not exist an order relationship among ICA outputs. Each independent component is treated equally by the single-subject ICA algorithm. Hence, after estimating the subject-level spatial maps by applying ICA on each subject, in order to obtain the group-level components from these subject-level ones we need to match these spatial maps across subjects. This is the most difficult task in developing GICA approaches under this group structure assumption. Many approaches on components matching either manually or automatically have been introduced in the literature. Calhoun et al. [7] used a method that manually matches these extracted subject-level spatial maps. Another class of approaches is based on some similarity measurements followed by automatic matching [e.g., 15, 12, 29]. The similarity metric can be the correlation coefficients of the spatial maps, the temporal courses, or a weighted sum of the two.

Under this type of group structure, Guo and Tang [19] recently proposed a *hierarchical PICA model*. The first-level model expresses each subject by a set of subject-specific temporal courses and spatial maps. The model of this level can be written as

$$X_i = A_i S_i + E_{1i},$$

where E_{1i} denotes the additive noise for the ith subject. The second-level model relates the subject-level spatial maps and the group-level components by

$$S_i = S + E_{2i},$$

where E_{2i} represents the difference between the subject-level spatial maps and the group-level components. Each group-level source spatial map is further assumed to follow a mixture of Gaussian distributions and the parameters of this hierarchical model are then estimated by the *EM algorithm*.

5.8 APPROACHES WITH MULTIPLE GROUP STRUCTURES

Most of the aforementioned approaches are developed based on a particular group structure assumption. Some other approaches can assume one or more group structures. Under the homogeneous in space setting assumption, Guo and Pagnoni [18] and Guo [17] proposed a general framework that can incorporate various assumptions of the temporal courses. For instance, the temporal courses can be assumed to be different across subjects, as in the homogeneous in space case, or to be the same but with subject-specific weights, as in the homogeneous in both space and time but with subject-specific weights case. The distributions of spatial maps are modeled by a mixture of Gaussian distributions. The temporal courses with a particular group structure assumption and the parameters in the mixture of Gaussian distributions are estimated via the EM algorithm.

5.9 SOFTWARE

There are a number of software packages that implement GICA approaches for neuroimaging data analysis. The most widely used software packages of GICA are *GIFT* (http://mialab.mrn.org/software/gift/) and *MELODIC* (http://fsl.fmrib.ox.ac.uk/fsl/fslwiki/MELODIC). Some other toolboxes implementing GICA approaches that we reviewed in this chapter include *AnalyzeFMRI* [6], *BrainVoyager* [16], *EEGIFT* [13], and so on.

5.10 CONCLUSION

Group ICA is a very popular and powerful approach for analyzing multiple or groups of neuroimaging samples. In this chapter, we gave a brief review of various group

ICA approaches. We organized the review according to the group structure assumptions: homogeneous in space, homogeneous in both space and time with or without subject-specific weights, inhomogeneous in space, and approaches with multiple group structures. For each GICA approach, we focused on the estimation of the spatial maps and temporal courses with particular application in fMRI.

Bibliography

1. S. Amari. Natural gradient works efficiently in learning. *Neural Computation*, 10(2):251–276, 1998.
2. S. Amari, A. Cichoki, and T.P. Chen. A new learning algorithm for blind source separation. *Advances in Neural Information Processing Systems*, 8:757–763.
3. C. F. Beckmann and S. M. Smith. Probabilistic independent component analysis for functional magnetic resonance imaging. *IEEE Transactions on Medical Imaging*, 23(2):137–152, 2004.
4. C. F. Beckmann and S. M. Smith. Tensorial extensions of independent component analysis for multisubject FMRI analysis. *NeuroImage*, 25:294–311, 2005.
5. A. J. Bell and T. J. Sejnowski. An information-maximization approach to blind separation and blind deconvolution. *Neural Computation*, 7(6):1129–1159, 1995.
6. C. Bordier, M. Dojat, and P. L. de Micheaux. Temporal and spatial independent component analysis for fMRI data sets embedded in the AnalyzeFMRI R package. *Journal of Statistical Software*, 44(9):1–24, 2011.
7. V. D. Calhoun, T. Adali, V. B. McGinty, J. J. Pekar, T. D. Watson, and G. D. Pearlson. fMRI activation in a visual-perception task: Network of areas detected using the general linear model and independent components analysis. *NeuroImage*, 14(5):1080 – 1088, 2001.
8. V. D. Calhoun, T. Adali, G. D. Pearlson, and J. J. Pekar. A method for making group inferences from functional MRI data using independent component analysis. *Human Brain Mapping*, 14:140–151, 2001.
9. A. Cichocki, R. Zdunek, A. H. Phan, and S. Amari. *Nonnegative Matrix and Tensor Factorizations: Applications to Exploratory Multi-Way Data Analysis and Blind Source Separation*. John Wiley & Sons, 2009.
10. P. Comon. Independent component analysis, a new concept? *Signal Processing*, 36(3):287–314, 1994.
11. P. Comon and C. Jutten. *Handbook of Blind Source Separation: Independent Component Analysis and Applications*. Academic Press, 2010.
12. M. De Luca, C. F. Beckmann, N. De Stefano, P. M. Matthews, and S. M. Smith. fMRI resting state networks define distinct modes of long-distance interactions in the human brain. *NeuroImage*, 29(4):1359–1367, 2006.
13. T. Eichele, S. Rachakonda, B. Brakedal, R. Eikeland, and V. D. Calhoun. EEGIFT: Group independent component analysis for event-related EEG data. *Computational Intelligence and Neuroscience*, 2011:1–9, 2011.

14. E. B. Erhardt, S. Rachakonda, E. J. Bedrick, E. A. Allen, T. Adali, and V. D. Calhoun. Comparison of multi-subject ICA methods for analysis of fMRI data. *Human Brain Mapping*, 32(12):2075–2095, 2011.

15. F. Esposito, T. Scarabino, A. Hyvarinen, J. Himberg, E. Formisano, S. Comani, G. Tedeschi, R. Goebel, E. Seifritz, and F. Di Salle. Independent component analysis of fMRI group studies by self-organizing clustering. *NeuroImage*, 25: 193–205, 2005.

16. R. Goebel, F. Esposito, and E. Formisano. Analysis of functional image analysis contest (FIAC) data with BrainVoyager QX: From single-subject to cortically aligned group general linear model analysis and self-organizing group independent component analysis. *Human Brain Mapping*, 27(5):392–401, 2006.

17. Y. Guo. A general probabilistic model for group independent component analysis and its estimation methods. *Biometrics*, 67:1532–1542, 2011.

18. Y. Guo and G. Pagnoni. A unified framework for group independent component analysis for multi-subject fMRI data. *NeuroImage*, 42(3):1078–1093, 2008.

19. Y. Guo and L. Tang. A hierarchical model for probabilistic independent component analysis of multi-subject fMRI studies. *Biometrics*, 69(4):970–981, 2013.

20. R. A. Harshman. Foundations of the PARAFAC procedure: Models and conditions for an "explanatory" multimodal factor analysis. *UCLA Working Papers in Phonetics*, 16:1–84, 1970.

21. J. Herault and C. Jutten. Space or time adaptive signal processing by neural network models. In *Neural Networks for Computing*, volume 151, pages 206–211. AIP Publishing, 1986.

22. A. Hyvärinen. Fast and robust fixed-point algorithms for independent component analysis. *Neural Networks, IEEE Transactions on*, 10(3):626–634, 1999.

23. A. Hyvärinen and E. Oja. A fast fixed-point algorithm for independent component analysis. *Neural Computation*, 9(7):1483–1492, 1997.

24. A. Hyvärinen and E. Oja. Independent component analysis: Algorithms and applications. *Neural Networks*, 13(4):411–430, 2000.

25. A. Hyvärinen, J. Karhunen, and E. Oja. *Independent Component Analysis*. Wiley, New York, 2001.

26. C. Jutten and J. Herault. Blind separation of sources, part i: An adaptive algorithm based on neuromimetic architecture. *Signal Processing*, 24(1):1–10, 1991.

27. V. J. Schmithorst and S. K. Holland. Multiple networks recruited during a story processing task found using group inferences across subjects from independent component analysis. In *Proceedings of the 10th Annual Meeting of ISMRM*, page 754, Honolulu, HI, 2002.

28. V. J. Schmithorst and S. K. Holland. Comparison of three methods for generating group statistical inferences from independent component analysis of functional magnetic resonance imaging data. *Journal of Magnetic Resonance Imaging*, 19:365–368, 2004.

29. V. Schopf, C. H. Kasess, R. Lanzenberger, F. Fischmeister, C. Windischberger, and E. Moser. Fully exploratory network ICA (FENICA) on resting-state fMRI

data. *Journal of Neuroscience Methods*, 192(2):207–213, 2010.

30. J. V. Stone. *Independent Component Analysis: A Tutorial Introduction.* MIT Press, Cambridge, MA, 2004.

31. M. Svensén, F. Kruggel, and H. Benali. ICA of fMRI group study data. *NeuroImage*, 16(3):551–563, 2002.

32. G. Varoquaux, S. Sadaghiani, P. Pinel, A. Kleinschmidt, J. B. Poline, and B. Thirion. A group model for stable multi-subject ICA on fMRI datasets. *NeuroImage*, 51(1):288–299, 2010.

33. X. Yu, D. Hu, and J. Xu. *Blind Source Separation: Theory and Applications.* John Wiley & Sons, 2014.

6 Polynomial Spline Independent Component Analysis with Application to fMRI Data

Atsushi Kawaguchi
Kyoto University Graduate School of Medicine

Young K. Truong
University of North Carolina at Chapel Hill

CONTENTS

An independent component analysis (ICA) algorithm was developed to capture the brain activity spatially and temporally. In order to improve the spatial map extraction, a mixture of density functions was employed to model the spatial distribution. Here, the sparse density (believed to be directly related to the experimental task) is estimated using a flexible nonparametric density procedure using polynomial splines, whereas the remainder is approximated by the logistic density function. The advantage of our method in spatial map extraction was demonstrated in a comparative study using several popular algorithms. The proposed procedure was applied to a medical study involving functional magnetic resonance imaging (fMRI) data for identifying activated spatial features that are correlated with various specific experimental tasks.

6.1 INTRODUCTION

In independent component analysis (ICA), it is assumed that the components of the observed k-dimensional random vector $\mathbf{x} = (x_1, \ldots, x_k)$ are linear combinations of the components of a latent k-vector $\mathbf{s} = (s_1, \ldots, s_k)$ such that $s_1, \ldots, s_k$ are mutually independent. This is denoted by

$$\mathbf{x} = \mathbf{As}, \tag{6.1}$$

where $\mathbf{A}$ is a $k \times k$ full-rank non-random mixing matrix. The main objective then is to extract the mixing matrix through a set of observations $\mathbf{x}_1, \mathbf{x}_2, \ldots, \mathbf{x}_n$. Chapter 5 provides an overview of this method, including its motivation, existence, and relationship with other well-known statistical methods such as principal component analysis, and factor analysis. See also the references cited there.

In signal processing applications, it will be convenient to view the observations as values recorded at k locations over time periods $1, \ldots, n$. The number of locations k varies depending on the application area. For instance, $k = 2$ in a blind source separation problem to $k \approx 10^5$ in a typical human neuroimaging data set. The number of time points n also varies and it ranges from $n \approx 10^2$ (fMRI) to $n \approx 10^6$ (EEG).

The spatial (location or space) and temporal (time) description of the data has generated a huge number of biomedical applications such as cognitive or genomic research. In this chapter, we will focus on human brain data acquired from functional magnetic resonance imaging (fMRI) technique where $k \approx 10^5$ and $n \approx 10^2$. This imaging technique has been used to effectively study brain activities in a non-invasive manner by detecting the associated changes in blood flow. Typically, fMRI data consists of a 3D grid of voxels; each voxels response signal over time reflects brain activity. However, response signals are often contaminated by other signals and noise, the magnitude of which may be as large as that of the response signal. Therefore, independent component analysis has been applied to extract the spatial and temporal features of fMRI data [23, 6].

For fMRI datasets, we remark that it is theoretically possible to search for signals that are independent over space (spatial ICA) or time (temporal ICA). In fact, the above ICA description involving the spatial k and temporal n scales should be called, more precisely, the *temporal ICA*, while in spatial ICA, k will be treated as time, and n as location. Thus one can see that temporal ICA is just the transpose of spatial ICA. However, in practice, it is very difficult to obtain accurate and meaningful results from the temporal ICA of fMRI data because of the correlation among the temporal physiological components. Therefore, the use of spatial ICA is preferred for fMRI analysis [23].

Our ICA on fMRI data is carried out by first reducing the number of independent components (IC) using tools such as principal component analysis (PCA) or singular value decomposition (SVD), followed with an algorithm for determining the ICs. The most commonly used ICA algorithms for analyzing fMRI data are Infomax [3], FastICA [18], and joint approximate diagonalization of eigenmatrices (JADE) [9]. The Infomax method was noted to yield consistent and reliable results in fMRI data analysis, followed closely by JADE and FastICA [6]. In this chapter, we introduce a novel ICA algorithm that is a modification of the *logspline ICA algorithm* (LICA) [19] with an illustration using human neuroimaging data.

In LICA, we employ a likelihood approach to search for ICs by estimating their probability distribution or density functions (pdf). This is equivalent to maximizing the independence among ICs, and it is realized by using polynomial splines to approximate the logarithmic pdf; we call this the *logspline* model. In the framework of blind source separation (BSS), it was shown by computer simulation that the method performed quite well compared to many existing methods [19]. Section 6.6 highlights some of these comparisons with a description of the methodology. The situation in fMRI data, however, is quite different from BSS. Probability density functions (pdfs) of spatial sources in fMRI data are known to be a mixture due to the fact that brain activation is, in general, sparse and highly localized [23, 26]. Such activated regions have very low frequency relative to the non-activation regions because of the localization. Thus it is difficult to acquire the information of activated regions in the density estimation. To account for the sparsity of spatial fMRI maps, we further treat the pdf as a mixture of a logspline and a logistic density function with the aim to model the activation and non-activation regions of the spatial sources. This approach has been shown to be very effective for handling *sparse features* in fMRI data. Using simulated and real data, we compared our method with several well-known, existing methods and demonstrated the relative advantage of our method in extracting ICs.

The remainder of this chapter is organized as follows. Section 6.2 describes the proposed method. Section 6.3 presents the simulation studies. Section 6.4 describes the application of the proposed method to real data. Finally, Section 6.5 presents some remarks on this study.

6.2 METHOD

Let $\mathbf{Y}$ denote a $T \times V$ data matrix: each column of this matrix corresponds to a voxel time series, and there are V voxels and T time points. We invoke singular value decomposition (SVD) to yield $\mathbf{Y} = \mathbf{UDX}$, where $\mathbf{U}$ is a $T \times M$ orthogonal matrix, $\mathbf{D} = diag(d_1, d_2, \ldots, d_M)$ with $d_1 \geq d_2 \geq \cdots \geq d_M$, and $\mathbf{X}$ is an $M \times V$ orthogonal matrix. Here, we selected (orthogonal) columns of $\mathbf{U}$ to consider some experimental task functions as well as the physiological components. In addition, the dimension of $\mathbf{D}$ has been reduced by discarding values below a certain threshold; in other words, these values are essentially treated as noise.

We determine the ICs based on the matrix $\mathbf{X}$ so that $\mathbf{X} = \mathbf{AS}$, where $\mathbf{A}$ is an $M \times M$ mixing matrix and S is an $M \times V$ source matrix. That is, the v-th column of $\mathbf{X}$ is equal to $\mathbf{A}$ multiplied by the v-th column of $\mathbf{S}$, where $v = 1, 2, \ldots, V$. Equivalently, each column of $\mathbf{X}$ is a mixture of M independent sources. Let $\mathbf{S}_v$ denote the source vector at voxel v so that $\mathbf{S}_v = (S_1, S_2, \ldots, S_M)$, $v = 1, 2, \ldots, V$. Suppose that each S_j has a density function f_j for $j = 1, 2, \ldots, M$. Then, the density function of $\mathbf{X}$ can be expressed as $f_{\mathbf{X}}(\mathbf{x}) = \det(\mathbf{W}) \prod_{j=1}^{M} f_j(\mathbf{w}_j \mathbf{x})$, where $\mathbf{W} = \mathbf{A}^{-1}$ and $\mathbf{w}_j$ is the j-th row of $\mathbf{W}$.

We now model each source density according to the mixture with unknown probability a:

$$f_j(x) = a f_{1j}(x) + (1-a) f_{2j}(x), \tag{6.2}$$

where the logarithm of $f_{1j}(x)$ is modeled by using polynomial splines

$$\log(f_{1j}(x)) = C(\beta_j) + \beta_{01j} x + \sum_{i=1}^{m_j} \beta_{1ij}(x - r_{ij})_+^3,$$

where $\beta_j = (\beta_{01j}, \beta_{11j}, \ldots, \beta_{1m_j j})$ is a vector of coefficients, $C(\beta_j)$ is a normalized constant, r_{ij} are the knots, and $(a)_+ = \max(a, 0)$; $f_{2j}(x) = \operatorname{sech}^2(x)/2$ is a logistic density function.

We denote the vector of parameters in the density function by $\theta = (a, \beta)$. The maximum likelihood estimate (MLE) of $(\mathbf{W}, \theta)$ is obtained by maximizing the likelihood of $\mathbf{X}$ with respect to $(\mathbf{W}, \theta)$:

$$\ell(\mathbf{W}, \theta) = \sum_{i=1}^{n} \sum_{j=1}^{k} \log(f_j(\mathbf{w}_j^T \mathbf{x}_i)).$$

We use a profile likelihood procedure to compute the MLE because a direct computation of the estimates is generally not feasible. The iterative algorithm is shown in Table 6.1.

Table 6.1

Algorithm

---

1. Initialize $\mathbf{W} = \mathbf{I}$.
2. Repeat until the convergence of $\mathbf{W}$, using the Amari metric.
   a. Given $\mathbf{W}$, estimate the log density $g_j = \log f_j$ for the jth element X_j of $\mathbf{X}$ (separately for each j) by using the stochastic EM algorithm described in Section 6.7
   b. Given g_j $(j = 1, 2, \ldots, p)$,

$$\mathbf{w}_j \leftarrow \operatorname{ave}[\mathbf{X} g_j{}'(\mathbf{w}_j^T \mathbf{X})] - \operatorname{ave}[g_j{}''(\mathbf{w}_j^T \mathbf{X})] \mathbf{w}_j$$

   where $\mathbf{w}_j$ is the jth column of $\mathbf{W}$ and ave is a sample average over $\mathbf{X}$.
   c. Orthogonalize $\mathbf{W}$

Note that the *Amari metric* [1] used in the algorithms is defined as

$$d(\mathbf{P},\mathbf{Q}) = \frac{1}{p(p-1)} \left\{ \sum_{i=1}^{p} \left(\frac{\sum_{j=1}^{p} |a_{ij}|}{\max_j |a_{ij}|} - 1 \right) + \sum_{j=1}^{p} \left(\frac{\sum_{i=1}^{p} |a_{ij}|}{\max_i |a_{ij}|} - 1 \right) \right\},$$

where $a_{ij} = (\mathbf{P}^{-1}\mathbf{Q})_{ij}$, $\mathbf{P}$, and $\mathbf{Q}$ are $p \times p$ matrices. This metric is normalized, and is between 0 and 1.

Several authors have discussed initial guesses for ICA algorithms. Instead of setting several initial guesses, as discussed in [19], we prewhitened $\mathbf{X}$ by multiplying $\widetilde{\mathbf{W}}$, which is the output of the algorithm when the log density function $g(x)$ is replaced with $g(x) = 1/\{2b^{1/b}\Gamma(1+1/b)\}\exp\{-|x|^b/b\}$, with $b = 3$. The final output is obtained in the form $\widehat{\mathbf{W}} = \widetilde{\mathbf{W}}\mathbf{W}_0$, where $\mathbf{W}_0$ is the output of the algorithm shown in Table 6.1.

The purpose of spatial ICA is to obtain independent spatial maps and the corresponding temporal activation profiles (time courses). Multiplying $\mathbf{X}$ by $\widehat{\mathbf{W}}$, we obtain the estimates of the spatial map $\widehat{\mathbf{S}}$ as $\widehat{\mathbf{S}} = \widehat{\mathbf{W}}\mathbf{X}$. On the other hand, the corresponding time courses are obtained in the form $\widehat{\mathbf{A}} = \widehat{\mathbf{W}}(\mathbf{UD})^{-1}$.

6.3 SIMULATION STUDY

We conducted a simulation study to compare the proposed method with existing methods such as Infomax [3], fastICA [18], and KDICA [12]. We designed our comparative study by using data that emulated the properties of fMRI data. The spatial sources $\mathbf{S}$ consisted of a set of 250×250 pixels. These spatial sources were modulated with four corresponding time courses $\mathbf{A}$ of length 128 to form a 62,500 $\times$ 128 dataset. The spatial source images $\mathbf{S}$ shown in the left-hand side of Figure 6.1 are created by generating random numbers from normal density functions with mean 0 and standard deviation 0.15 for a non-activation region, and mean 1 and standard deviation 0.15 for an activation region. The activated regions consist of squares of d_i pixels on a side, for $i = 1, 2, 3, 4$, that are located at different corners. We consider two situations: d_i's are the same among the four components ($d_1 = d_2 = d_3 = d_4 = d$) and d_i's are different. For the former, we used $d = 20, 30, 40,$ and 50. For the latter, we generated uniform random numbers between 20 and 50 for each d_i. The temporal source signals in the right-hand side of Figure 6.1 are the stimulus sequences convolved with an ideal hemodynamic response function as a task-related component, and sin curves with frequencies of 2, 17, and 32 as other sources. We generated the task-related component by using the R package `fmri` with onset times (11,75) and a duration of 11. We repeated the above procedure 10 times for the case in which d_i's were the same and 50 times for the case in which d_i's were different.

Both the spatial and the temporal accuracies of ICA were assessed by R-square fitting of a linear regression model. The evaluation was carried out as follows. For every estimated time course, the R-square is computed from the linear regression model with the response being each of the estimates and the predictor being true, that is, the stimulus sequence (Comp 1 on the right-hand side in Figure 6.1). The

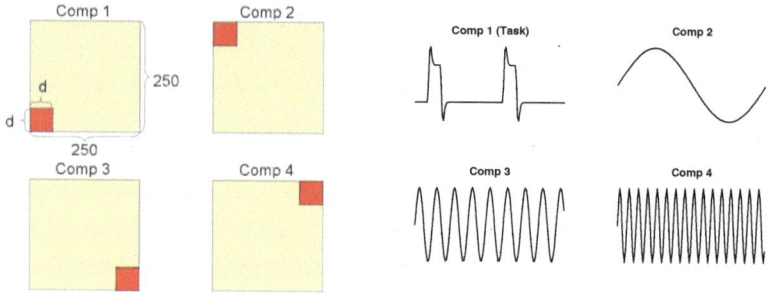

Figure 6.1: Spatial and temporal simulation data

component that has the maximum R-square is considered to be task related. We used the R-square value of this component for the comparison with the existing methods with respect to temporal accuracy and to determine the corresponding spatial map. The intensities of the spatial map are vectorized and used in the linear regression model as the response with the vectorized true (Comp 1 on the left-hand side in Figure 6.1) as the predictor to compute R-square for the spatial accuracy.

Table 6.2

Temporal R-square for simulation data. The mean over $d = 20, 30, 40,$ and 50 is calculated in the row labeled as average. The rand row shows the average over 50 replications when d_i's were chosen randomly from the range 20 to 50.

	Infomax	fastICA	KDICA	LICA
$d=50$	0.627	0.852	0.679	0.843
$d=40$	0.456	0.460	0.472	0.735
$d=30$	0.408	0.463	0.424	0.586
$d=20$	0.358	0.270	0.709	0.518
average	0.462	0.511	0.571	0.670
rand	0.623	0.651	0.529	0.699

Table 6.3

Spatial R-square for simulation data. The mean over $d = $ **20, 30, 40, and 50 is calculated in the row labeled as average. The rand row shows the average over 50 replications when** d_i**'s were chosen randomly from the range 20 to 50.**

	Infomax	fastICA	KDICA	LICA
d=50	0.801	0.765	0.641	0.761
d=40	0.462	0.502	0.545	0.726
d=30	0.409	0.528	0.552	0.680
d=20	0.323	0.478	0.624	0.587
average	0.499	0.568	0.591	0.688
rand	0.537	0.607	0.579	0.643

The averaged R-squares over simulations are summarized in Tables 6.2 and 6.3 for the temporal and spatial data, respectively. When the sizes of the activation region were the same among all components, R-squares of the proposed method were significantly larger than those of others for moderate sizes ($d = 40$ and 30) for both temporal and spatial data. For $d = 50$, fastICA had the largest R-square for both temporal and spatial data, with the difference from the result of the proposed method being small. For $d = 20$, KDICA had the largest R-square for both temporal and spatial data, with the difference from the result of the proposed method being significant for temporal data but not for spatial data. With respect to the average for $d = 50, 40,$ 30, and 20, the proposed method had larger values than did the others. When d_i was determined randomly, which might be more practical, we observed that the proposed method had larger R-squares than did the others in the rand row in the table.

6.4 APPLICATION

To demonstrate the applicability of the proposed method to real data, we separate fMRI data into independent spatial components that can be used to determine three-dimensional brain maps. To study brain regions that are related to different finger tapping movements, fMRI data were obtained from two normal twin subjects (Twin 1 and Twin 2) performing different tasks alternately. The paradigm shown in Figure 6.2 consisted of externally guided (EG) or internally guided (IG) movements based on three different finger sequencing movements performed alternately by either the right or the left hand.

The fMRI dataset has 128 scans that were acquired using a modified 3T Siemens MAGNETOM Vision system. Each acquisition consists of 49 contiguous slices. Each slice contains 64×64 voxels. Hence, each scan produces 64×64×49 voxels. The size of each voxel is 3 mm×3 mm×3 mm. Each acquisition took 2.9388 seconds, with the scan-to-scan repetition time (TR) set to 3 seconds. The dataset was pre-processed using SPM5 [15]. The preprocessing included slice timing, realignment, and smoothing. We masked the image outside the human head using the GIFT

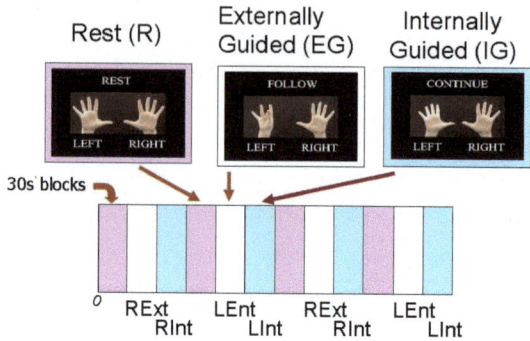

Figure 6.2: Experimental paradigm.

software package (Group ICA of fMRI Toolbox, 7). We used 21 components for Twin 1 and 30 for Twin 2; these were estimated using the minimum description length (MDL) criteria.

We applied four ICA algorithms—Informax [3], fastICA [18], KDICA [12], and the proposed method LICA—to the twins' data. The R-square statistic was calculated from the fitted multiple linear regression model with the estimated time course as the response. The predictors were the right EG, right IG, left EG, and left IG, which consists of the expected BOLD response for the task indicator function given by the argument as a convolution with the hemodynamic response function modeled by the difference between two gamma functions. Table 6.4 shows the corresponding R-square statistics. From this table, we can see that the proposed method extracted more correlated components for a task than did the other methods for both twins.

Table 6.4
R-square statistics for the twin data.

	Infomax	fastICA	KDICA	LICA
Twin 1	0.640	0.666	0.655	0.680
Twin 2	0.847	0.661	0.805	0.862

Figures 6.3 and 6.4 show the resulting spatial maps of LICA for Twin 1 and Twin 2, respectively, in which the right motor area is highly activated and the corresponding time course shows a fit to the left-hand blocks.

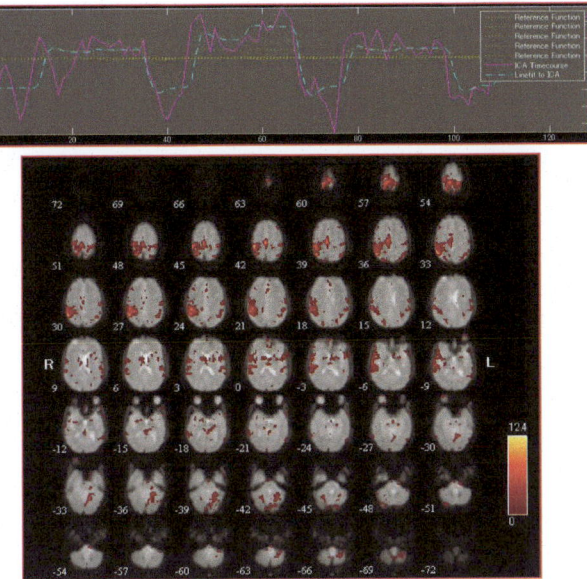

Figure 6.3: Spatial image for Twin 1.

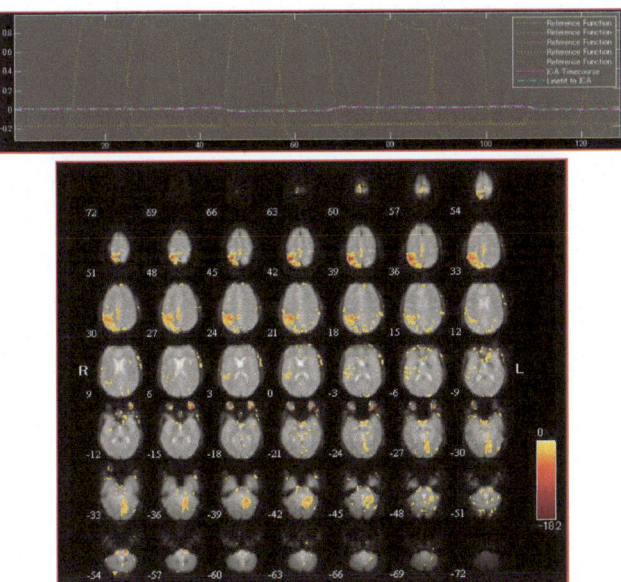

Figure 6.4: Spatial image for Twin 2

We mention a few important observations in this real human brain analysis:

1. After the analysis, it was revealed to us that Twin 1 had shown signs and symptoms (tremors and slowed movements) of Parkinson's disease (PD), while Twin 2 was considered normal at the time the data were collected. This may help to explain why Twin 2, the normal subject, has higher R-squares in three of the four methods (Table 4). In these methods, FastICA shows practically no difference between the twins.

2. In interpreting results from ICA, one should note that ICA is ambiguous about the sign: $\mathbf{x} = \mathbf{As} = (-\mathbf{A})(\mathbf{s})$. This fact has produced different color scales in the spatial maps (located in the lower right corner). With this in mind, one can say that Twin 2, or the normal subject, has a higher intensity or activation level in the right motor area, (because of the left-hand task paradigm).

3. Further examination of the spatial maps indicates that the normal subject (on the right panel) has a more focused location of the motor area, see particularly the red region in slices 51, 48, 45, 42, 39, 36 and 33. The activated motor area of the PD twin (the left panel) is not as sharply defined.

6.5 DISCUSSIONS AND CONCLUSIONS

In this study, we developed an ICA algorithm based on a maximum likelihood approach using a mixture of logspline and logistic density models with adaptive knot locations. The first concern about this approach is that its model dimension seems to be much higher than those of its peers. Here, model dimensionality is defined as the number of model parameters including possibly the spline knot locations. Depending on how noisy the data are, the built-in model selection procedure (which is based on AIC or BIC) works in a sensible adaptive way: there is constantly a trade-off in balancing the bias and variance of the estimate of the parameter since the optimal strategy is to minimize the mean square error loss at the expense of the model dimension. Moreover, the logistic component is included to reduce the model dimension from the spline part in handling the sparsity of the spatial map. The main issue then is the time required to extract the ICs this way. It is considerably more time consuming, but the accuracy is very rewarding. The improvement over its peers performance was demonstrated numerically in Tables 6.2 and 6.3 using the R-square as a criterion.

It is important to point out that we should also provide a sensitivity and specificity analysis of the activated spatial locations as described in [21], where popular methods such as Infomax and fastICA were shown to have a higher false-positive/negative rate. This implies that brain activation should be studied more carefully, and one should avoid using methods that tend to yield false activation.

As in our previous approaches to ICA, the key feature has always been the flexibility in modeling the source. The marginal distribution of the temporal source component was modeled by the logspline methodology and we noted the improvement over its peers. The comparative study was based on a wide variety of density functions, some of which are known to be very challenging to estimate [19]. In pursuing spatial ICA for fMRI based on human brain data, we observed that simply taking the transpose of the temporal ICA approach mentioned in the introduction did not always work. This is due to the fact that the spatial activation maps are very sparse: density estimation using the logspline approach in the presence of sparsity has never been investigated before. One of our findings is that the logspline estimate of the spatial distribution is too noisy, and perhaps the model dimension is too high. Thus the logistic component is added to our previous temporal ICA procedure in order to address this issue. The advantage over the simple tranposition of the temporal approach has been clearly shown in this chapter.

The mixture modeling has been used previously for the detection of brain activation in fMRI data [14, 16, 25]. In fMRI data, the density functions of spatial sources are known to be supergaussion with heavy tails due to the fact that brain activation is sparse and highly localized [23], and often skewed due to larger signal amplitudes in activated regions [26]. Our spline approach is more flexible than some of the parametric procedures such as the improved exponential power family [13].

In addition, the method may have some important extensions. For example, assessing the variability of ICA has been an important problem, especially how best to display the variance in the spatial map. One way to examine the variation of the mixing coefficient estimates is to use the bootstrap method while preserving information about the spatial structure. For example, in spatial ICA, one can generate bootstrap random samples from the logspline density estimates of the source over space. Mix these samples using the estimate mixing coefficients to yield the observed fMRI (BOLD) signals, which will then pass through ICA to produce the so-called bootstrapped spatial maps and mixing coefficients [20]. The algorithm can be described similarly for temporal ICA. Thus it is feasible to develop the statistical inference framework for assessing the variability of the estimator of the mixing matrix via the bootstrap method while preserving information about the spatial or temporal structure.

In extending our temporal ICA to spatial ICA, we merely added the logistic component to the logspline piece, which is essentially a one-dimensional density estimation, or marginal density estimation procedure. Alternatively, in order to capture the actual spatial feature of the three-dimensional brain, or the two-dimensional map, one can incorporate the spatial correlation structure of the spatial map by introducing tensor products of spline functions or the interaction terms in the logspline formulation. For temporal ICA, this can be implemented by using time series models to account for the source serial correlations. Indeed, it has been observed that there is noticeable improvement over the marginal density–based ICA procedures [21]. It will be important to see if the same will hold for the above spatial ICA approach using tensor products of splines.

Another issue that we have not addressed is how to extend our method to compare groups of subjects. This is known as the group ICA problem. In principle, we can follow the Group ICA of the fMRI Toolbox [7] by simply concatenating the observed data matrices. This will certainly increase the computational complexity and one has to address the efficiency problem as well.

Finally, we recall that prior to applying any of the ICA algorithms, one must carry out a dimension reduction step on the observed data matrix first. In temporal ICA with T and V as time and space scales, V will be reduced by, typically, employing the PCA, while the time factor T will be reduced in the spatial ICA. We have found that even greater improvement can be achieved by using informative dimension reduction methods such as SVD by choosing the eigenvectors to closely relate to the experimental task paradigm. This is being referred to as a *supervised SVD dimension reduction procedure* (see Chapter 10) and has been used effectively in fMRI data analysis [21].

In conclusion, the results presented in this chapter can be viewed as a tool for setting up a new framework for addressing some of the known issues in applying ICA to fMRI or other brain imaging modalities such as EEG or neural spike sorting problems. We have demonstrated that the key element here is the flexibility in modeling the source distribution and that was achieved by using polynomial splines as an approximation tool. We also used a mixture of distribution approaches to account for the spatial distribution in ICA for fMRI data analysis. Although there are still many issues to be addressed, we have illustrated the usefulness of our approach to fMRI brain activation detection in both simulated and real data analysis.

6.6 LOGSPLINE DENSITY ESTIMATION

6.6.1 METHODOLOGY

Let X be a random variable having a continuous and positive density function. The log density of X is modeled by

$$g(x) = \log(f(x)) = C(\beta) + \beta_{01}x + \sum_{i=1}^{m} \beta_{1i}(x - r_i)_{+}^{3},$$

where $\beta = (\beta_{01}, \beta_{11}, \ldots, \beta_{1m})$ is a vector of coefficients, $C(\beta)$ is a normalized constant, and r_{ji} are the knots. Let $X_1, \ldots, X_n$ be independent random variables having the same distribution as X. The log-likelihood function corresponding to the logspline family, given by $\ell(\beta) = \sum_{i=1}^{n} g(X_i)$. The maximum likelihood estimate $\hat{\beta}$ is obtained by maximizing the log-likelihood function.

The knot selection methodology involves initial knot placement, stepwise knot addition, stepwise knot deletion, and final model selection based on the information criterion. We set the initial knot placement to be minimum, median, and maximum values of the distribution of data. At each addition step, we first find a good location for a new knot in each of the intervals $(L, r_1), (r_1, r_2), \ldots, (r_{K-1}, r_K), (r_K, U)$ determined by the existing knots $r_1, r_2, \ldots, r_K$ and some constants L and U. Let $X_{(1)}, \ldots, X_{(n)}$ be the data written in nondecreasing order. Set $l_1 = 0$ and $u_K = n$. Define l_i and u_i by

$$l_i = d_{\min} + \max\{j : 1 \leq j \leq n \text{ and } X_{(j)} \leq r_i\}, \quad i = 2, \ldots, K$$

and

$$u_i = -d_{\min} + \max\{j : 1 \leq j \leq n \text{ and } X_{(j)} \geq r_i\}, \quad i = 1, \ldots, K - 1,$$

where $d_{\min}$ is the minimum distance between consecutive knots in order statistics.

For $i = 0, \ldots, K$ and for the model with X_{j_i} as a new knot where $j_i = [(l_i + u_i)/2]$ with $[x]$ being the integer part of x, we compute the Rao statistics R_i defined by

$$R_i = \frac{[S(\hat{\beta})]_i}{\sqrt{[I^{-1}(\hat{\beta})]_{ii}}},$$

where $S(\hat{\beta})$ is the score function; that is, the vector with entries $\partial \ell(\hat{\beta})/\partial \beta_j$, and $I(\hat{\beta})$ is the matrix whose entry in row j and column k is given by $-\partial^2 \ell(\hat{\beta})/\partial \beta_j \partial \beta_k$. We place the potential new knot in the interval $[X_{l_{i^*}}, X_{u_{i^*}}]$ where $i^* = \arg\max R_i$. Within this interval we further optimize the location of the new knot. To do this, we proceed by computing the Rao statistics R_l for the model with $X_{(l)}$ as the knot with $l = [(l_{i^*} + j_{i^*})/2]$ and R_u for the model with $X_{(u)}$ as the knot with $u = [(j_{i^*} + u_{i^*})/2]$. If $R_{i^*} \geq R_l$ and $R_{i^*} \geq R_u$, we place the new knot at $X_{(i^*)}$. If $R_{i^*} < R_l$ and $R_l \geq R_u$, we continue searching for a knot location in the interval $[X_{(l_{i^*})}, X_{(j_{i^*})}]$. If $R_{i^*} < R_u$ and $R_l < R_u$, we continue searching for a knot location in the interval $[X_{(j_{i^*})}, X_{(u_{i^*})}]$.

After a maximum number of knots $K_{\max} = \min(4n^{1/5}, n/4, N, 30)$ where N is the number of distinct X_i's, we continue with stepwise knot deletion. During knot deletion we successively remove the knot, that has minimum Wald statistics defined by

$$W_i = \frac{\hat{\beta}_i}{\sqrt{[I^{-1}(\hat{\beta})]_{ii}}}$$

among the existing knots.

Among all models that are fit during the sequence of knot addition and knot deletion, we choose the model that minimizes the Bayesian information criterion (BIC) defined by $BIC = -2\ell(\hat{\beta}) + m \log(n)$.

6.6.2 NUMERICAL RESULTS

Figure 6.5 contains 18 distributions to be used for a numerical comparative study [2]. The true mixing matrix is chosen at random with bounded condition number between 1 and 2 to simulate our data. The leading ICA algorithms to be compared include FastICA [17]; the JADE algorithm [8]; the extended infomax algorithm: ex-info [22]; one of two versions of KernelICA: KGV [2], the RADICAL algorithm [24]; the NPICA algorithm [5]; the KDICA [11]. Since we got the consistent result with [2] that the resulting Amari metric of KGV are smaller than that of KCCA, we picked KGV over KCCA for the comparative study. The initial values set by KGV is described in [2]. The settings in other algorithms followed the original ones. The data was prewhitened before invoking the algorithm. The difference between the true matrix **A** and the estimated **W** was measured by the Amari metric after the both matrices were adequately orthogonalized.

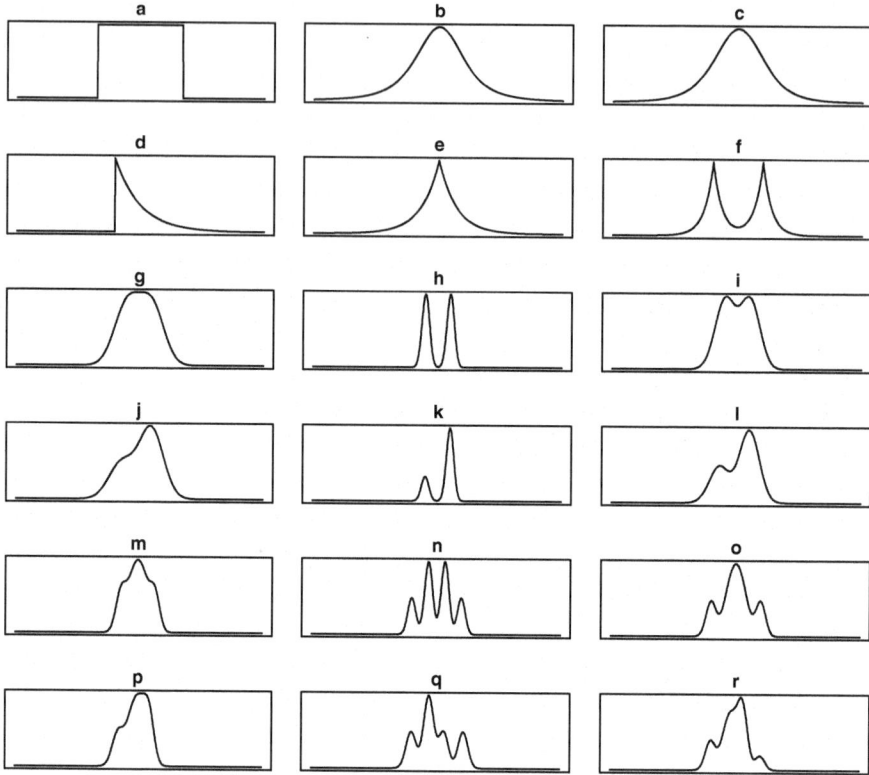

Figure 6.5: Probability density functions of sources with their kurtoses. (a) uniform; (b) student with three degrees of freedom; (c) student with five degrees of freedom; (d) exponential; (e) double exponential; (f) mixture of two double exponentials; (g)-(h)-(i) symmetric mixtures of two Gaussians: unimodal, multimodal and transitional; (m)-(n)-(o) symmetric mixtures of four Gaussians: unimodal, multimodal and transitional; (p)-(q)-(r) nonsymmetrical mixtures of four Gaussians: unimodal, multimodal and transitional. The original MATLAB® code to sample from these distributions was provided from the website of Francis Bach, UC Berkeley. We translated it into R [19].

The first study is for 2 components. Table 6.5 shows the mean results for each source density on each row, with $N = 250$, the number of input points, and 100 replications of each experiment. The bottom row of the table presents the average over replications for which two sources were chosen at random among the 18 densities. KGV and KDICA have the smallest average on the mean row. KGV also has the smallest average for the rand row. LICAs perform better than the other algorithms except for KGV and KDICA, which are (especially for KGV) only marginally better than our algorithm. NPICA seems to have the largest average, but when the initialization employed by KGV is used, the value became smaller. This means that the accuracy of this algorithm depends crucially on the setting of initial values.

Table 6.5

Amari metric (multiplied by 100) for two-component study with 250 samples. For each probability density function (from a to r), averages over 100 replicates are presented. The overall mean is calculated in the row labeled mean. The rand row presents the average over 1000 replications when two (generally different) densities were chosen randomly from the 18 possible densities.

	fastICA	JADE	exinfo	KGV	RADICAL	NPICA	KDICA	LICA
a	3.75	2.34	10.83	2.93	5.00	10.49	2.47	2.46
b	7.69	6.43	13.22	7.04	5.99	10.39	4.84	5.06
c	10.13	12.27	10.54	15.36	13.96	28.13	11.53	13.32
d	9.78	9.65	11.03	1.84	2.15	1.66	1.58	1.99
e	7.08	10.87	10.12	5.91	6.45	15.67	5.87	5.45
f	4.52	4.89	9.18	1.57	2.05	76.04	1.49	1.63
g	20.51	15.25	11.41	25.48	38.98	41.99	18.49	23.20
h	2.17	1.45	11.30	1.20	1.93	91.69	1.19	1.32
i	11.71	8.58	11.40	10.18	21.21	23.91	11.17	13.80
j	23.37	19.71	11.71	9.74	20.08	21.16	15.30	14.62
k	50.65	12.60	11.67	1.29	2.24	77.50	2.22	1.41
l	13.93	10.10	11.71	4.96	8.00	9.11	5.64	6.16
m	10.43	7.26	11.57	9.11	21.08	19.24	8.03	10.99
n	10.30	6.47	11.60	2.64	2.21	17.95	1.50	1.70
o	26.21	10.64	11.64	4.89	7.61	25.92	4.78	4.45
p	10.57	9.80	11.55	7.53	13.70	14.47	8.87	9.99
q	11.70	5.69	11.70	3.16	3.04	4.03	2.18	2.55
r	43.09	25.96	11.67	5.64	14.25	26.97	11.81	7.46
mean	15.42	10.00	11.33	6.69	10.55	28.68	6.61	7.09
rand	11.85	9.11	19.43	4.98	7.76	11.48	5.30	5.08

Table 6.6 shows the mean results for each source density on each row, with $N = 1000$, the number of input points, and 100 replications of each experiment. The table shows that LICA has the smallest average on both the mean and rand rows, clearly demonstrating its superior performance.

Table 6.6

Amari metric (multiplied by 100) for two-component study with 1000 samples. For each probability density function (from a to r), averages over 100 replicates are presented. The overall mean is calculated in the row labeled mean. The rand row presents the average over 1000 replications when two (generally different) densities were chosen randomly from the 18 possible densities.

	fastICA	JADE	exinfo	KGV	RADICAL	NPICA	KDICA	LICA
a	4.49	1.34	1.84	0.86	1.63	6.74	0.85	0.68
b	2.96	3.89	4.12	2.70	3.92	9.22	2.59	2.34
c	6.35	4.86	11.00	5.05	6.33	17.55	5.55	4.77
d	3.07	4.18	7.47	0.56	0.74	0.70	0.53	0.97
e	5.23	4.63	11.09	2.76	3.16	6.42	2.23	2.22
f	2.25	2.32	1.07	0.73	1.01	89.98	0.70	0.76
g	8.54	6.92	6.22	11.27	15.59	29.12	9.16	8.37
h	1.55	0.62	0.54	0.56	0.90	93.01	0.52	0.57
i	3.56	3.91	3.47	4.99	6.84	9.10	4.34	4.34
j	7.93	6.80	5.86	5.85	6.71	6.22	5.51	4.19
k	59.45	4.55	7.02	0.61	1.01	87.41	0.58	0.64
l	5.32	3.97	3.44	2.67	3.46	2.83	2.41	2.01
m	3.44	2.81	3.88	3.78	6.20	5.96	3.20	3.16
n	4.21	2.17	3.47	0.57	0.81	50.16	0.56	0.61
o	14.44	3.39	15.98	1.42	2.26	28.40	1.36	1.45
0	6.63	3.73	4.49	4.32	5.44	2.92	3.54	3.68
q	6.97	3.01	7.46	0.85	1.43	1.72	0.86	0.87
r	44.00	12.34	23.80	2.20	3.16	9.86	2.94	2.32
mean	10.58	4.19	6.79	2.87	3.92	25.41	2.64	2.44
rand	5.47	4.19	5.91	2.11	2.97	4.58	1.99	1.86

The second study is for 2, 4, 8 and 16 components. The source densities were selected at random from the 18 densities. Results are presented in Table 6.7. The bottom row presents the average over each column. LICA has the smallest average among all algorithms.

In summary, LICA performance is similar to the best algorithms for smaller data sets while it is the best for moderate data sets. This may be explained by the fact that LICA is based on the highly adaptive logspline density estimation methodology having the knots selected adaptively. For smaller data sets, the variability of the logspline estimates seems to dominate the bias. The variability apparently subsides for moderate data sets.

Table 6.7

Results for experiments in higher dimensions (again, mean Amari error multiplied by 100). The table shows experiments for dimensions 2 through 16. The number of points used for each experiment is shown in the second column and the number of experiment replications is shown in the third column [19]. (Reprint with permission from *Bulletin of Informatics and Cybernetics*)

dims	N	#repl	fastICA	JADE	exinfo	KGV
2	250	1000	11.85	9.11	19.43	4.98
2	1000	1000	5.47	4.19	5.91	2.11
4	1000	100	4.27	4.56	11.06	3.43
4	4000	100	1.91	2.21	4.21	1.29
8	2000	50	2.97	3.21	11.32	3.53
8	4000	50	1.97	2.22	6.04	2.79
16	4000	25	1.95	2.50	13.52	17.83
		mean	4.34	4.00	10.21	5.13

dims	N	#repl	RADICAL	NPICA	KDICA	LICA
2	250	1000	7.76	11.48	5.30	5.08
2	1000	1000	2.97	4.58	1.99	1.86
4	1000	100	3.12	2.30	2.36	2.09
4	4000	100	1.49	1.36	1.22	0.94
8	2000	50	1.90	1.36	2.07	1.33
8	4000	50	1.19	0.99	1.48	2.91
16	4000	25	1.44	0.91	3.05	1.08
		mean	2.84	3.28	2.50	2.19

6.7 STOCHASTIC EM ALGORITHM

We apply the stochastic EM algorithm to estimate parameters a and β in (6.2). The stochastic EM algorithm was introduced in [10] to avoid stabilization on saddle points in parametric mixture models, incorporating the stochastic step (S-step) into the EM algorithm. In [4], semiparametric mixture models were generalized by using kernel density estimation.

Suppose we have observations $x_1, x_2, \ldots, x_n$, and the observations are grouping by the k-means clustering method with $k = n/10$. Denote the number of members in each group by n_g ($g = 1, 2, \ldots, n/10$) and $x_g = (x_{i_{g1}}, x_{i_{g2}}, \ldots, x_{i_{gn_g}})$. The algorithm used in this chapter is given below.

(1) E-step: Compute $\tau(j|x_g)$ ($g = 1, 2, \ldots, n/10$, $j = 1, 2$) using

$$\tau(j|x_g) = \frac{1}{n_g} \sum_{h=1}^{n_g} \widetilde{\tau}(j|x_{i_{gh}})$$

where $\widetilde{\tau}(j|x) = af_j(x)/f(x)$.

(2) S-step: Draw randomly $z(x_g)$ from a Bernoulli distribution with probability of $\tau(1|x_g)$ and define $z(x_{i_{gh}}) = 1$ if $z(x_g) = 1$ and $z(x_{i_{gh}}) = 1$, otherwise for $g = 1, 2, \ldots, n/10$ and $h = 1, 2, \ldots, n_g$.

(3) M-step: The estimator of a is given by

$$\hat{a} = \frac{1}{n} \sum_{i=1}^{n} z(x_i).$$

f_1 is estimated by maximizing the likelihood described in Section 6.6 based on x_i for $i \in \{i; z(x_i) = 1\}$.

These steps alternate until convergence. For the log-spline density f_1, the maximum likelihood estimation is applied. The data-driven knot locations in f_1 are optimized as in Section 6.6. We use the k-means method to initialize a and f_1. Of separated observations by the k-means method, the ones that have larger mean are used to initialize f_1. It is possible that the stochastic EM algorithm may not converge but be stable [10, 4]. Therefore, we use a large number of iterations so as to stabilize the estimate of f. We then pick up the $\hat{f}$ as the final estimate, whose likelihood $\sum_{i=1}^{n} \log \hat{f}(x_i)$ is maximum among the iterations.

6.8 SOFTWARE: R

The R functions used in the chapter are:

```
Example.R
MLICA.r
a24random.r
amarimetric.r
logsplinederivative.r
pdfunction.r
UniOrthoMat4multi.r
```

The code of each function is listed in the following subsections.

6.8.1 EXAMPLE.R

```
####################################################################
# Example
#  you need to specify the working directory
####################################################################

##################
### Load Files ###
##################

source("amarimetric.r")
```

```
source("logsplinederivative.r")
source("MLICA.r")
source("pdfunction.r")
source("a24random.r")
source("UniOrthoMat4multi.r")


###############
### Setting ###
###############

pdfname <- letters[1:18]
   #pdfname (a)-(r) corresponding to Bach and Jordan (2003)

### parameter in MLICA ###
maxitri <- 10 #the number of iteration for global maximization
penai <- 1 #alpha in penalty of AIC
tailp <- 0 #initial knot locations (min, median max)

n.comp<-2 #number of components

### set initial matrices ###
Wcands <- uni.ortho.mat(n.comp, maxitri)


#################
### Example 1 ###
#################

pdfi <- 1 #number of pdf's
n.data <- 1000 #number of samples

### Data ###
set.seed(1)
S <- matrix(myrandom(pdfi, n.data*n.comp), n.data, n.comp)
A <- matrix(runif(n.comp^2, 1, 2), n.comp, n.comp)
X<-S%*%A

### Run algorithm ###
set.seed(1)
system.time(
fit1<-MLICA(X=X, white.on=1, Wcands=Wcands,
        max.itr=maxitri, penai, tailp))[3]

### RESULT ###
print(paste("###############", pdfname[pdfi], "###############"))
```

```
print("### True Mixing Matrix")
print(A)
print("### Target")
print(solve(A%*%fit1$K))
print("### Estimated DeMixing Matirx")
print(fit1$W)
oWO <- solve(svd(A%*%fit1$K)$u%*%t(svd(A%*%fit1$K)$v))
    #orthogonalization
print(paste("Amari =", round(amari(fit1$W, oWO),3)))
print("### White x Estimated")
print(fit1$K%*%fit1$W)



#################
### Example 2 ###
#################

n.data <- 1000 #number of samples

### Data (randm select) ###
set.seed(1)
pdfis<-rep(0, n.comp)
S <- matrix(0, n.data, n.comp)
for(i in 1:n.comp)
{
pdfi<-ceiling(runif(1)*18)
pdfis[i]<-pdfi
S[,i] <- myrandom(pdfi, n.data)
}
A <- matrix(runif(n.comp^2, 1, 2), n.comp, n.comp)
X<-S%*%A

### Run algorithm ###
set.seed(1)
system.time(
fit1<-MLICA(X=X, white.on=1, Wcands=Wcands,
        max.itr=maxitri, penai, tailp))[3]

### RESULT ###
print(paste("#############", pdfname[pdfis], "#############"))
print("### True Mixing Matrix")
print(A)
print("### Target")
print(solve(A%*%fit1$K))
```

```
print("### Estimated DeMixing Matirx")
print(fit1$W)
oW0 <- solve(svd(A%*%fit1$K)$u%*%t(svd(A%*%fit1$K)$v))
    #orthogonalization
print(paste("Amari =", round(amari(fit1$W, oW0),3)))
print("### White x Estimated")
print(fit1$K%*%fit1$W)
```

6.8.2 MLICA.R

```
################################################################
#logspline Independent Component Analysis
################################################################


library(polspline)

# load source file of amari metric
source("amarimetric.r")
# load source file of derivative of density
source("logsplinederivative.r")
pdfname <- letters[1:18]


#############################################################
#####                Main Function                  #####
##### X : data
##### white.on: 1=whitening data, else=do not
##### Wcands: the initial matrixes
##### max.itr :  the number of iteration
##### pena: the penalty term of AIC on density estimation
#############################################################
MLICA<-function(X, white.on, Wcands, max.itr, penai, tailp)
{ N <- length(X[,1])
  P <- length(X[1,])

if(white.on==1) {
  ###Whitening(by principal component package)###
  pcX   <- princomp(X)
  # Whitening_mat (sweep: similar to "apply")
  White <- sweep(pcX$loadings[,1:P], 2, pcX$sdev[1:P], FUN="/");
  Xw  <- sweep(X, 2, pcX$center) %*% White
}
else if(white.on==2) {
  ###Whitening(by SVD)###
  X2 <- sweep(X, 2, apply(X,2,mean))#centering
```

```
  tXX <- t(X2) %*% X2#sum of square (Covariance of X2)
  Xu <- svd(tXX)$u
  Xd <- svd(tXX)$d/(N-1)
  White2<-Xu%*%diag(1/sqrt(Xd))
  Xw2 <- X2%*%White2
}
else {
  Xw<-X
}


glob1 <- global_maxi(Xw, N, P, max.itr, Wcands, penai, tailp)
W.final <- glob1$W

##### Final output #####
#X pre-processed data matrix
#K pre-whitening matrix that projects data onto
#                                    the first n.comp PC.
#W estimated un-mixing matrix (see definition in details)
#A estimated mixing matrix
#S estimated source matrix
if(white.on==0) {
  list(X=Xw, W=W.final, S=Xw%*%t(W.final))
} else {
   list(X=Xw, K=White, A=solve(White%*%W.final), W=W.final,
     S=Xw%*%W.final, ahis=glob1$ahis, cotrhis=glob1$cotrhis,
     chist=glob1$chist, wchist=glob1$wchist,
     Wihist=glob1$Wihist)
  }
}
################################################################


############################
### function : local_maxi ###
############################
local_maxi <- function(Xw, N, P, ini.a, max.itr, penai, tailp)
{

###Initialize ###

den<-matrix(0, N, P)
upA<-ini.a
oldA<-ini.a
```

```
ad<-rep(0, 101)
ai<-1
########## Algorithm 1 ##########
repeat
{
s<-Xw %*% oldA
########## Algorithm 2 ##########

### update A ###
for(j in 1:P)
{
  s0<-s[,j]
  qs<-quantile(s0, prob=c(tailp, .5, 1-tailp))
  fit <- logspline(s0, knots=qs, penalty=penai*log(N))
  g1<-mydlogspline(s0, fit)$y1
  g2<-mydlogspline(s0, fit)$y2
  for(i in 1:P)
  {upA[i,j]<-mean(Xw[,i]*g1)-mean(g2)*oldA[i,j]}
}

### Orthogonalize updated A ###
upA <- svd(upA)$u%*%t(svd(upA)$v)

########## Algorithm 3 ##########

### contrast function ###
tmp.s <- Xw %*% upA
for(j in 1:P)
{
  s1<-tmp.s[,j]
  qs<-quantile(s1, prob=c(tailp, .5, 1-tailp))
  fit <- logspline(s1, knots=qs, penalty=penai*log(N))
  den[,j]<-mydlogspline(s1, fit)$y
}
contr<-sum(den)

### Compute Amari Metric ###
ad[ai]<-amari(upA, ini.a)

###Convergence###
if((abs(ad[ai]-ad[ai-1])<0.01 && ai>1) || ai>20)
{
break}else
{
```

```
oldA<-upA
ai<-ai+1}

}### end of algorithm


### output ###
list(W=upA, contr=contr)
}



###############################
### function : global_maxi   ###
###############################
global_maxi<-function(Xw, N, P, max.itr, Wcands, penai, tailp)
{
### Initialization ###
local.a.histry<-matrix(0, P^2, max.itr)
cotr.histry<-rep(0, max.itr)
change.histry<-rep(0, max.itr)

for(itr in 1:max.itr)
{

  w1st <- Wcands[1,itr]
  wfin <- Wcands[length(Wcands[,itr]),itr]

  Winit<-matrix(Wcands[,itr], ncol=P)

  ### local maximumization ###
  local.max <- local_maxi(Xw, N, P, Winit, max.itr,
                              penai, tailp)
  local.a<-local.max$W
  local.contr<-local.max$contr

  if(itr==1)
  {
    Wmax<-local.a
    Cmax<-local.contr
  }else if (local.contr>Cmax){
    Wmax<-local.a
    Cmax<-local.contr
    change.histry[itr]<-1
  }
```

```
  local.a.histry[,itr]<-c(local.a)
  cotr.histry[itr]<-local.contr
}

### output ###
list(W=Wmax, ahis=local.a.histry, cotrhis=cotr.histry,
      chist=change.histry)
}
```

6.8.3   A24RANDOM.R

```
####################
### random number ###
####################
myrandom <- function(pdfi, n) {
  x<-numeric(0)
  if(pdfi==1){x<-rt(n, 3)}
  if(pdfi==2){x<-sign(runif(n)-0.5)*rexp(n,sqrt(2))}
  if(pdfi==3){x<-runif(n, -sqrt(3), sqrt(3))}
  if(pdfi==4){x<-rt(n, 5)}
  if(pdfi==5){x<--1+rexp(n,1)}
  if(pdfi==6) {
    prop=c(.5, .5);mus=c(-1, 1);covs=c(.5, .5);
    for(j in 1:n) {
      i<-sample_discrete(prop, 1, 1);
      x[j]<-sign(runif(1)-0.5)*rexp(1,sqrt(2))*covs[i]+mus[i]
    }
  }
  if(pdfi==7) {
    prop=c(.5, .5);mus=c(-.5, .5);covs=c(.15, .15);
    for(j in 1:n) {
      i<-sample_discrete(prop, 1, 1);
      x[j]<-rnorm(1, mus[i], covs[i]);
    }
  }
  if(pdfi==8){
    prop=c(.5, .5);mus=c(-.5, .5);covs=c(.4, .4);
    for(j in 1:n) {
      i<-sample_discrete(prop, 1, 1);
      x[j]<-rnorm(1, mus[i], covs[i]);
    }
```

```
}

if(pdfi==9) {
  prop=c(.5, .5)
  mus=c(-.5, .5)
  covs=c(.5, .5)
  outp=0
  for(j in 1:n) {
    i <- sample_discrete(prop, 1, 1);
    x[j] <- rnorm(1, mus[i], covs[i]);
  }
}

if(pdfi==10) {
  prop=c(1, 3)
  prop=prop/sum(prop)
  mus=c(-.5, .5)
  covs=c(.15, .15)
  outp=0;
  for(j in 1:n) {
    i <- sample_discrete(prop, 1, 1);
    x[j] <- rnorm(1, mus[i], covs[i]);
  }
}

if(pdfi==11) {
  prop=c(1, 2)
  prop=prop/sum(prop)
  mus=c(-.7, .5)
  covs=c(.4, .4)
  for(j in 1:n) {
    i <- sample_discrete(prop, 1, 1);
    x[j] <- rnorm(1, mus[i], covs[i]);
  }
}

if(pdfi==12) {
  prop=c(1, 2)
  prop=prop/sum(prop)
  mus=c(-.7, .5)
  covs=c(.5, .5)
  for(j in 1:n) {
    i <- sample_discrete(prop, 1, 1);
    x[j] <- rnorm(1, mus[i], covs[i]);
```

```
  }
}

if(pdfi==13) {
  prop=c(1,2,2,1)
  prop=prop/sum(prop)
  mus=c(-1, -.33, .33,1)
  covs=c(.16, .16, .16, .16)
  for(j in 1:n) {
    i <- sample_discrete(prop, 1, 1);
    x[j] <- rnorm(1, mus[i], covs[i]);
  }
}

if(pdfi==14) {
  prop=c(1,2,2,1)
  prop=prop/sum(prop)
  mus=c(-1, -.2, .2,1)
  covs=c(.2, .3, .3, .2)
  for(j in 1:n) {
    i <- sample_discrete(prop, 1, 1)
    x[j] <- rnorm(1, mus[i], covs[i])
  }
}

if(pdfi==15) {
  prop=c(1,2,2,1)
  prop=prop/sum(prop)
  mus=c(-.7, -.2, .2,.7)
  covs=c(.2, .3, .3, .2)
  for(j in 1:n) {
    i <- sample_discrete(prop, 1, 1);
    x[j] <- rnorm(1, mus[i], covs[i]);
  }
}

if(pdfi==16) {
  prop=c(1,1,2,1)
  prop=prop/sum(prop)
  mus=c(-1, .3 ,-.3 ,1.1)
  covs=c(.2, .2, .2, .2)
  for(j in 1:n) {
    i <- sample_discrete(prop, 1, 1);
    x[j] <- rnorm(1, mus[i], covs[i]);
```

```
    }
  }

  if(pdfi==17) {
    prop=c(1,3,2,.5)
    prop=prop/sum(prop)
    mus=c(-1, -.2, .3,1)
    covs=c(.2, .3, .2, .2)
    for(j in 1:n) {
      i <- sample_discrete(prop, 1, 1);
      x[j] <- rnorm(1, mus[i], covs[i]);
    }
  }

  if(pdfi==18) {
    prop=c(1, 2, 2, 1)
    prop=prop/sum(prop)
    mus=c(-.8, -.2, .2, .5)
    covs=c(.22, .3, .3, .2)
    for(j in 1:n) {
      i <- sample_discrete(prop, 1, 1);
      x[j] <- rnorm(1, mus[i], covs[i]);
    }
  }

  x
}


sample_discrete <- function(prob, r, c) {
  cumprob = cumsum(prob);
  n1 = length(cumprob);
  R = matrix(runif(1), r, c);
  M = matrix(1, r, c);
  for(i in 1:(n1-1)) {
    M = M + ifelse(R > cumprob[i], 1, 0);
  }
  M
}
```

6.8.4 AMARIMETRIC.R

```
###Amari Metric###
```

```r
amari <- function(A1, A2) {
  p <- length(A1[1,])
  r <- solve(A1)%*%A2
  sr1 <- apply(abs(r),1,sum)
  mr1 <- apply(abs(r),1,max)
  sr2 <- apply(abs(r),2,sum)
  mr2 <- apply(abs(r),2,max)
  d <- sum(sr1/mr1-1)/(2*p*(p-1))+sum(sr2/mr2-1)/(2*p*(p-1))
  d
}
```

6.8.5 LOGSPLINEDERIVATIVE.R

```r
### Logspline, 1st, 2nd Derivative ###
mydlogspline <- function(q, fit) {
  if(class(fit)!="logspline")
      stop("fit is not a logspline object")
  if(!missing(q))q <- unstrip(q)
  x <- q
  y <- fit$coef.pol[1] + x * fit$coef.pol[2]
  y1 <- fit$coef.pol[2]
  y2 <- 0
  for(i in 1:length(fit$knots)) {
    y <- y + fit$coef.kts[i]
        * ifelse((x- fit$knots[i])>0, (x- fit$knots[i]), 0)^3
    y1 <- y1 + fit$coef.kts[i] * 3
        * ifelse((x- fit$knots[i])>0, (x- fit$knots[i]), 0)^2
    y2 <- y2 + fit$coef.kts[i] * 6
        * ifelse((x- fit$knots[i])>0, (x- fit$knots[i]), 0)
  }
  if(fit$bound[1] > 0) {
     y[x < fit$bound[2]] <- -1000
     y1[x < fit$bound[2]] <- -1000
     y2[x < fit$bound[2]] <- -1000
  }
  if(fit$bound[3] > 0) {
    y[x > fit$bound[4]] <- -1000
    y1[x > fit$bound[4]] <- -1000
    y2[x > fit$bound[4]] <- -1000
  }

  list(y = y, y1=y1, y2=y2)
}
```

6.8.6   PDFUNCTION.R

```
####################
### pdf function ###
####################
mypdflines<-function(pdfi)
{
x<-seq(-5, 5, by=0.01)
if(pdfi==1){lines(x, dt(x, 3), col=2, lty=2)}
if(pdfi==2){lines(x, exp(-sqrt(2)*abs(x))/sqrt(2),
    col=2, lty=2)}

if(pdfi==3){lines(x, 1/2/sqrt(3)*(x<sqrt(3))*(x>-sqrt(3))
, col=2, lty=2)}

if(pdfi==4){lines(x, dt(x, 5), col=2, lty=2)}

if(pdfi==5){lines(x, exp(-(x+1))*(x>-1), col=2, lty=2)}

if(pdfi==6){
prop=c(.5, .5);
mus=c(-1, 1);
covs=c(.5, .5);
outp=0;
for(i in 1:2)
{outp=outp+prop[i]/covs[i] *
        exp(-sqrt(2)*abs(x-mus[i])/covs[i])/sqrt(2);}
lines(x, outp, col=2, lty=2)
}

if(pdfi==6){
prop=c(.5, .5);mus=c(-1, 1);covs=c(.5, .5);
outp=0;
for(i in 1:2)
{outp=outp+prop[i]/covs[i]*
        exp(-sqrt(2)*abs(x-mus[i])/covs[i])/sqrt(2);}
lines(x, outp, col=2, lty=2)
}

if(pdfi==7){
prop=c(.5, .5);mus=c(-.5, .5);covs=c(.15, .15);
outp=0;
```

```
for(i in 1:2)
{outp=outp+prop[i]*dnorm(x, mus[i], covs[i]);}
lines(x, outp, col=2, lty=2)
}


if(pdfi==8){
prop=c(.5, .5);mus=c(-.5, .5);covs=c(.4, .4);
outp=0;
for(i in 1:2)
{outp=outp+prop[i]*dnorm(x, mus[i], covs[i]);}
lines(x, outp, col=2, lty=2)
}


if(pdfi==9){
prop=c(.5, .5);mus=c(-.5, .5);covs=c(.5, .5);
outp=0;
for(i in 1:2)
{outp=outp+prop[i]*dnorm(x, mus[i], covs[i]);}
lines(x, outp, col=2, lty=2)
}


if(pdfi==10){
prop=c(1, 3);prop=prop/sum(prop);
mus=c(-.5, .5);covs=c(.15, .15);
outp=0;
for(i in 1:2)
{outp=outp+prop[i]*dnorm(x, mus[i], covs[i]);}
lines(x, outp, col=2, lty=2)
}


if(pdfi==11){
prop=c(1, 2);prop=prop/sum(prop);mus=c(-.7, .5);
covs=c(.4, .4); outp=0;
for(i in 1:2)
{outp=outp+prop[i]*dnorm(x, mus[i], covs[i]);}
lines(x, outp, col=2, lty=2)
}


if(pdfi==12){
prop=c(1, 2);prop=prop/sum(prop);
mus=c(-.7, .5);covs=c(.5, .5);
outp=0;
for(i in 1:2)
{outp=outp+prop[i]*dnorm(x, mus[i], covs[i]);}
```

```
lines(x, outp, col=2, lty=2)
}


if(pdfi==13){
prop=c(1,2,2,1);prop=prop/sum(prop);
mus=c(-1, -.33, .33,1);covs=c(.16, .16, .16, .16);
outp=0;
for(i in 1:4)
{outp=outp+prop[i]*dnorm(x, mus[i], covs[i]);}
lines(x, outp, col=2, lty=2)
}


if(pdfi==14){
prop=c(1,2,2,1);prop=prop/sum(prop);
mus=c(-1, -.2, .2,1);covs=c(.2, .3, .3, .2);
outp=0;
for(i in 1:4)
{outp=outp+prop[i]*dnorm(x, mus[i], covs[i]);}
lines(x, outp, col=2, lty=2)
}


if(pdfi==15){
prop=c(1,2,2,1);prop=prop/sum(prop);
mus=c(-.7, -.2, .2,.7);covs=c(.2, .3, .3, .2);
outp=0;
for(i in 1:4)
{outp=outp+prop[i]*dnorm(x, mus[i], covs[i]);}
lines(x, outp, col=2, lty=2)
}


if(pdfi==16){
prop=c(1,1,2,1);prop=prop/sum(prop);
mus=c(-1, .3 ,-.3 ,1.1);covs=c(.2, .2, .2, .2);
outp=0;
for(i in 1:4)
{outp=outp+prop[i]*dnorm(x, mus[i], covs[i]);}
lines(x, outp, col=2, lty=2)
}


if(pdfi==17){
prop=c(1,3,2,.5);prop=prop/sum(prop);
mus=c(-1, -.2, .3,1);covs=c(.2, .3, .2, .2);
outp=0;
for(i in 1:4)
```

```
{outp=outp+prop[i]*dnorm(x, mus[i], covs[i]);}
lines(x, outp, col=2, lty=2)
}

if(pdfi==18){
prop=c(1, 2, 2, 1);prop=prop/sum(prop);
mus=c(-.8, -.2, .2, .5);covs=c(.22, .3, .3, .2);
outp=0;
for(i in 1:4)
{outp=outp+prop[i]*dnorm(x, mus[i], covs[i]);}
lines(x, outp, col=2, lty=2)
}

}
```

6.8.7   UNIORTHOMAT4MULTI.R

```
source("amarimetric.r")

uni.ortho.mat <- function(P, n.ini) {
  set.seed(1)
  w3 <- diag(P)

  ta <- rep(0, 10000)

  w31<-matrix(0, P^2, 10000)
  for(i in 1:10000) {
    a30 <- matrix(rnorm(P^2),P,P)
    a30svd <- svd(a30)
    a3 <- a30svd$u%*%t(a30svd$v)
    ta[i] <- amari(a3, w3)
    w31[,i] <- c(a3)
  }

  sta <- sort(ta)
  qsta <- sta[seq(500, 9500, length=n.ini)]
  print(qsta)
  exind <- rep(0, n.ini)
  for(i in 1:n.ini) {
    exind[i] <- which(ta==qsta[i])
  }
  w32 <- matrix(0, P^2, n.ini)
  w32 <- w31[, exind]
```

```
for(i in 1:n.ini) {
  print(matrix(w32[,i], ncol=P))
  print(paste("Amari = ", amari(matrix(w32[,i], ncol=P), w3)))
}
w32
}
```

Acknowledgment

We are grateful to Dr. Aiyou Chen for providing the KDICA programming code. We are also deeply grateful to Dr. Mechelle Lewis for her insight about the Twin data set, and her fruitful discussion of our analysis. This research was supported in part by the Banyu Fellowship Program sponsored by the Banyu Life Science Foundation International and by Grants-in-Aid from the Ministry of Education, Culture, Sport, Science and Technology of Japan (21700312) to AK, and by NSF DMS-0707090 and DMS-1106962 to YT.

Bibliography

1. S. Amari, A. Cichocki, and H. H. Yang. A new learning algorithm for blind signal separation. *Advances in Neural Information Processing Systems*, 8:757–763, 1996.
2. F. R. Bach and M. I. Jordan. Kernel independent component analysis. *Journal of Machine Learning Research*, 3(1):1–48, 2003.
3. A. J. Bell and T. J. Sejnowski. An information maximisation approach to blind separation and blind deconvolution. *Neural Computation*, 7:1129–1159, 1995.
4. L. Bordes, D. Chauveau, and P. Vandekerkhove. A stochastic EM algorithm for a semiparametric mixture model. *Comput. Stat. Data Anal.*, 51:5429–5443, July 2007. ISSN 0167-9473. doi: 10.1016/j.csda.2006.08.015.
5. R. Boscolo, H. Pan, and V. P. Roychowdhury. Independent component analysis based on nonparametric density estimation. *IEEE Transactions on Neural Networks*, 15(1):55–65, 2004.
6. V. D. Calhoun and T. Adali. Unmixing fMRI with independent component analysis. *Engineering in Medicine and Biology Magazine, IEEE*, 25:79–90, 2006.
7. V. D. Calhoun, T. Adali, G. D. Pearlson, and J. J. Pekar. A method for making group inferences from functional MRI data using independent component analysis. *Human Brain Mapping*, 14:140–151, 2001.
8. J. F. Cardoso. High-order contrasts for independent component analysis. *Neural Computation*, 11(1):157–192, 1999.
9. J. F. Cardoso and A. Souloumiac. Blind beamforming for non-Gaussian signals. *IEE-Proc.-F*, 140:362–370, 1993.
10. G. Celeux and J. Diebolt. A stochastic approximation type EM algorithm for the mixture problem. *Stochastics Stochastics Rep.*, 41(1-2):119–134, 1992.

11. A. Chen. Fast kernel density independent component analysis. *Independent Component Analysis and Blind Signal Separation*, 3889:24–31, 2006.

12. A. Chen and P. J. Bickel. Efficient independent component analysis. *Annals of Statistics*, 34:2825–2855, 2006.

13. D. Cordes and R. Nandy. Independent component analysis in the presence of noise in fMRI. *Magnetic Resonance Imaging*, 25(9):1237–1248, 2007.

14. B. S. Everitt and E. T. Bullmore. Mixture model mapping of brain activation in functional magnetic resonance images. *Human Brain Mapping*, 7(1):1–14, 1999.

15. K. J. Friston, A. P. Holmes, K. J. Worsley, J. B. Poline, C. Frith, and R. S. J. Frackowiak. Statistical parametric maps in functional imaging: A general linear approach. *Human Brain Mapping*, 2:189–210, 1995.

16. N. V. Hartvig and J. L. Jensen. Spatial mixture modeling of fMRI data. *Human Brain Mapping*, 11(4):233–248, 2000.

17. A. Hyvärinen and E. Oja. A fast fixed-point algorithm for independent component analysis. *Neural Computation*, 9(7):1483–1492, 1997.

18. A. Hyvärinen and E. Oja. A fast fixed point algorithm for independent component analysis. *Neural Computation*, 9:1483–1492, 1997.

19. A. Kawaguchi and Y. K. Truong. Logspline independent component analysis. *Bulletin of Informatics and Cybernetics*, 43:83–94, 2011.

20. A. Kawaguchi, X. Huang, and Y. K. Truong. Application of polynomial spline independent component analysis to fMRI data. In G. R. Naik, editor, *Independent Component Analysis for Audio and Biosignal Applications*. INTECH Open Access Publisher.

21. S. Lee, H. Shen, Y. K. Truong, M. M. Lewis, and X. Huang. Independent component analysis involving autocorrelated sources with an application to functional magnetic resonance imaging. *Journal of the American Statistical Association*, 106(495):1009–1024, 2011.

22. T. W. Lee, M. Girolami, and T. J. Sejnowski. Independent component analysis using an extended infomax algorithm for mixed subgaussian and supergaussian sources. *Neural Computation*, 11(2):417–441, 1999. ISSN 0899-7667.

23. M. J. McKeown, S. Makeig, G. Brown, T-P. Jung, S. Kindermann, T. Bell, V. Iragui, and T. J. Sejnowski. Analysis of fMRI by blind separation into independent spatial components. *Human Brain Mapping*, 6:160–188, 1998.

24. E. G. Miller and J. W. Fisher III. ICA using spacings estimates of entropy. In *Proceedings of the Fourth Symposium on Independent Component Analysis and Blind Source Separation*, 2003.

25. J. Neumann, D. Cramon, and G. Lohmann. Model-based clustering of meta-analytic functional imaging data. *Human Brain Mapping*, 29(2):177–192, 2008.

26. J. V. Stone, J. Porrill, N. R. Porter, and I. D. Wilkinson. Spatiotemporal independent component analysis of event-related fMRI data using skewed probability density functions. *Neuroimage*, 15:407–421, 2002.

7 Colored Independent Component Analysis

Seonjoo Lee
Columbia University, New York

Haipeng Shen
University of Hong Kong, China

Young K. Truong
University of North Carolina at Chapel Hill

CONTENTS

As indicated in the last two chapters, instantaneous ICA has been developed to extract latent sources by modeling and estimate the marginal probability density functions. In this chapter, an alternative viewpoint of ICA will be described by relying on some biological or physiological characteristics of the sources. An example taken from electroencephalogram (EEG) will be used as a leading case for this approach.

7.1 INTRODUCTION

Independent component analysis (ICA) is an effective data-driven technique for extracting the source signals from their mixtures [19, 32]. It aims to solve the *blind source separation* problem by expressing a set of observed mixed signals as linear combinations of independent latent random variables (or source signals or components). It has many important applications, especially in *functional magnetic resonance imaging* (fMRI) analysis [25, 32, 17, 8].

The ICA problem can be formally expressed as follows. Suppose there are M mixed signals of length T each, which are stored in the observed signal matrix $\mathbf{X}$. ICA then allows one to decompose $\mathbf{X}$ as

$$\mathbf{X}_{M \times T} = \mathbf{A}_{M \times M} \mathbf{S}_{M \times T}, \tag{7.1}$$

where $\mathbf{A}$ is a non-random mixing matrix and $\mathbf{S}$ is a matrix of *independent* source signals. The goal of ICA is to recover the latent source signals (rows of $\mathbf{S}$) as

$$\mathbf{S} = \mathbf{W}\mathbf{X} \quad \text{with} \quad \mathbf{W} = \mathbf{A}^{-1}. \tag{7.2}$$

Many ICA procedures have been developed over the last fifteen years. The majority of the methods are based on estimates of the marginal densities of the sources, either parametrically or nonparametrically. Parametric approaches include *Infomax* [3], which estimates the density parameters via minimization of mutual information, and is equivalent to maximum likelihood estimation using high-order cumulants [13, 9]; or *FastICA* [19] that maximizes non-Gaussianity as measured by the approximated negative-entropy. (Note that the sources should not have a Gaussian distribution in order for the mixing matrix to be identifiable.)

Nonparametric methods include estimating the score function using kernel approximation [33], kernel density estimation [2, 4, 11, 31, 10], smoothing splines [16], B-spline approximation [12], and logsplines [20]. Two other relaxations of the basic ICA model are subspace ICA [18, 30], which allows the sources to form mutually independent subgroups and does not require the sources within the same subgroup to be independent; and AMICA [26], which uses Gaussian scale mixtures to model the sources, which was extended to include mixtures of linear processes [27]. Note that when all the sources are mutually independent, subspace ICA reduces to ordinary ICA [18].

All the above methods, however, only make use of the marginal densities (with the exception of the recent extension of AMICA), which do not contain information about the correlation structures within the source signals. For example, in fMRI studies the experiment-stimulus-related signals or physiological signals such as heart beat or breathing are usually periodic, and therefore embedded in the fMRI data are some autocorrelation or colored noise structures within the signals [7]. Such information is not incorporated when using the marginal-density-based ICA methods.

Reference [28] is the first ICA procedure that takes into account the autocorrelation structures within the sources. The authors imposed certain parametric correlation assumptions on the sources. Specifically, they assumed that the spectral density of each source was known up to some scale parameters, and proposed a quasi-maximum likelihood method to recover the sources. Their formulation is in the spectral domain, and builds upon the asymptotic independence and normality of the discrete Fourier Transform (DFT). Since the spectral densities are known except the scale parameters, the authors used the corresponding (known) separating filters in the quasi-likelihood, and only needed to maximize the likelihood to get estimates for the scale parameters and the mixing matrix. Later, [22] relaxed that assumption and proposed a new spectral domain ICA procedure for sources with autocorrelation, i.e., colored sources, and this approach was extended to spatially correlated cases [35].

7.2 COLORED INDEPENDENT COMPONENT ANALYSIS

We start with several definitions in spectral analysis. Consider a vector-valued stationary process $\mathbf{X}(t) = (X_1(t), \ldots, X_M(t))^\top$, $t = 0, \pm 1, \pm 2, \ldots$, with mean zero and the covariance function $\mathbf{c}_{\mathbf{XX}}(u) = \text{Cov}(\mathbf{X}(t), \mathbf{X}(t+u))$. If $\sum_{u=-\infty}^{\infty} |\mathbf{c}_{\mathbf{XX}}(u)| < \infty$, we define the $M \times M$ spectral density matrix of the series $\mathbf{X}(t)$ as

$$\mathbf{f}_{\mathbf{XX}}(r) = \frac{1}{2\pi} \sum_{u=-\infty}^{\infty} \mathbf{c}_{\mathbf{XX}}(u) \exp\{-iru\}, \quad r \in \mathbb{R},$$

where r is the angular frequency per unit time, or simply the frequency. The jth diagonal element of $\mathbf{f}_{\mathbf{XX}}$, $f_{X_j X_j}$, is the spectral density of the univariate time series $X_j(t)$. For a more detailed discussion of spectral density, see [5].

In practice, we suppose that $\mathbf{X}(t)$, $t = 0, 1, \ldots, T-1$, are observed. Consider the Fourier frequencies $r_k = (2\pi k)/T$, $k = 0, \ldots, T-1$. Then, we can define the discrete Fourier transform (DFT) of $\mathbf{X}(t)$, $t = 0, 1, \ldots, T-1$, as

$$\varphi(r_k, \mathbf{X}) = \sum_{t=0}^{T-1} \mathbf{X}(t) \exp\{-ir_k t\},$$

and its second-order periodogram as

$$\tilde{\mathbf{f}}(r_k, \mathbf{X}) = \frac{1}{2\pi T} \varphi(r_k, \mathbf{X}) \varphi^*(r_k, \mathbf{X}),$$

where φ^* is the conjugate transpose of the vector φ.

Considering the source signals $\mathbf{S}(t) = (S_1(t), \ldots, S_M(t))^\top, t = 0, 1, \ldots, T-1$, we can similarly define their spectral density matrix $\mathbf{f}_{SS}$, DFT, and periodogram. Since the sources are mutually independent, we have that $\mathbf{f}_{SS} = \text{diag}\{f_{11}, \ldots, f_{MM}\}$, where f_{jj} is the spectral density of the jth source.

To minimize the bias in estimation associated with the misspecification of the time series distributions, we will be using the *Whittle likelihood* [34], which was formulated based on the asymptotic distributional properties of DFT (see Theorem 4.4.1 of [5]). Specifically, suppose that we were able to observe the source signals and compute their periodograms: $\tilde{f}(r_k, S_1), \ldots, \tilde{f}(r_k, S_M)$. In addition, if the sources are stationary with finite moments, then it can be shown that $\tilde{f}(r_k, S_j)$ is asymptotically $f_{jj}(r_k)\chi_2^2/2$, independently of the other variates for $k = 0, 1, \ldots, T-1$ and $j = 1, \ldots, M$ (see Theorem 5.2.6 of [5]). Using the mixing relationship (7.2) and the linearity of DFT, the log-likelihood can be written as

$$L(\mathbf{W}, \mathbf{f}_{SS}; \mathbf{X}) = -\frac{1}{2} \sum_{j=1}^{M} \sum_{k=0}^{T-1} \left\{ \frac{\mathbf{e}_j^\top \mathbf{W}_j^\top \tilde{\mathbf{f}}(r_k, \mathbf{X}) \mathbf{W}_j \mathbf{e}_j}{f_{jj}(r_k)} + \ln f_{jj}(r_k) \right\} + T \ln |\det(\mathbf{W})|,$$

(7.3)

where $\mathbf{W}_j$ is the j-th column vector of $\mathbf{W}$.

We can model spectral densities by parameterizing source signals. This will be illustrated in the following sections.

7.3   STATIONARY TIME SERIES MODELS

Our discussion will be based on a general stationary time series Y_t, $t \in \mathbb{Z} \equiv \{0, \pm 1, \pm 2, \ldots\}$. See also Chapter 3 and books on time series analysis [5, 6]. In the case of the source signal, simply replace Y by S.

7.3.1   WHITE NOISE

The time series Y is said to be a *white noise* series if Y_t, $t \in \mathbb{Z}$, are stochastically independent. If each Y_t has a Guassian distribution, we say Y is a Gaussian white noise.

In defining more general time series models, it is convenient to denote the white noise by $\varepsilon_t, t \in \mathbb{Z}$. This is also referred to as the *innovation*. Moreover, we will assume the innovation ε_t is stationary with mean zero and variance σ^2.

The *spectral density function* or simply *power spectrum* of the white noise series is

$$f_{ee}(\lambda) = \frac{\sigma^2}{2\pi}, \qquad \lambda \in \mathbb{R}.$$

(7.4)

7.3.2 MOVING-AVERAGE PROCESSES

The time series Y is said to be a *moving-average process* of order q, abbreviated by MA(q), if

$$Y_t = \varepsilon_t + \theta_1 \varepsilon_{t-1} + \cdots + \theta_q \varepsilon_{t-q}, \quad t = 0, \pm 1, \ldots, \tag{7.5}$$

where

$$q : \text{is a non-negative integer,}$$
$$\theta_1, \theta_2, \ldots, \theta_q : \text{are real numbers,}$$
$$\varepsilon_t : \text{is the innovation.}$$

The power spectrum of the MA(q) process is

$$f(\lambda) = \frac{\sigma^2}{2\pi} |1 + \theta_1 e^{-i\lambda} + \cdots + \theta_q e^{-iq\lambda}|^2, \qquad \lambda \in \mathbb{R}. \tag{7.6}$$

7.3.3 AUTOREGRESSIVE PROCESSES

The time series Y is said to be an *autoregressive process* of order p, abbreviated by AR(p), if Y can be written as

$$Y_t - \phi_1 Y_{t-1} - \cdots - \phi_p Y_{t-p} = \varepsilon_t, \quad t = 0, \pm 1, \ldots, \tag{7.7}$$

where

$$p : \text{is a non-negative integer,}$$
$$\phi_1, \phi_2, \ldots, \phi_p : \text{are real numbers,}$$
$$\varepsilon_t : \text{is the innovation.}$$

The power spectrum of the AR(p) process is

$$f(\lambda) = \frac{\sigma^2}{2\pi} |1 - \phi_1 e^{-i\lambda} + \cdots + \phi_q e^{-iq\lambda}|^{-2}, \qquad \lambda \in \mathbb{R}. \tag{7.8}$$

Before describing the next process, it will be convenient to introduce a more general way to express the above processes. Let $\theta(\cdot)$ and $\phi(\cdot)$ denote, respectively qth and pth polynomials

$$\theta(z) = 1 + \theta_1 z + \theta_2 z^2 + \cdots + \theta_q z^q, \tag{7.9}$$

and

$$\phi(z) = 1 - \phi_1 z - \phi_2 z^2 - \cdots - \phi_p z^p, \qquad z \in \mathbb{C}. \tag{7.10}$$

Let B be the backward shift operator defined by

$$B^j Y_t = Y_{t-j}, \qquad j \in \mathbb{Z}. \tag{7.11}$$

Thus the MA(q) process can be represented by

$$Y_t = \theta(B)\varepsilon_t, \qquad t \in \mathbb{Z}. \tag{7.12}$$

The power spectrum is

$$f(\lambda) = \frac{\sigma^2}{2\pi} |\theta(e^{-i\lambda})|^2, \qquad \lambda \in \mathbb{R}. \tag{7.13}$$

Similarly, the AR(p) process is given by

$$\phi(B)Y_t = \varepsilon_t, \qquad t \in \mathbb{Z}. \tag{7.14}$$

The power spectrum is

$$f(\lambda) = \frac{\sigma^2}{2\pi} |\phi(e^{-i\lambda})|^{-2}, \qquad \lambda \in \mathbb{R}. \tag{7.15}$$

7.3.4   AUTOREGRESSIVE AND MOVING-AVERAGE PROCESSES

The time series Y is said to be an *autoregressive and moving-average process of orders p and q*, abbreviated as ARMA(p), if

$$\begin{aligned} \phi(B)Y_t &= \theta(B)\varepsilon_t, \\ Y_t - \phi_1 Y_{t-1} - \cdots - \phi_p Y_{t-p} &= \varepsilon_t + \theta_1 \varepsilon_{t-1} + \cdots + \theta_q \varepsilon_{t-q}, \quad t \in \mathbb{Z}, \end{aligned} \tag{7.16}$$

where p, q, ϕ's, θ's and ε_t are given as before. Also, Y_t is an ARMA(p,q) with mean $\mu \in \mathbb{R}$ if $Y_t - \mu$ is ARMA(p,q).

An ARMA(p,q) process is *causal* iff the polynomial $\phi(z)$ does not admit any roots inside the unit disc of the complex plane [6]:

$$\phi(z) \neq 0, \qquad |z| \leq 1. \tag{7.17}$$

The power spectrum of the ARMA(p,q) is

$$f(\lambda) = \frac{\sigma^2}{2\pi} \left| \frac{\theta(e^{-i\lambda})}{\phi(e^{-i\lambda})} \right|^2, \qquad \lambda \in \mathbb{R}. \tag{7.18}$$

7.3.5   HARMONIC PROCESSES

The time series Y_t, $t \in \mathbb{Z}$ is a harmonic process if

$$Y_t = A\cos(\lambda t + \phi) + \varepsilon_t, \qquad t \in \mathbb{Z}, \tag{7.19}$$

where A is the amplitude, λ is the frequency, and ϕ is the phase.

Let $\eta(\cdot)$ denote the Delta comb function. Then the power spectrum of the harmonic process is

$$f(\omega) = \frac{A^2}{4}[\eta(\omega - \lambda) + \eta(\omega + \lambda)], \qquad \omega \in \mathbb{R}. \tag{7.20}$$

This type of power spectra is called the *line spectra* and is very useful for modeling brain rhythms in EEG data analysis.

7.4   STATIONARY COLORED SOURCE MODELS

It will now be easy to model the latent colored sources. For example, if the jth source follows some stationary ARMA(p_j, q_j) model so that $\Phi_j(B)S_j(t) = \Theta_j(B)\varepsilon_j(t)$, $\varepsilon_j(t) \sim WN(0, \sigma_j^2)$, where B is the backshift operator described above, $\Phi_j(z) = 1 - \phi_{j,1}z - \cdots - \phi_{j,p_j}z^{p_j}$, and $\Theta_j(z) = 1 + \theta_{j,1}z + \cdots + \theta_{j,q_j}z^{q_j}$. Then the power spectrum of this source is given by

$$f_{jj}(r) = \frac{\sigma_j^2}{2\pi} \frac{|\Theta_j(e^{-ir})|^2}{|\Phi_j(e^{-ir})|^2}, \qquad r \in \mathbb{R}. \tag{7.21}$$

A very useful model for the sources will the AR(p) processes described above. In this case, the polynomial $\theta \equiv 1$. The Whittle likelihood function has fewer parameters to estimate, and that is the leading case in our software implementation. Another important feature is the automatic order p selection for each hidden source. Thus, this approach is flexible in modeling the sources, auto-correlation structures. Moreover, in our EEG analysis, it will be extremely useful to model the periodic activity as part of the latent sources by adding the line spectra to the AR ones, which will be further discussed in the application later in the chapter.

7.5   MAXIMUM LIKELIHOOD ESTIMATION

The unmixing matrix and nuisance parameters related to spectral densities are estimated iteratively by maximizing using the Whittle likelihood (7.3) [22]. The orthogonality of the unmixing matrix $\mathbf{W}$ can be imposed in two different ways: 1) performing the minimization of the objective function according a Newton–Raphson method with Lagrange multiplier [22]; and 2) performing eigenvalue decomposition on $\sum_{k=0}^{T-1} \left\{ \frac{\hat{\mathbf{f}}(r_k, \mathbf{X})}{f_{jj}(r_k)} \right\}$ and the eigenvector corresponds to the smallest eigenvalue is the estimates of $\mathbf{W}_j$. Both methods have been implemented in the `coloredICA` R-package [29, 24]. To determine convergence, Amari's distance [1] is used as the convergence criterion due to this scale and permutation invariance.

7.6   COLOREDICA R-PACKAGE

In this section, we discuss how a colorICA analysis can be conducted using
`coloredICA` package in R [24]. A similar version written in MATLAB will be described in Section 7.8. First, we load the `coloredICA` package by using the `library`
function in R:

```
R> library(coloredICA)
```

In this example, we generate two sources from MA(2) with parameters $\theta_1 = 1, \theta_2 = 0.25$, and AR(1) with parameter $\phi_1 = -0.5$. For the error term, the first source
is generated from the standard normal distribution, and the second source is gener-
ated from uniform distribution. The length of time series (i.e., sample size) is set to
be $T = 256$.

```
R> set.seed(1234)
R> T=256
R> S1 = arima.sim(list(order=c(0,0,2),ma=c(1,0.25)),T)
R> S2 = arima.sim(list(order=c(1,0,0), ar=-0.5), T, +
          rand.gen = function(n, ...) (runif(n)-0.5)*sqrt(3))
```

We then randomly generate an unmixing matrix **W** from uniform distribution and
make it identifiable using the `rerow` function. Note that the `rerow` function first
scales each row of the matrix to have norm one, and orders the rows by their diagonal
elements. Then, the mixing matrix **A** is the inverse of **W**.

```
R> M=2
R> W = rerow(matrix(runif(M^2)-0.5,M,M))
R> A = solve(W)
```

The mixed or observed matrix **X** is now generated by multiplying the mixing
matrix with the source matrix **S**. Using the function `cICA` from `coloredICA`, you
can recover the true sources up to permutation and scale. Amari's distance can also
be computed to measure how close the estimated matrix is to the true unmixing
matrix.

```
R> S=rbind(S1,S2)
R> X = A %*% S
R> cica = cICA(X,tol=0.001)
R> amari_distance(W,cica$W %*% cica$K)
```

The `cICA` function outputs the following:

Parameter	Explanation
W	Estimate of the M x M unmixing matrix in the whitened space.
K	Pre-whitening matrix that projects data onto the first M principal components. Dimensions are M x p.
A	Estimate of the p x M mixing matrix.
S	Estimate of the M x T source matrix.
X	Original p x T data matrix.
iter	Number of iterations.
NInv	Number of times the algorithm is rerun after it does not achieve convergence.
den	Estimate of the spectral density of the sources.

Besides `coloredICA`, several ICA methods have been implemented in R or `matlab`. Several authors have provided their source codes through their private weibstes.

Software Package	Language	Reference
coloredICA	R	[22]
FastICA	R	[19]
Infomax	MATLAB	[3]
JADE	R	[9]
KICA	MATLAB	[2]
AMICA	MATLAB	[26]
PFICA	MATLAB	[11]

7.7 RESTING-STATE EEG DATA ANALYSIS

In this section, we will further illustrate the colored ICA methodology by applying the R package `coloredICA` [24] to an EEG data set. The electroencephalogram (EEG) (also known as *brain waves*) represents the electrical activity of the brain [14]. In clinical practice, several channels of the EEG are recorded simultaneously from various locations on the scalp for comparative analysis of activities in different regions of the brain. Our data is based on the recording of a resting-state run experiment. During the run, the subject is asked to relax and sit quietly with their eyes closed (for the alpha rhythm). The EEG cap contains 32 tin electrodes (AFZ ground right mastoid (M2) reference). Vertical and horizontal electro-oculograms (VEOG/HEOG) using right mastoid (M2) reference are measured bipolarly and recorded at tin electrodes placed at the outer canthi of each eye and above and below the right eye. The sampling rate was 500 Hz and the recording filters were set to 0.15-70 Hz bandpass. The acquired dataset is preprocessed using EEGLAB [15]. The data were filtered with 0.1 high-pass filter and noise artifacts are manually rejected. We applied `cICA` from `coloredICA` to the raw data and extracted sixteen IC components. This was carried out by using *principal component analysis* (PCA) to reduce dimensionality, and the first 16 components, explaining at least 90% of total variations, were chosen. Figures 7.1–7.4 depict the results. The sixteen spatial or scalp maps are constructed from the sixteen columns of the estimated mixing matrices **A**, and time course data are the sixteen rows of the estimated source matrix **S**. The

Table 7.1

Noise artifacts among the estimated sources

cICA-LSP	Main Frequency Band (approximated)	Spatial Pattern
7	<3Hz	Eye movement
9	<3Hz	Right eyes
10	<3Hz and 8–12Hz	Eye movement
13	<3Hz	Left frontal lobe

detected artifacts are listed in Table 7.1 with their frequency band and summary of spatial patterns [23]. The time series features of each independent component can be extracted by passing the data to one of the spectral density estimation procedures. An illustration is carried out by downsampling the data using 250 Hz, resulting in 40,000 time points. Eight components were selected after the dimension reduction via PCA mentioned above. We then applied cICA to the dimension-reduced data. Figures 7.5 and 7.6 display the topography maps and the reconstructed time courses using cICA. The spectra of the recovered sources were estimated by a logspline spectral density procedure [21, 23].

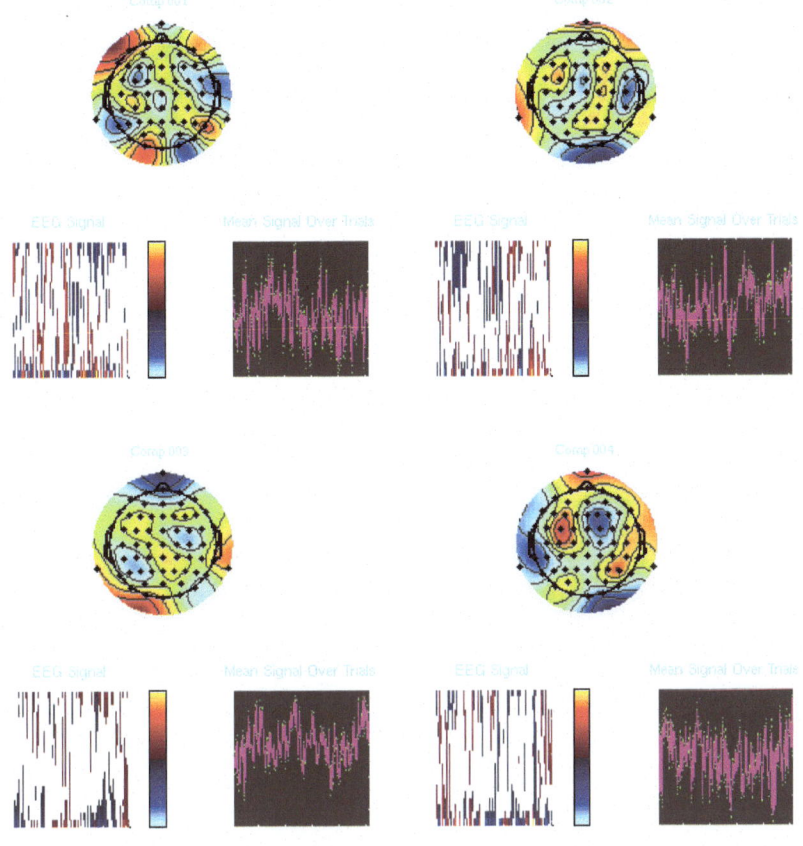

Figure 7.1: Topography and reconstructed source time series by cICA: Components 1–4.

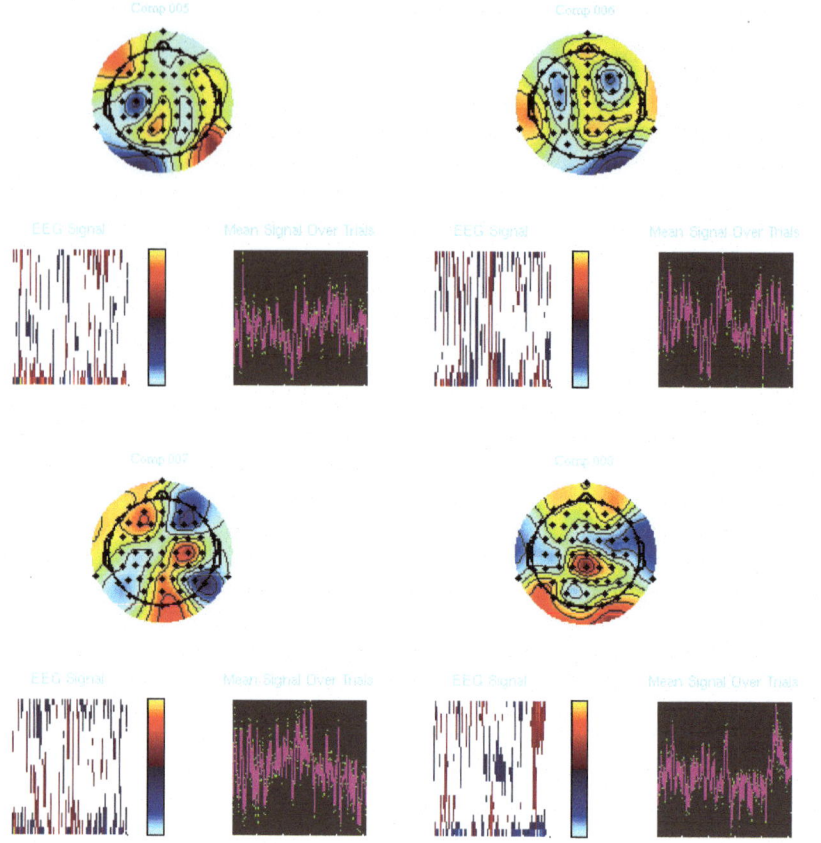

Figure 7.2: Topography and reconstructed source time series by cICA: Components 5–8.

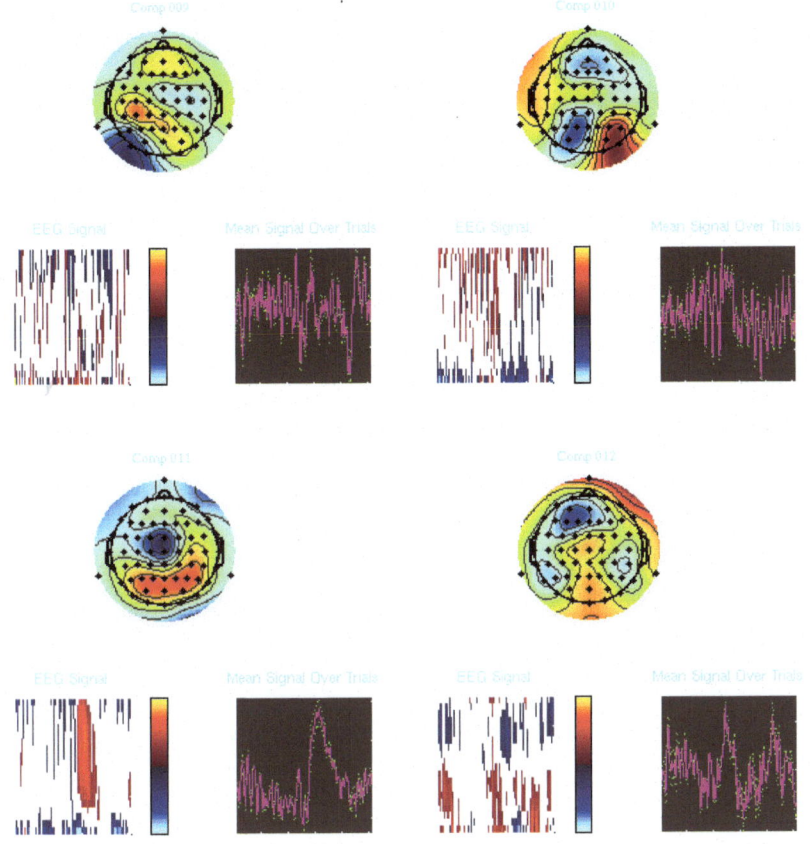

Figure 7.3: Topography and reconstructed source time series by cICA: Components 9–12.

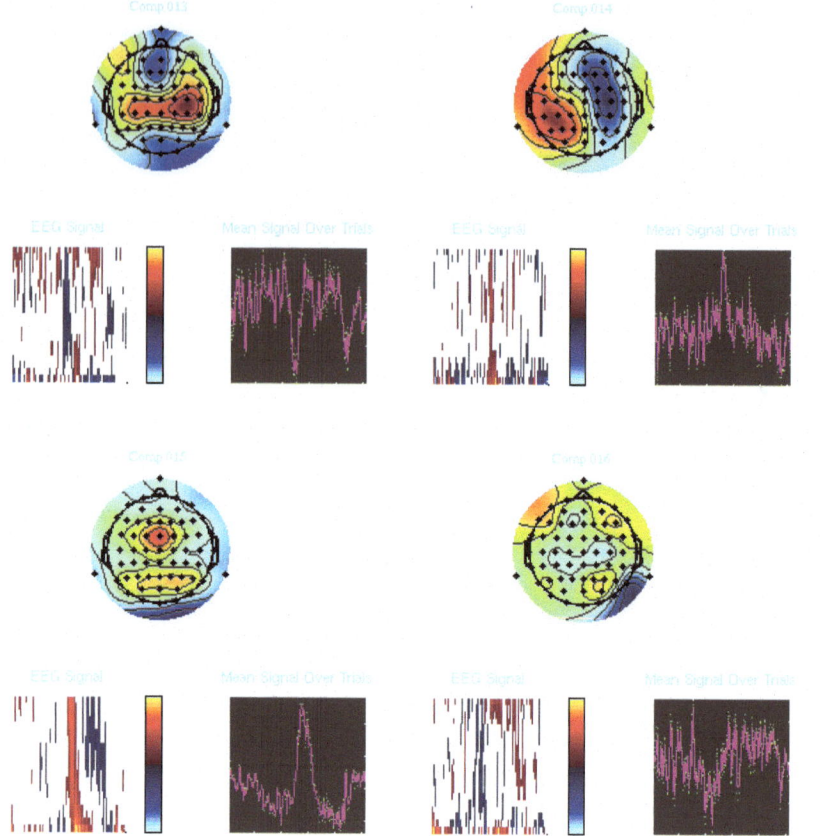

Figure 7.4: Topography and reconstructed source time series by cICA: Components 13–16.

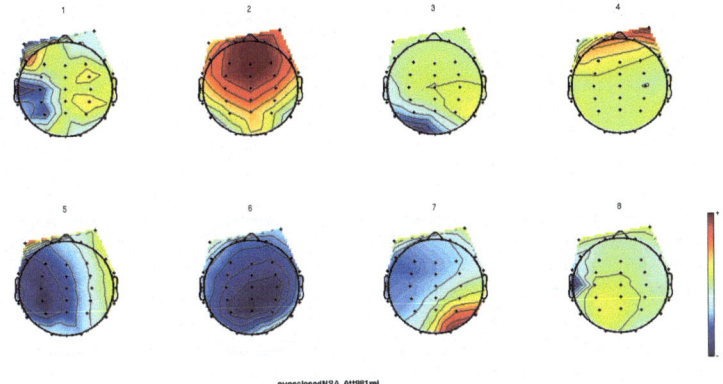

Figure 7.5: Topography maps of the eight independent components.

7.8 M-FILES

The MATLAB functions similar to those in the R package `colorICA` discussed previously in the chapter are:

```
aic_unit.m
aic_unit2.m
amari_distanceW.m
assignspec.m
cICA_xarma2.m
calculateAIC.m
colorICA_pmm.m
colorICA_pmm4.m
dexprand.m
estimateAR.m
getspec2.m
model_select_MM.m
specAR.m
whiteICA.m
```

The code of each function is listed in the following subsections.

7.8.1 AIC UNIT

```
% aic_unit.m
function [outscore] = aic_unit(ts, phi,sigmasq, method)
%% Description
%%%returns AIC for model selection in AR
```

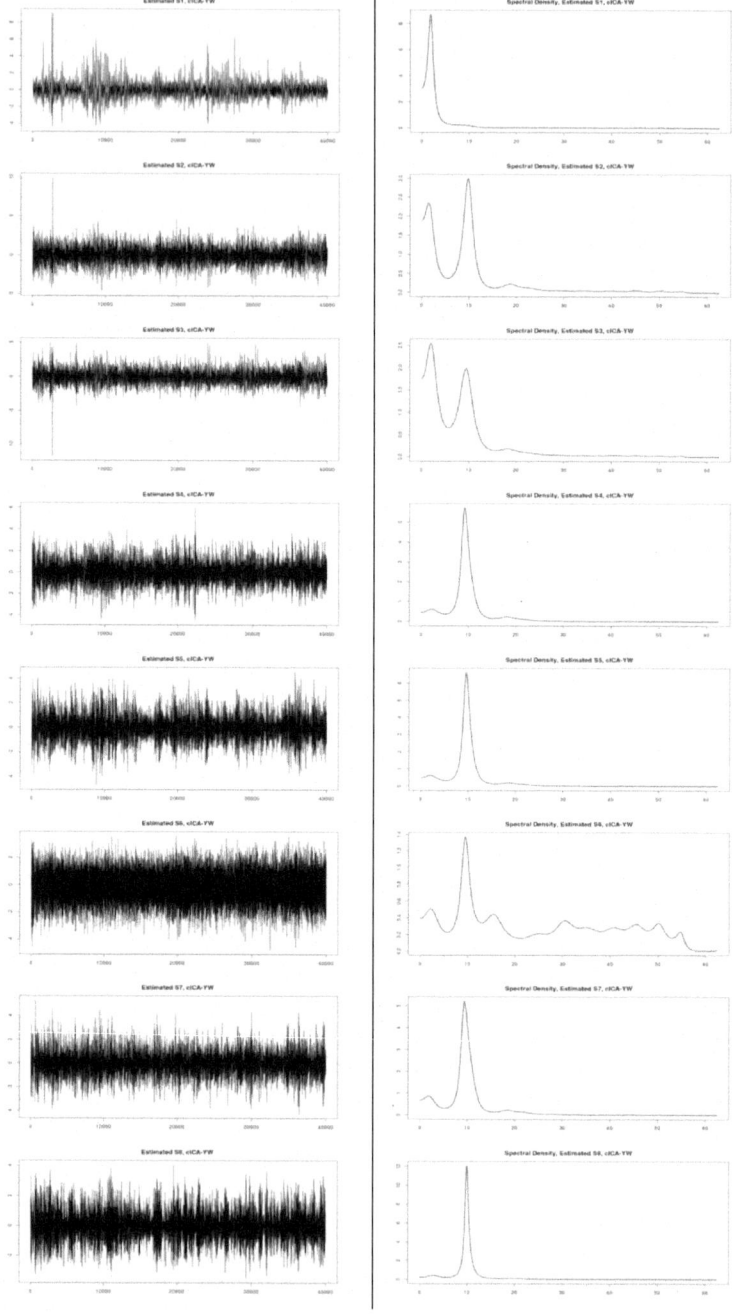

Figure 7.6: Topography and estimated source spectra by cICA-LSP.

```
%%% ts : DFT of [(N+1)/2]-by-1 univariate time series
%%% This is specifically for colorICA_pml
%%% Copyright (c) Seonjoo Lee, Jan 2009 .

    if (size(ts,1)==1)
        ts=ts'; end;
    if (size(phi,1)==1)
        phi=phi'; end;
    N = size(ts,1);
    freq = (0:(floor(N/2)))'/N*2*pi;

    if (phi == 0)
        p = 0;
        phi_l = ones((floor((N-1)/2))+1,1);
    else
        p = size(phi,1);
        phi_l = 1-(exp(-1i * freq * (1:p) ) * phi);
    end
        h =1./(abs(phi_l).^2);

        per = (abs(ts).^2);
        per = per(1:(floor((N-1)/2)+1));
    if strcmp(method,'aic'), outscore = ...
        2*sum(per./(sigmasq*h) + log(sigmasq * h)) + 2*p;
        elseif strcmp(method,'aicc')
            outscore = 2*sum( per./(sigmasq*h) + ...
            log(sigmasq * h)) + 2*(p+1)*N/(N-p-1);
        elseif strcmp(method,'bic')
            outscore = 2*sum( per./(sigmasq*h) + ...
             log(sigmasq * h)) +p*log(N);
    end
end
```

7.8.2   AIC UNIT2

% aic_unit2.m

```
function [outscore,per,pxx] = aic_unit(ts, phi,sigmasq, method)
%% Description
%%%returns AIC for model selection in AR
%%% ts : DFT of [(N+1)/2]-by-1 univariate time series
%%% This is specifically for colorICA_pml
%%% Copyright (c) Seonjoo Lee, Jan 2009 .
```

```
     if (size(ts,1)==1)
         ts=ts'; end;
     if (size(phi,1)==1)
         phi=phi'; end;

     N = size(ts,1);
     freq = (0:(floor((N-1)/2)))'/N*2*pi;

     if (phi == 0)
         p = 0;
         pxx =sigmasq/(2*pi)*ones((floor((N-1)/2))+1,1);
     else
         p = size(phi,1);
         pxx = sigmasq/(2*pi)./(abs(1-(exp(-1i * ...
                 freq * (1:p) ) * phi)).^2);
          %phi_l = 1-(exp(-1i * freq * (1:p) ) * phi);
     end
%        f =sigmasq/(2*pi)./(abs(phi_l).^2);

     per = (abs(fft(ts)).^2)/(2*pi*N);
     per = per(1:(floor((N-1)/2)+1));
     if strcmp(method,'aic'), outscore = ...
                     2*sum(per./pxx + log(pxx)) + 2*p;
         elseif strcmp(method,'aicc')
             outscore = 2*sum( per./(pxx) + log(pxx)) + ...
                     2*(p+1)*N/(N-p-1);
         elseif strcmp(method,'bic')
             outscore = 2*sum( per./(pxx) + log(pxx)) +p*log(N);
     end
end
```

7.8.3 AMARI DISTANCEW

```
%amari_distanceW.m

function Perf=amari_distanceW(Q1,Q2);
% AMARI_DISTANCE - distance between two matrices.
% Beware: it does not verify the axioms
%         of a distance. It is always between 0 and 1.
% Copyright (c) Francis R. Bach, 2002.

m=size(Q1,1);
%Per=inv(Q1)*Q2;
Per = Q1/Q2;
```

```
Perf=[sum((sum(abs(Per))./max(abs(Per))-1)/(m-1))/m;
sum((sum(abs(Per'))./max(abs(Per'))-1)/(m-1))/m];
Perf=mean(Perf);
```

7.8.4 ASSIGNSPEC.M

```
function [g] = assignspec(X, W2)
%%%%%%%%%%%%%%%%%%%%%%%%%%%%%%%%%%%%%%%%%%%%%%%%%%%%%
%%% STEP2 : Estimate Spectral Density Part
%%%%%%%%%%%%%%%%%%%%%%%%%%%%%%%%%%%%%%%%%%%%%%%%%%%%%
        [M N] = size(X);
        g = [];
        wX = W2 * X;
        putRdata('N',N);
    for  j = 1:M
            period1 = abs(fft(wX(j,:))).^2/(N * 2 * pi);
            period1 = period1(2:floor((length(period1) + 2)/2));
            period1(find(period1==0)) = 0.0001;
            putRdata('period1',period1);
            evalR('fit = lspec( period = period1, odd=FALSE)');
            evalR('sde = dlspec((0:(N-1))*2*pi/N, fit)');
            evalR('sde$m[is.na(sde$m)] = 0');
            evalR('e_spec = (sde$m + sde$d)*2*pi');
            e_spec = getRdata('e_spec');
            g = [g;e_spec];
        end
end
```

7.8.5 CICA-XARMA2

```
% cICA_xarma2.m

function [W, phi, p, sigma, iter, NInv, wlscore] = ...
                cICA_xarma2(X, varargin)
%%% X is M by N matrix, M signals,
    %%% N samples, it will be centered.
%%% tol : convergence number : Default-0.0001
%%% maxit : maximum iteration number : Default-200
%%% initial_W is initial matrix of unmixing matrix :
    %%%             Default : I_M
```

```
%%% prewhite :   0 - No prewhitening : Default
%%%              1 = Prewhitening
%%% Modified 110504: Modified the algorithm of colorICA_pmm
%%%     in the optimization
%%% ( get rid of log(det(W)) in the optimization step)
%%% Modified 110608: the second optimization step using
%%% eigen decomposition.


%%% Copyright (c) Seonjoo Lee, Jan 2009 .
%%%%%%%%%%%%%%%%%%%%%%%%%%%%%%%%%%%%%%%%%%%%%%%%%%%%%%%%%
%% Defaults
%%%%%%%%%%%%%%%%%%%%%%%%%%%%%%%%%%%%%%%%%%%%%%%%%%%%%%%%%
    [M N] = size(X);         % M signals, N samples
    tol = 0.0001;
    maxit = 200;
    initialw = eye(M); %    initial_W=rand_orth(M);
    prewhite = 1;
    method='bic';
    verbose=0;
    wlscore = -inf;
    maxnmodels=100;
    W2 = zeros(M);
    tau=0.5;
    if N>10000,
        maxnmodels = 35;
    end
%%%%%%%%%%%%%%%%%%%%%%%%%%%%%%%%%%%%%%%%%%%%%%%%%%%%%%%%%%%%5
%% Reading options
%%%%%%%%%%%%%%%%%%%%%%%%%%%%%%%%%%%%%%%%%%%%%%%%%%%%%%%%%%%%5
    for k=1:2:(length(varargin)-1)
        switch lower(varargin{k})
            case 'tol'
                tol= varargin{k+1};
            case 'maxit'
                maxit= varargin{k+1};
            case 'initialw'
                initialw= varargin{k+1};
            case 'prewhite'
                prewhite= varargin{k+1};
            case 'method'
                method = varargin{k+1};
            case 'verbose'
                verbose = varargin{k+1};
            case 'maxnmodels'
```

```
                    maxnmodels = varargin{k+1};
            end
        end

        W1 = initialw;
        Xc=X-repmat(mean(X,2),1,N);  % center data
        clear X;
        if (prewhite == 1)
            %[Us,Ss,Vs] = svd(Xc*Xc'/N);%
            invsqcovmat = eye(M)/(sqrtm(Xc*Xc'/N));
                %Us * diag(1./sqrt(diag(Ss))) * Us';
            Xc = invsqcovmat * Xc;
%           display(invsqcovmat);pause;
        end

%%%Discrete Fourier Transformation
        X_dft = zeros(M,N);
        for k = 1:M
            X_dft(k,:) = fft(Xc(k,:))/sqrt(2*pi*N);
        end

%%%%%%%%%%%%%%%%%%%%%%%%%%%%%%%%%%%%%%%%%%%%%%%%%%%%%%%%%%%%%%%%%%%%%%%%
%%% Initial Step : Initialize W, Spectral density g = sigma * h
%%%%%%%%%%%%%%%%%%%%%%%%%%%%%%%%%%%%%%%%%%%%%%%%%%%%%%%%%%%%%%%%%%%%%%%%
phi = cell(M,1);
    g = zeros(M,floor(N/2));
sigma = zeros(M,1);
p = cell(M,1);
indx = 2:(floor(N/2)+1);
freq = (1:floor(N/2))/N*2*pi;
    wX = W1 * Xc;

    for j = 1:M
        [phi_t,sigma_t,p_t]=model_select_MA(wX(j,:),...
                    'method',method, 'maxnmodels', maxnmodels);
        if p_t == 0, g(j,:) = sigma_t/(2*pi)*ones(1,floor(N/2));
        else
            g(j,:) = (sigma_t/(2*pi)).*(abs(phi_t * ...
                        exp(-1i*(0:p_t)'*freq)).^2);
        end
        p{j} = p_t;
        phi{j} =  phi_t;
sigma(j) = sigma_t;
    end
```

```
%     plot(g');pause;

    clear wX;
lim = 1;
iter = 0;
NInv = 0;
    eigenval = zeros(M,1);

%'Start!'
index1 = reshape(reshape(repmat(1:M,1,M),M,M)', 1,M^2);
index2 = repmat(1:M,1,M); % 123123123

    while   ((lim>tol) && (iter<maxit) && (NInv <maxit))
        iter = iter+1;
%%%%%%%%%%%%%%%%%%%%%%%%%%%%%%%%%%%%%%%%%%%%%%%%%%
%%% STEP1 : Estimate W
%%%%%%%%%%%%%%%%%%%%%%%%%%%%%%%%%%%%%%%%%%%%%%%%%%
%%%Score function of Loglikelihood funct
        tmp = real(X_dft(index1,indx) .* conj(X_dft(index2,indx)));
        taucount=1;
        err=1;orthoerror=1;
        W2 = W1;
        tau=0.5;
        while (taucount<60 && err>0.0001 && orthoerror>0.0001)
            for j=1:M
                Gam=0;
                for k = 1:M
                    if k ~= j
                        nu = W2(k,:)' * W2(k,:);
                        Gam = Gam + nu;
                    end
                end
                tmpmat = reshape(tmp*(1./g(j,:))',M,M)';
                tmpV = tmpmat+tau*Gam;
                [u,v] = eig(tmpV);
                eigenval(j)=v(1,1);
                W2(j,:) = u(:,1)'/(norm(u(:,1)));
            end
            orthoerror = sum(sum((W2*W2' - eye(M)).^2));
            err = sqrt(sum(sum( (rerow(W1)-rerow(W2)).^2)));
            taucount = taucount+1;
            tau = 2*tau;
        end
```

```
        if (iter == maxit), %% When the iteration reaches maximum.
           disp('colorICA_mme: iteration reaches to maximum. \n');
        end
        if (NInv == maxit) %% When the iteration reaches maximum.
           disp('No convergence');
        end

        if (any( abs(sigma) == Inf) || (iter == maxit))
           display('Either sigma is infinity or ... \n');
           display('the iteration reaches to maximum.\n');
W2 = rand_orth(M);
iter = 0;
NInv = NInv +1;
        end

        lim = amari_distanceW(rerow(W2), rerow(W1));
     W1 = W2;

%%%%%%%%%%%%%%%%%%%%%%%%%%%%%%%%%%%%%%%%%%%%%%%
%%% STEP2 : Estimate Spectral Density Part
%%%%%%%%%%%%%%%%%%%%%%%%%%%%%%%%%%%%%%%%%%%%%%%
%      h = zeros(M, N);
     wX = rerow(W2) * Xc;
     for  j = 1:M
        [phi_t,sigma_t, p_t]=model_select_MA(wX(j,:), ...
                'method', method, 'maxnmodels', maxnmodels);
        if p_t == 0, g(j,:) = ...
                       sigma_t/(2*pi)*ones(1,floor(N/2));
        else
            g(j,:) = (sigma_t/(2*pi)).*(abs(phi_t * ...
                    exp(-1i*(0:p_t)'*freq)).^2);
        end
        p{j} = p_t;
        phi{j} =  phi_t;
        sigma(j) = sigma_t;
     end
     plot(g');pause;
     clear wX;
     wltmp = -1*(sum(eigenval) + sum(sum(log(g))));

     if (wltmp<wlscore)
      fprintf('Previous steps whittle score: %0.7g:\n', ...
                  wlscore);
      fprintf('Whittle Likelihood (%.7g) is decreasing.\n',...
```

```
                        wltmp);
        end
        wlscore = wltmp;
        if verbose==1,
            fprintf('rcondh: %0.5g \n', rcondh);
            fprintf('lim: %0.5g \n', lim);
            fprintf('iter: %0.5g \n', iter);
            fprintf('Whittle Score: %0.8g \n', wlscore);
        end
    end

    if (prewhite==1)
        W=W2*invsqcovmat;
    else W = rerow(W2);
    end
end
```

7.8.6   CALCULATEAIC.M

```
% calculateAIC.m
function [aic] = calculateAIC(X, W)
    [M N] = size(X);
        X_dft = [];
    for k = 1:M
        X_dft = vertcat(X_dft, (fft(X(k,:))/sqrt(N)) );
    end
g = [];
    wX_dft = W * X_dft;
    openR;
    evalR('library(polspline)');
    putRdata('N',N);
    [g] = assignspec(X, W);

    aic = 2 * ( sum(sum(abs(X_dft).^2./g + log(g))) ) + M*(M+1);
    closeR;
end
```

7.8.7   COLORICA-PMM

```
% colorICA_pmm.m
```

```
function [W, phi,sigma, iter, NInv,p_t] = colorICA_pmm(X, ...
                                        varargin)
%%% X is M by N matrix, M signals, N samples, centered.
%%% tol : convergence number : Default-0.0001
%%% maxit : maximum iteration number : Default-200
%%% initial_W is initial matrix of unmixing matrix :
    %%%         Default : I_M
%%% prewhite :  0 - No prewhitening : Default
    %%%             1 = Prewhitening

%%% Copyright (c) Seonjoo Lee, Jan 2009 .

% Default
[M N] = size(X);            % M signals, N samples
tol = 0.0001;
maxit = 200;
initialw = eye(M); %    initial_W=rand_orth(M);
arma=0;
prewhite = 0;
method='aic';
for k=1:2:(length(varargin)-1)
    switch lower(varargin{k})
        case 'tol'
            tol= varargin{k+1};
        case 'maxit'
            maxit= varargin{k+1};
        case 'initialw'
            initialw= varargin{k+1};
        case 'prewhite'
            prewhite= varargin{k+1};
        case 'method'
            method = varargin{k+1};
    end
end

W1 = initialw;
Xc=X-repmat(mean(X,2),1,N);  % center data
if (prewhite == 1)
    %[Us,Ss,Vs] = svd(Xc*Xc'/N);%
    invsqcovmat = eye(M)/(sqrtm(Xc*Xc'/N));
            %Us * diag(1./sqrt(diag(Ss))) * Us';
    Xc = invsqcovmat * Xc;
end
```

```
%%%Discrete Fourier Transformation
    X_dft = zeros(M,N);
    for k = 1:M
        X_dft(k,:) = fft(Xc(k,:))/sqrt(2*pi*N);
    end

%%%%%%%%%%%%%%%%%%%%%%%%%%%%%%%%%%%%%%%%%%%%%%%%%%%%%%%%%%%%%%%%%%%%
%% Initial Step : Initialize W, Spectral density g = sigma * h
%%%%%%%%%%%%%%%%%%%%%%%%%%%%%%%%%%%%%%%%%%%%%%%%%%%%%%%%%%%%%%%%%%%%
phi = cell(M,1);
        g = zeros(M,N);
        sigma = zeros(M,1);
%        h = zeros(M, N);
        p = cell(M,1);

freq = (0:(N-1))/N*2*pi;
    wX = W1 * Xc;
for  j = 1:M
        if arma==1, [phi_t,theta_t,sigma_t, p_t] ...
                =model_select_ARMA(wX(j,:),'method',method);
            if (~all(p_t))
                g(j,:)=sigma_t^2/(2*pi)*ones(1,N);
            elseif (p_t(1) == 0 && p_t(2)>0)
                g(j,:)=sigma_t^2/(2*pi)*(abs(1+(exp(-1i ...
                    * freq' * (0:(p_t(2))) ) * theta_t')).^2)';
            elseif (p_t(2) == 0 && p_t(1)>0)
                g(j,:)=sigma_t^2/(2*pi)./(abs(1+(exp(-1i ...
                    * freq' * (0:(p_t(1))) ) * phi_t')).^2)';
            elseif (all(p_t))
                g(j,:)=sigma_t^2/(2*pi)*((abs(1+(exp(-1i ...
                    * freq' * (0:(p_t(2))))*theta_t')).^2 ...
                    ./(abs(1+(exp(-1i * freq' ...
                        * (0:(p_t(1))) * phi_t')).^2)))';
            end

        elseif arma==0, [phi_t,sigma_t, p_t]= ...
                model_select_MM(wX(j,:), 'method',method);
            if p_t == 0, g(j,:) = sigma_t/(2*pi)*ones(1,N);
            else
                g(j,:) = (sigma_t/(2*pi))...
                    ./(abs(1- phi_t * exp(-1i*(1:p_t)'*freq)).^2);
            end
        end
        p{j} = p_t;
```

```
          phi{j} =  phi_t;
sigma(j) = sigma_t;
 %        g(j,:) = sigma_t*h(j,:);
     end

lim = 1;
iter = 0;
NInv = 0;
lambda = ones(M*(M+1)/2,1);
%'Start!'
          index1 = reshape(reshape(repmat(1:M,1,M),M,M)',1,M^2);
                       % 111222333
index2 = repmat(1:M,1,M); % 123123123
tempmx1 = reshape(index2,M,M);
tempmx2 = reshape(index1,M,M);
tempmx1 = tempmx1(index1<=index2);
                      %Upper Triangular elements...column
tempmx2 = tempmx2(index1<=index2);
                      %Upper Triangular elements...row
          indx_temp = [ eye(M*(M+1)/2), ...
                      zeros(M*(M+1)/2,M*(M-1)/2)];
while   ((lim>tol) && (iter<maxit) && (NInv <maxit))
          iter = iter+1;
X_wdft = W1 * X_dft;
%%%%%%%%%%%%%%%%%%%%%%%%%%%%%%%%%%%%%%%%%%%%%%%
%%% STEP1 : Estimate W
%%%%%%%%%%%%%%%%%%%%%%%%%%%%%%%%%%%%%%%%%%%%%%%
%%%Score function of Loglikelihood function

W1_inv = inv(W1);
Score_l = 2*sum(real(conj(X_wdft(index1,:)) .* ...
          X_dft(index2,:))./g(index1,:),2)- N * ...
              reshape( W1_inv, M^2, 1);


%%%Score function of constraint
c_e = W1 * W1' - eye(M);
c_e= c_e(index1<=index2);
              %%Lower triangular parts including diag

J_e1 = W1(tempmx1,index2).* indx_temp(tempmx2,index1);
J_e2 = W1(tempmx2,index2).* indx_temp(tempmx1,index1);
J_e = J_e1 + J_e2;
%%%Hassian Matrix of Loglikelihood function
temp_l = zeros(M^2);
```

```
for  o = 1:M
          for  j = 1:M
            for  k = 1:M
              for  l = 1:M
                if ( k==o && o==j && j==l) ...
                    temp_l(((o-1)*M+j),((k-1)*M+l)) = ...
                        lambda((o*(o+1)/2));
                  end
                end
              end
            end
          end

temp_h = zeros(M^2);
        tmpm = real(conj(X_dft(index1,:)) .* X_dft(index2,:));
        for j = 1:M
temp_h(((j-1)*M+1):(j*M),((j-1)*M+1):(j*M) )...
              =2*reshape(tmpm*(1./g(j,:))', M,M)';
        end


        clear tmpm;%1 tmpm2;
Hassian_l = temp_h - temp_l + N * ...
            (W1_inv(index2,index1).*W1_inv(index2,index1)')')';
H =[Hassian_l, -J_e';
-J_e, zeros(M*(M+1)/2) ];

%[U,D] = svd(H,0);
        rcondh=rcond(H);
        % D = diag(D);
if (iter == maxit)
          fprintf('colorICA_mme: iteration reaches maximum \n');
        end
if  rcondh< 10^(-15) || any( abs(sigma) == Inf) || ...
                  (iter == maxit)
W2 = rand_orth(M);
lambda = ones(M*(M+1)/2,1);
iter = 0;
NInv = NInv +1;
W1 = W2;
end

if  (rcondh>= 10^(-15)) && (all(abs(sigma) < Inf))
V_W = vertcat(reshape(W1', M^2,1),lambda) ...
```

```
                   - H\vertcat( Score_l- (J_e' * lambda), -c_e');
W2 = reshape(V_W(1:M^2),M,M)';
lambda = V_W((M^2+1) : (M*(3*M+1)/2));
lim = amari_distanceW(rerow(W2), rerow(W1));
W1 = W2;
%%              fprintf('%0.5g',lim);
        end
if (NInv == maxit) fprintf('No convergence');
        end
%%%%%%%%%%%%%%%%%%%%%%%%%%%%%%%%%%%%%%%%%%%%%%%%%%%%
%%% STEP2 : Estimate Spectral Density Part
%%%%%%%%%%%%%%%%%%%%%%%%%%%%%%%%%%%%%%%%%%%%%%%%%%%%
%        h = zeros(M, N);
        wX = W2 * Xc;
        for  j = 1:M
        if arma==1, ...
            [phi_t,theta_t,sigma_t, p_t]=...
                model_select_ARMA(wX(j,:), 'method',method);
            if (~all(p_t))
                g(j,:)=sigma_t^2/(2*pi)*ones(1,N);
            elseif (p_t(1) == 0 && p_t(2)>0)
                g(j,:)=sigma_t^2/(2*pi) *(abs(1+(exp(-1i * ...
                    freq' * (0:(p_t(2))) ) * theta_t')).^2)';
            elseif (p_t(2) == 0 && p_t(1)>0)
                g(j,:)=sigma_t^2/(2*pi)...
                    ./(abs(1+(exp(-1i * freq' * ...
                    (0:(p_t(1))) ) * phi_t')).^2)';
            elseif (all(p_t))
              g(j,:)=sigma_t^2/(2*pi)*((abs(1+(exp(-1i * ...
              freq' * (0:(p_t(2))))*theta_t')).^2)./...
              (abs(1+(exp(-1i * freq' * (0:(p_t(1))) * ...
              phi_t')).^2)))';
            end

        elseif arma==0, [phi_t,sigma_t, p_t]= ...
                model_select_MM(wX(j,:), 'method',method);
            if p_t == 0, g(j,:) = sigma_t/(2*pi)*ones(1,N);
            else
                g(j,:) = (sigma_t/(2*pi))./(abs(1- phi_t ...
                    * exp(-1i*(1:p_t)'*freq)).^2);
            end
        end
        p{j} = p_t;
        phi{j} =  phi_t;
```

```
sigma(j) = sigma_t;
 %         g(j,:) = sigma_t*h(j,:);
     end
end
    if (prewhite==1) W=W1*invsqcovmat;
    else W = rerow(W1);
    end
end
```

7.8.8 COLORICA-PMM4

```
% colorICA_pmm4.m

function [W, W2, phi, sigma, iter, NInv, p_t, ...
    wlscore] = colorICA_pmm4(X, varargin)
%%% X is M by N matrix, M signals, N samples, centered.
%%% tol : convergence number : Default-0.0001
%%% maxit : maximum iteration number : Default-200
%%% initial_W is initial matrix of unmixing matrix :
%%%         Default : I_M
%%% prewhite :  0 - No prewhitening : Default
%%%             1 = Prewhitening
%%% Modified 110608: the second optimization step using
%%% eigen decomposition.

%%% Copyright (c) Seonjoo Lee, Jan 2009 .
%%%%%%%%%%%%%%%%%%%%%%%%%%%%%%%%%%%%%%%%%%%%%%%%%%%%%%%%%%
%% Defaults
%%%%%%%%%%%%%%%%%%%%%%%%%%%%%%%%%%%%%%%%%%%%%%%%%%%%%%%%%%
    [M N] = size(X);           % M signals, N samples
    tol = 0.0001;
    maxit = 200;
    initialw = eye(M); %    initial_W=rand_orth(M);
    prewhite = 1;
    method='bic';
    verbose=0;
    wlscore = -inf;
    maxnmodels=100;
    W2 = zeros(M);
    tau=0.5;
    if N>10000,
        maxnmodels = 35;
    end
```

```
%%%%%%%%%%%%%%%%%%%%%%%%%%%%%%%%%%%%%%%%%%%%%%%%%%%%%%%%%%%%%%5
%% Reading options
%%%%%%%%%%%%%%%%%%%%%%%%%%%%%%%%%%%%%%%%%%%%%%%%%%%%%%%%%%%%%%5
    for k=1:2:(length(varargin)-1)
        switch lower(varargin{k})
            case 'tol'
                tol= varargin{k+1};
            case 'maxit'
                maxit= varargin{k+1};
            case 'initialw'
                initialw= varargin{k+1};
            case 'prewhite'
                prewhite= varargin{k+1};
            case 'method'
                method = varargin{k+1};
            case 'verbose'
                verbose = varargin{k+1};
            case 'maxnmodels'
                maxnmodels = varargin{k+1};
        end
    end

    W1 = initialw;
    Xc=X-repmat(mean(X,2),1,N);  % center data
    clear X;
    if (prewhite == 1)
        %[Us,Ss,Vs] = svd(Xc*Xc'/N);%
        invsqcovmat = eye(M)...
            /(sqrtm(Xc*Xc'/N));%Us * diag(1./sqrt(diag(Ss))) * Us';
        Xc = invsqcovmat * Xc;
    end

%%%Discrete Fourier Transformation
    X_dft = zeros(M,N);
    for k = 1:M
        X_dft(k,:) = fft(Xc(k,:))/sqrt(2*pi*N);
    end

%%%%%%%%%%%%%%%%%%%%%%%%%%%%%%%%%%%%%%%%%%%%%%%%%%%%%%%%%%%%%%%
%% Initial Step : Initialize W, Spectral density g = sigma * h
%%%%%%%%%%%%%%%%%%%%%%%%%%%%%%%%%%%%%%%%%%%%%%%%%%%%%%%%%%%%%%%
phi = cell(M,1);
    g = zeros(M,floor(N/2));
sigma = zeros(M,1);
```

```
p = cell(M,1);
indx = 2:(floor(N/2)+1);
freq = (1:floor(N/2))/N*2*pi;
    wX = W1 * Xc;

    for j = 1:M
        [phi_t,sigma_t,p_t]=model_select_MM(wX(j,:),...
            'method',method, 'maxnmodels',maxnmodels);
        if p_t == 0, g(j,:) = sigma_t/(2*pi)*ones(1,floor(N/2));
        else
            g(j,:) = (sigma_t/(2*pi))...
                ./(abs(1- phi_t * exp(-1i*(1:p_t)'*freq)).^2);
        end
        p{j} = p_t;
        phi{j} =  phi_t;
sigma(j) = sigma_t;
    end

%    plot(g');pause;

    clear wX;
lim = 1;
iter = 0;
NInv = 0;
    eigenval = zeros(M,1);

%'Start!'
index1 = reshape(reshape(repmat(1:M,1,M),M,M)', 1,M^2);
index2 = repmat(1:M,1,M);

    while   ((lim>tol) && (iter<maxit) && (NInv <maxit))
        iter = iter+1;
%%%%%%%%%%%%%%%%%%%%%%%%%%%%%%%%%%%%%%%%%%%%%%%%%%%%%%
%%% STEP1 : Estimate W
%%%%%%%%%%%%%%%%%%%%%%%%%%%%%%%%%%%%%%%%%%%%%%%%%%%%%%
%%%Score function of Loglikelihood funct
        tmp = real(X_dft(index1,indx) .* ...
            conj(X_dft(index2,indx)));
        taucount=1;
        err=1;orthoerror=1;
        W2 = W1;
        tau=0.5;
        while (taucount<60 && err>0.0001 && orthoerror>0.00001)
```

```
            for j=1:M
                Gam=0;
                for k = 1:M
                    if k ~= j
                        nu = W2(k,:)' * W2(k,:);
                        Gam = Gam + nu;
                    end
                end
                tmpmat = reshape(tmp*(1./g(j,:))',M,M)';
                tmpV = tmpmat+tau*Gam;
                [u,v] = eig(tmpV);
                eigenval(j)=v(1,1);
                W2(j,:) = u(:,1)'/(norm(u(:,1)));
            end
            orthoerror = sum(sum((W2*W2' - eye(M)).^2));
            err = sqrt(sum(sum( (rerow(W1)-rerow(W2)).^2)));
            taucount = taucount+1;
            tau = 2*tau;
        end
        if (iter == maxit),
                %% When the iteration reaches maximum.
                disp('colorICA_mme: iteration reaches maximum.\n');
        end
        if (NInv == maxit)
                %% When the iteration reaches maximum.
                disp('No convergence');
        end

        if (any( abs(sigma) == Inf) || (iter == maxit))
                display('Eigher sigma is infinity or ...
                    iteration reaches to maxium.\n');
W2 = rand_orth(M);
iter = 0;
NInv = NInv +1;
        end

        lim = amari_distanceW(rerow(W2), rerow(W1));
    W1 = W2;

%%%%%%%%%%%%%%%%%%%%%%%%%%%%%%%%%%%%%%%%%%%%%%%%%%%%%%%%%%%%%
%%% STEP2 : Estimate Spectral Density Part
%%%%%%%%%%%%%%%%%%%%%%%%%%%%%%%%%%%%%%%%%%%%%%%%%%%%%%%%%%%%%
        wX = rerow(W2) * Xc;
        for  j = 1:M
```

```
            [phi_t,sigma_t, p_t]=model_select_MM(wX(j,:), ...
                'method', method, 'maxnmodels', maxnmodels);
            if p_t == 0, g(j,:) = ...
                            sigma_t/(2*pi)*ones(1,floor(N/2));
            else
                g(j,:) = (sigma_t/(2*pi))./(abs(1- phi_t ...
                    * exp(-1i*(1:p_t)'*freq)).^2);
            end
            p{j} = p_t;
            phi{j} =  phi_t;
            sigma(j) = sigma_t;
        end
%        plot(g');pause;
        clear wX;
        wltmp = -1*(sum(eigenval) + sum(sum(log(g))));

        if (wltmp<wlscore)
            fprintf('Previous steps whittle score: %0.7g; ...
                Whittle Likelihood is (%.7g)is
                Decreasing.\n',wlscore,wltmp);
        end
        wlscore = wltmp;
        if verbose==1, fprintf('rcondh: %0.5g, lim: %0.5g, ...
                iter:%0.5g, Whittle Score: %0.8g.\n',...
                    rcondh,lim,iter,wlscore);
        end
    end

    if (prewhite==1)
        W=W2*invsqcovmat;
    else W = rerow(W2);
    end
end
```

7.8.9   DEXPRAND.M

```
% dexprand.m

function [p] = dexprand(N,m)
    news1 = sign(rand(1,N)-0.5);
    news2=exprnd(m,1,N);
    p = news1 .* news2;
end
```

7.8.10 ESTIMATEAR.M

```
% estimateAR.m
function [phi,sigma] = estimateAR(x,p)
%%% x is a column vector
    n = length(x);
    y = x-mean(x);
    if p>0
        T = zeros((p+1),2*n);
        for k = 1:(p+1)
            T(k,(n+2-k):(2*n+1-k)) = y;
        end
        GammaHat = T * T'/n;
        phi = inv(GammaHat(1:p,1:p)) * GammaHat(1,2:(p+1))' ;
        sigma = GammaHat(1,:) * [1 -phi']';
    else
        phi = 0;
        sigma = var(y);
    end;
end


%%%Example
%%%n=1000
%%x = arima.sim(n =1000, list(ar = c(0.8897, -0.4858)), sd = 1)
%%%arima(x,order = c(2,0,0))
%%%p=2
%%%estimateAR(x,2)
%%%estimateAR(x,1)
%%%estimateAR(x,5)
```

7.8.11 GETSPEC2.M

```
% getspec2.m

function [g] = getspec2(X, W2)
%%%%%%%%%%%%%%%%%%%%%%%%%%%%%%%%%%%%%%%%%%%%%%%%%%%%%%%%%%
%%% STEP2 : Estimate Spectral Density Part
%%%%%%%%%%%%%%%%%%%%%%%%%%%%%%%%%%%%%%%%%%%%%%%%%%%%%%%%%%
        [M N] = size(X);
        g = zeros(M,floor(N/2));
        wX = W2 * X;
        for n = 1:M
            spec = callR('dlspec.sj', wX(n,:),...
```

```
                        (1:floor(N/2))*2*pi/N);
            linespec = spec.m';
            linespec(isnan(linespec))=0;
            cspec = spec.d';

            %%%test pure harmonic
            thresh_0 = 10^(-5);
        if max(linespec)>0 ...
          && (max(cspec)/min(linespec(linespec>0))) < thresh_0,
                g(n,:) = linespec;
                g(n,linespec==0) = inf;
            else
                g(n,:) = linespec+cspec;
            end
        end
end
```

7.8.12 MODEL-SELECT-MM

```
% model_select_MM.m

function [phi, sigmasq, p, msscore] = model_select_MM(ts, ...
                                       varargin)
%%% ts : time series data on real-line
%%% maxNmodels : maximum number of the models to be tested :
%%%     default is 10.
%%% method :    'aic' : default
%%%             'aicc'
%%%             'bic'

%%% Copyright (c) Seonjoo Lee, Jan 2009 .

    if ( size(ts,1)==1)
        ts = ts';
    end

    maxnmodels = 100;
    if size(ts,1)>10000,
        maxnmodels = 15;
    end
    method='aic'; scoreplot = 0;

    for k=1:2:(length(varargin)-1)
```

```
            switch lower(varargin{k})
                case 'maxnmodels'
                    maxnmodels = varargin{k+1};
                case 'method'
                    method = varargin{k+1};
                case 'scoreplot'
                    scoreplot = varargin{k+1};
            end
    end
    N = size(ts,1);
    ar = 1;
    stopping = 0;
    score = cell(maxnmodels,1);
    phi1 = cell(maxnmodels,1);
    sigmasq1 = cell(maxnmodels,1);
    p1 = cell(maxnmodels,1);
    freq = (0:(floor((N-1)/2)))'/N*2*pi;

    while ((ar < (maxnmodels+1)) && (stopping == 0))
%        [ar (ar < (maxnmodels+1)) && (stopping == 0)]
        [phi_t,sigma_t] = aryule( ts, (ar-1));
        if ar==1
            phi_t = 0;
        else phi_t = -phi_t(2:ar);
%%%            [phi_t,sigma_t] = estimateAR( ts, (ar-1));
        end

        if (phi_t == 0)
            pxx =sigma_t/(2*pi)*ones((floor((N-1)/2))+1,1);
        else
            pxx =sigma_t/(2*pi)./(abs(1-(exp(-1i ...
                * freq * (1:(ar-1)) ) * phi_t')).^2);
        end
        per = (abs(fft(ts)).^2)/(2*pi*N);
        per = per(1:(floor((N-1)/2)+1));
        if strcmp(method,'aic'), score{ar} = 2*sum(per./pxx ...
                + log(pxx)) + 2*(ar-1);
        elseif strcmp(method,'aicc')
            score{ar} = 2*sum( per./(pxx) + log(pxx)) ...
                + 2*((ar-1)+1)*N/(N-(ar-1)-1);
        elseif strcmp(method,'bic')
            score{ar} = 2*sum( per./(pxx) + log(pxx)) + ...
                        (ar-1)*log(N);
        end
```

```
        phi1{ar} = phi_t;
        sigmasq1{ar} = sigma_t; p1{ar} = ar-1;
        if ar>20,
            if (score{ar} > min(cell2mat(score(1:(ar-1)))) ...
                    || (ar>maxnmodels)),
                stopping = 1;
            end
        end
        ar = ar+1;
    end
%plot(score)
    [minscore idx] = min(cell2mat(score));
    phi = phi1{idx};
    sigmasq=sigmasq1{idx};
    p = p1{idx};
%    [cell2mat(p1),cell2mat(sigmasq1)]
    msscore = minscore;%score(idx);

    if scoreplot == 1
        plot(cell2mat(p1), cell2mat(score));
    end
end
```

7.8.13   SPECAR.M

```
% specAR.m

function [h] = specAR(ts, phi, freq)
%% Description
%%% returns spectral density without
%%%        sigmasq( phi_l(AR(p))/sigmasq)
%%%        at different frequency
%%% ts : N-by-1 univariate time series vector
%%% phi : p-by-1 vector for AR(p) model
if (size(ts,1)==1) ts=ts'; end;
if (size(phi,1)==1) phi=phi'; end;
p = size(phi,1);
h = 1./abs(1-phi' * exp(i * (1:p)' * freq ));
h = h';
end;
```

7.8.14   WHITEICA.M

```
% whiteICA.m

function [W1, iter, NInv, res] = ...
       whiteICA(X, tol, maxit, initial_W)
%%% Required functions
%%% X is Stationary Time series with mean zero (optional)
%%% tol : convergence number
%%% maxit : maximum iteration number
%%% initial_W is initial matrix of unmixing matrix

W1 = initial_W;
    [M N] = size(X);           % M signals, N samples
    % center signal
X = X- mean(X, 2) * ones (1, N);
%%%Discrete Fourier Transformation
    X_dft = [];
    for k = 1:M
        X_dft = vertcat(X_dft, (fft(X(k,:))/sqrt(N)) );
    end

lim = 1;
iter = 0;
res = [];
NInv = 0;
lambda = ones(M*(M+1)/2,1);

while   ((lim>tol) && (iter<maxit) && (NInv <maxit)),
iter = iter+1;
X_wdft = W1 * X_dft;

%% Initialize+Estimate Sigma^2
        g = sum(abs(W1 * X_dft).^2,2)/N;

%% Score function of Loglikelihood function
        index1 =  reshape(reshape(repmat(1:M,1,M),M,M)',1,M^2);
                    % 111222333
index2 = repmat(1:M,1,M); % 123123123
W1_inv = inv(W1);
        Score_1 = sum(real(conj(X_wdft(index1,:)) ...
            .* X_dft(index2,:)),2)./g(index1) - N ...
              * reshape( W1_inv, M^2, 1);

%%%Score function of constraint
```

```
c_e = W1 * W1' - eye(M);
c_e= c_e(find(index1<=index2));
          %%Lower triangular parts including diag

tempmx1 = reshape(index2,M,M);
tempmx2 = reshape(index1,M,M);
tempmx1 = tempmx1(find(index1<=index2));
tempmx2 = tempmx2(find(index1<=index2));
          indx_temp=[eye(M*(M+1)/2),zeros(M*(M+1)/2,M*(M-1)/2)];
J_e1 = W1(tempmx1,index2).* indx_temp(tempmx2,index1);
J_e2 = W1(tempmx2,index2).* indx_temp(tempmx1,index1);
J_e = J_e1 + J_e2;
%%%Hassian Matrix of Loglikelihood function
temp_l = zeros(M^2);

for  o = 1:M
        for  j = 1:M
          for  k = 1:M
            for  l = 1:M
              if ( k==o && o==j && j==l) ...
                      temp_l(((o-1)*M+j),((k-1)*M+l)) ...
                          = lambda((o*(o+1)/2));
              end
            end
          end
        end
      end

      Hassian_l = kron(diag(1./g),...
        (reshape(sum(real(conj(X_dft(index1,:)) ...
          .* X_dft(index2,:)),2),M,M))')+ N ...
            *(W1_inv(index2,index1).* ...
            W1_inv(index2,index1)')')'- temp_l;

H =[Hassian_l, -J_e';
-J_e, zeros(M*(M+1)/2) ];

[U,D,V] = svd(H);

      D = diag(D );

      if (iter == maxit)
          fprintf('whiteICA : iteration reached to maximum \n');
      end
```

```
if (min(D)/max(D))< 10^(-15) || any( abs(g) == Inf)...
                 || (iter == maxit)
W2 = rand_orth(M);
lambda = ones(M*(M+1)/2,1);
iter = 0;
NInv = NInv +1;
W1 = W2;
  end

if  ((min(D)/max(D))>= 10^(-15)) && (all(abs(g) < Inf))
 V_W = vertcat(reshape(W1', M^2,1),lambda) - inv(H) ...
                *vertcat( Score_1- (J_e' * lambda), -c_e');
W2 = reshape(V_W(1:M^2),M,M)';
lambda = V_W((M^2+1) : (M*(3*M+1)/2));
lim = amari_distance(W2, W1);
W1 = W2;
    res = [res, [V_W;g]];
 %%            fprintf('%0.5g',lim);
        end
if (NInv == maxit)
            fprintf('whiteICA : No convergence \n');
        end
    end
end
```

Bibliography

1. S. Amari, A. Cichocki, H.H. Yang, et al. A New Learning Algorithm for Blind Signal Separation. *Advances in Neural Information Processing Systems*, 8: 757–763, 1996.
2. F.R. Bach and M.I. Jordan. Kernel Independent Component Analysis. *Journal of Machine Learning Research*, 3(1):1–48, 2003.
3. A.J. Bell and T.J. Sejnowski. An Information-Maximization Approach to Blind Separation and Blind Deconvolution. *Neural Computation*, 7(6):1129–1159, 1995.
4. R. Boscolo, H. Pan, and V.P. Roychowdhury. Independent Component Analysis Based on Nonparametric Density Estimation. *IEEE Transactions on Neural Networks*, 15(1):55–65, 2004.
5. D.R. Brillinger. *Time Series: Data Analysis and Theory*. Society for Industrial and Applied Mathematics, 2001.
6. P.J. Brockwell and R.A. Davis. *Time Series: Theory and Methods*. Springer, second edition, 1991.
7. E. Bullmore, C. Long, J. Suckling, J. Fadili, G. Calvert, F. Zelaya, T.A. Carpenter, and M. Brammer. Colored Noise and Computational Inference in Neu-

rophysiological (fMRI) Time Series Analysis: Resampling Methods In Time and Wavelet Domains. *Human Brain Mapping*, 12(2):61–78, 2001.

8. V.D. Calhoun, J. Liu, and T. Adali. A Review of Group ICA for fMRI Data and ICA for Joint Inference of Imaging, Genetic, and ERP Data. *Neuroimage*, 45(1S1):163–172, 2009.

9. J.F. Cardoso. High-Order Contrasts for Independent Component Analysis. *Neural Computation*, 11(1):157–192, 1999.

10. A. Chen. Fast Kernel Density Independent Component Analysis. *Independent Component Analysis and Blind Signal Separation*, 3889:24–31, 2006.

11. A. Chen and P.J. Bickel. Consistent Independent Component Analysis and Prewhitening. *IEEE Transactions on Signal Processing*, 53(10 Part 1):3625–3632, 2005.

12. A. Chen and P.J. Bickel. Efficient Independent Component Analysis. *Annals of Statistics*, 34(6):2825–2855, 2006.

13. P. Comon. Independent Component Analysis: A New Concept? *Signal Processing*, 36(3):287–314, 1994.

14. R. Cooper, J.W. Osselton, and J.C. Shaw. *EEG Technology*. London, Butterworth & Co, 1980.

15. A. Delorme and S. Makeig. EEGLAB: An Open Source Toolbox for Analysis of Single-Trial EEG Dynamics Including Independent Component Analysis. *Journal of Neuroscience Methods*, 134(1):9–21, 2004. ISSN 0165-0270.

16. T. Hastie and R. Tibshirani. Independent Components Analysis through Product Density Estimation. *Advances in Neural Information Processing Systems*, pages 665–672, 2003.

17. S.A. Huettel, A.W. Song, and G. McCarthy. *Functional Magnetic Resonance Imaging*. Sinauer Associates Sunderland, second edition, 2008.

18. A. Hyvärinen and P. Hoyer. Emergence of Phase- and Shift-Invariant Features by Decomposition of Natural Images into Independent Feature Subspaces. *Neural Computation*, 12(7):1705–1720, 2000.

19. A. Hyvärinen, J Karhunen, and E. Oja. *Independent Component Analysis*. John Wiley & Sons, 2001.

20. A. Kawaguchi and Y. K. Truong. Logspline independent component analysis. *Bulletin of Informatics and Cybernetics*, 43:83–94, 2011.

21. C. Kooperberg, C.J. Stone, and Y.K. Truong. Logspline Estimation of a Possibly Mixed Spectral Distribution. *Journal of Time Series Analysis*, 16(4):359–388, 1995.

22. S. Lee, H. Shen, Y.K. Truong, M.M. Lewis, and X. Huang. Independent Component Analysis Involving Autocorrelated Sources with an Application to Functional Magnetic Resonance Imaging. *Journal of the American Statistical Association*, 106(495):1009–1024, 2011.

23. S. Lee, H. Shen, and Y.K. Truong. Nonparametric Independent Component Analysis for the Sources with Mixed Spectra. *Technical Report*, 2015.

24. Shen H. Truong Y. Lee, S. and P. Zanini. *coloredICA: Implementation of Colored Independent Component Analysis and Spatial Colored Independent Com-*

ponent Analysis, 2015. R package version 1.0.0.

25. M.J. McKeown, L.K. Hansen, and T.J. Sejnowski. Independent Component Analysis of functional MRI: What Is Signal and What Is Noise? *Current Opinion in Neurobiology*, 13(5):620–629, 2003.

26. J.A. Palmer, S. Makeig, K.K. Delgado, and B.D. Rao. Newton Method for the ICA Mixture Model. *Acoustics, Speech and Signal Processing, 2008. ICASSP 2008. IEEE International Conference on*, pages 1805–1808, 2008.

27. J.A. Palmer, K. Kreutz-Delgado, and S. Makeig. *An Adaptive Mixture of Independent Component Analyzers with Shared Components*. Technical report, 2010. Available as http://sccn.ucsd.edu/~jason/amica̓r.pdf.

28. D.T. Pham and P. Garat. Blind Separation of Mixture of Independent Sources through a Quasi-Maximum Likelihood Approach. *IEEE Transactions on Signal Processing*, 45(7):1712–1725, 1997.

29. R Core Team. *R: A Language and Environment for Statistical Computing*. R Foundation for Statistical Computing, Vienna, Austria, 2013. URL `http://www.R-project.org/`. ISBN 3-900051-07-0.

30. A. Sharma and K.K. Paliwal. Subspace Independent Component Analysis Using Vector Kurtosis. *Pattern Recognition*, 39(11):2227–2232, 2006.

31. Hao Shen, Knut Hüper, and Alexander Smola. Newton-Like Methods for Nonparametric Independent Component Analysis. *Neural Information Processing*, 4232:1068–1077, 2006.

32. J.V. Stone. *Independent Component Analysis*. MIT Press, 2004.

33. N. Vlassis and Y. Motomura. Efficient Source Adaptivity in Independent Component Analysis. *IEEE Transactions on Neural Networks*, 12(3):559–566, 2001.

34. P. Whittle. Some Results in Time Series Analysis. *Skand. Aktuarietidskr*, 35: 48–60, 1952.

35. P. Zanini, H. Shen, and Y. K. Truong. Independent component analysis for spatial stochastic processes on a lattice. *Ann. Applied. Statist.*, 2015.

8 Group Blind Source Separation (GBSS)

Dong Wang
University of North Carolina at Chapel Hill

Haipeng Shen
University of Hong Kong, China

Young K. Truong
University of North Carolina at Chapel Hill

CONTENTS

A new group blind source separation (GBSS) method is introduced and is based on the group structure of the spectral properties of the latent sources. The hidden sources are assumed to be stationary time series. Their spectral density functions will be estimated directly from the sensors or mixed signals according to the specified or known group structures, which can be further described by space or time or both. Our results will be compared with a group independent component analysis method

(GroupICA) based on some popular ICA algorithms. This will be carried out using simulated and real human brain data.

8.1 INTRODUCTION

Independent Component Analysis (ICA) is an important *blind source separation* (BSS) approach which separates hidden source signals from linear mixtures [10, 14]. It has become a common approach to analyze *functional magnetic resonance imaging* (fMRI) data ever since the introduction by McKeown et al. [12]. ICA recovers source signals by assuming observations as a linear mixture of sources (or independent components (ICs)) which are independent of each other and at most one source signal is Gaussian [10]. Many ICA algorithms have been developed to estimate sources. For instance, *Infomax* [2], *fastICA* [10], *Kernel ICA* [1], and *ProDenICA* [8]. Most of these methods only depend upon marginal density estimations of source signals. However, for fMRI data and other functional type neuroimaging data, usually there exist temporal autocorrelation structures for each source [16]. Incorporating the temporal information into ICA would increase estimation accuracy.

Recently, Lee et al. [11] proposed a novel ICA algorithm by taking into account the joint density of the sources; the algorithm is named *colorICA* or *parametric independent colored sources* (PICS). The algorithm assumes that each temporal source has its own parametric autocorrelation structure either in *autoregressive* (AR), *moving average* (MA), or *autoregressive moving average* (ARMA) forms. The approach is carried out in the spectral domain via the Whittle likelihood [15], which is expressed as a function of observations, time series parameters (both the correlation coefficients and noise level), and a matrix reflecting linear mixing operations (mixing matrix). The estimates of time series parameters and the mixing matrix are obtained by minimizing the negative Whittle log-likelihood.

The application of ICA to fMRI data decomposes the observation into a set of spatial maps and the associated temporal courses. According to the independent source assumption on temporal courses or spatial maps, ICA can be categorized into *temporal ICA* (tICA) or *spatial ICA* (sICA) . PICS is a tICA approach because the algorithm assumes independent temporal sources.

ICA has been successfully applied to single fMRI observations and it is called *single subject ICA*. Here, the general term *single subject* means that the 4D measurement is for one run, one session, or one subject. However, in many cases, we record various observations. For instance, multiple runs/sessions/subjects from one session/subject/group. Motivated by these situations, Calhoun et al. [6] proposed the *group ICA* (GICA) approach, which applies ICA to multiple subjects or groups. GICA has enjoyed an increasing popularity in the analysis of multiple neuroimaging observations and many GICA procedures have been developed over the last ten years. See Calhoun et al. [7], Calhoun and Adali [5], and Hyvärinen [9] for recent reviews on this topic.

Multiple fMRI observations can be organized in a hierarchical way such that one observation from the upper level consists of various samples from one level below. For instance, the dataset would have many groups and each group contains multiple subjects (i.e., a two-level organization). The two most commonly studied hierarchical organizations are one-level (i.e., multiple subjects) and two-level (i.e., various groups with each group consisting of multiple subjects). An organization with more than two levels can be analyzed in a similar way. Hence, in this chapter, we focus on these two types. GICA on one-level and two-level organizations are named *multisubject GICA* and *multigroup GICA*, respectively. Multisubject GICA usually assumes heterogeneity across subjects while multigroup GICA assumes homogeneity among subjects within the same group and heterogeneity across groups. In the following, we only review multisubject GICA. Multigroup GICA can be extended by additional homogeneity subject constraints.

One of the goals of multisubject GICA is to identify features shared by subjects. Recall that single subject ICA decomposes fMRI data into spatial maps and temporal courses. Based on prior knowledge of the data, we are interested in estimating common spatial maps, temporal courses, or both. These different types of priors are named *group structures*. Each GICA approach assumes one or more types of group structures as a prior and the estimation procedure is mainly directed by the assumption. Calhoun et al. [6] imposes common spatial maps and subject-specific temporal course assumptions and conducts analysis via sICA. Note that the temporal autocorrelation structures have been ignored in these GICA approaches. Also, only limited group structures have been incorporated and most of them are based on sICA for smaller datasets. In this chapter, we propose a new tICA-type GICA approach, which is an extension of PICS to group inferences. We name our approach *group parametric independent colored sources* (GPICS). GPICS models each temporal source by a parametric time series model and hence automatically takes the autocorrelation structures into account. Moreover, if subjects share the same temporal sources, previous GICA approaches would assume individuals have exactly the same set of temporal courses. However, GPICS would assume the sources of each subject possess the same time series models and allows temporal courses to vary across subjects. Thus GPICS will allow subject-level noise. Furthermore, the parameters are estimated through maximizing log-likelihood in the spectral domain by an iterative procedure. The log-likelihood is a function of data samples, time series parameters and mixing matrices. Hence, various group structures can be accommodated into GPICS. In addition, the procedure only needs one step dimension-reduction via PCA on the spatial direction for single-subjects. Since the temporal direction of single-subject fMRI data is usually small, after the spatial direction dimension reduction step, GPICS is scalable to a large number of subjects.

The remainder of this chapter is organized as follows. We give a brief overview of ICA and PICS in Section 8.2. The GPICS approach and its optimization procedure are described in Section 8.3. The numerical performances are studied through simulations in Section 8.4. A real data analysis example is presented in Section 8.5. We further discuss our GPICS approach in Section 8.6.

8.2 BACKGROUND ON ICA AND PICS

We first introduce some basic definitions and notations for time series analysis in Section 8.2.1 and review the ICA model in Section 8.2.2. We provide details of PICS procedure in Section 8.2.3.

8.2.1 PRELIMINARIES

This section describes some basic notions and tools for analyzing time series considered in this chapter.

Let $Y(t), t = 0, \pm 1, \pm 2, \ldots$ denote a real-valued stationary time series in the strict sense. That is, the finite dimensional distribution of $Y(t_1 + u), Y(t_2 + u), \ldots, Y(t_k + u)$ is the same as $Y(t_1), Y(t_2), \ldots, Y(t_k)$ for $k = 1, 2, \ldots$ and $u \in \mathbb{R}$ [3, 4]. The autocovariance function is defined by $\gamma(u) = \text{cov}(Y(t), Y(t+u)), u \in \mathbb{R}$.

Suppose the condition $\sum_u |\gamma(u)| < \infty$ holds. Then the spectral density function or simply spectrum of Y is defined by

$$f(r) = \frac{1}{2\pi} \sum_{u=-\infty}^{\infty} \gamma(u) \exp\{-iru\}, \quad r \in \mathbb{R}.$$

Statistical inference will be focused mainly on estimating the spectral density function f. Here r is referred to as the frequency. This will lead to an estimate of the autocovariance function $\gamma(\cdot)$ via the inverse Fourier transformation [3].

A popular approach to estimate the spectral density function is based on the discrete Fourier transform (DFT), which can be described as follows. Let T denote a positive integer. Suppose a finite sample of length T given by $Y(0), Y(1), \ldots, Y(T-1)$ is observed. The DFT of the finite sample Y is defined by

$$\varphi(r_k, Y) = \sum_{t=0}^{T-1} Y(t) \exp(-ir_k t), \quad r_k = 2\pi k/T, \quad k = 0, \ldots, T-1.$$

The second-order periodogram is given by

$$\tilde{f}(r_k, Y) = \frac{1}{2\pi T} |\varphi(r_k, Y)|^2, \quad k = 0, \ldots, T-1.$$

Under appropriate conditions, this function can be shown to possess a chi-square distribution, which can be used to construct a more efficient estimate of the spectral density function $f(\cdot)$ [3].

Alternatively, the spectrum f can be estimated by fitting parametric time series models such as the AR process of order p (AR(p)). For these processes, the spectral density functions have a parametric form $f(r, \phi)$, where ϕ is a vector of parameters specified by the AR models above. Consequently, the statistical estimation will concentrate on the estimation of ϕ using some likelihood approaches [4].

The Whittle likelihood [15] is one of such approaches. The basic idea of this approach is to compare the periodogram and the parametric model-based estimates.

This is given by

$$L(f;Y) = -\frac{1}{2}\sum_{k=0}^{T-1}\left\{\frac{\tilde{f}(r_k,Y)}{f(r_k,\phi)} + \ln f(r_k,\phi)\right\}. \tag{8.1}$$

Before describing ICA, it is necessary to note that the above description for the univariate time series Y can be extended to vector-valued series. More specifically, let $\mathbf{x}(t) \in \mathbb{R}^M$ for $t = 0, 1, \ldots, T-1$ denote T observations of a real-valued vector stationary process with mean $\mathbf{0}$ and autocovariance function $\mathbf{c_{xx}}(u) = \mathrm{cov}(\mathbf{x}(t), \mathbf{x}(t+u))$ satisfying the condition $\sum_{t=-\infty}^{\infty}|\mathbf{c_{xx}}(u)| < \infty$. Then the second-order periodogram is given as

$$\tilde{\mathbf{f}}(r_k,\mathbf{X}) = \frac{1}{2\pi T}\varphi(r_k,\mathbf{X})\varphi^*(r_k,\mathbf{X}),$$

where $\varphi(r_k,\mathbf{X}) = \sum_{t=0}^{T-1}\mathbf{x}(t)\exp\{-ir_kt\}$ is the DFT and * denotes the conjugate transpose operator.

8.2.2 INDEPENDENT COMPONENT ANALYSIS (ICA)

ICA assumes observed multivariate signals as a linear mixture of hidden independent sources such that at most one source can have Gaussian distribution. Let $\mathbf{x}(t) = [x_1(t), x_2(t), \ldots, x_M(t)]^T$ be an M dimensional random vector representing the observation signal at time point t for $t = 0, 1, \ldots, T-1$ and let the vector $\mathbf{s}(t) = [s_1(t), s_2(t), \ldots, s_M(t)]^T$ denote the corresponding random multivariate source signal. The *noise-free ICA* model can be mathematically expressed as:

$$\mathbf{x}(t) = \mathbf{As}(t), \quad t \in \{0, 1, 2, \ldots, T-1\}. \tag{8.2}$$

Here $\mathbf{A}$ is a nonsingular deterministic matrix of size $M \times M$ representing the linear mixing operator and it is called the *mixing matrix*. The inverse of A is denoted as $\mathbf{W} = \mathbf{A}^{-1}$ and is called the *unmixing matrix*.

Given T observations $\mathbf{x}(0), \mathbf{x}(1), \ldots, \mathbf{x}(T-1)$, ICA aims at recovering the mixing matrix $\mathbf{A}$ and sources $\mathbf{s}(0), \mathbf{s}(1), \ldots, \mathbf{s}(T-1)$. Let $\mathbf{X} = [\mathbf{x}(0), \ldots, \mathbf{x}(T-1)]$ and $\mathbf{S} = [\mathbf{s}(1), \ldots, \mathbf{s}(T-1)]$ be the row-wise concatenation of $\mathbf{x}(t)$ and $\mathbf{s}(t)$, respectively. Then, for $t = 0, 1, \ldots, T-1$, Equation (8.2) can be combined and written in a concise form as:

$$\mathbf{X}_{M \times T} = \mathbf{A}_{M \times M}\mathbf{S}_{M \times T}. \tag{8.3}$$

The source signals $\mathbf{S}$ can be recovered by $\hat{\mathbf{S}} = \hat{\mathbf{W}}\mathbf{X}$ where $\hat{\mathbf{W}}$ is an estimate of the unmixing matrix $\mathbf{W}$.

ICA has been widely used in fMRI data analysis. Let $\mathbf{Y}$ of size $V \times T$ denote a single observation where V and T are the numbers of voxels and time points respectively. Each row of $\mathbf{Y}$ is a temporal course at a specific voxel and each column represents a vectorized 3D image at a time point. Single subject ICA aims at decomposing $\mathbf{Y}$ as outer products of M spatial maps (represented by the columns of $\mathbf{H}_{V \times M}$)

and the associated temporal courses (denoted by the corresponding row of $\mathbf{S}_{M \times T}$). Then ICA decomposition on the observation $\mathbf{Y}$ can be written as

$$\mathbf{Y}_{V \times T} = \mathbf{H}_{V \times M} \mathbf{S}_{M \times T}.$$

For fMRI data, V is usually very large. Hence, there is usually a pre-processing step to reduce the dimension of $\mathbf{Y}$ to $M \times T$ and ICA decomposition is conducted via Equation (8.3).

8.2.3   PARAMETRIC INDEPENDENT COLORED SOURCES (PICS)

Suppose the sources $s_j(0), s_j(1), \ldots, s_j(T-1)$ were available. Then each source can be fitted by a parametric model using the Whittle likelihood (8.1) given more specifically by

$$L(\phi_j; s_j) = -\frac{1}{2} \sum_{k=0}^{T-1} \left\{ \frac{\tilde{f}(r_k, s_j)}{f_{jj}(r_k, \phi_j)} + \ln f_{jj}(r_k, \phi_j) \right\}.$$

Since sources are mutually independent, the joint Whittle log-likelihood for all sources is the sum of individual ones and can be written as

$$L(\phi; \mathbf{s}) = -\frac{1}{2} \sum_{j=1}^{M} \sum_{k=0}^{T-1} \left\{ \frac{\tilde{f}(r_k, s_j)}{f_{jj}(r_k, \phi_j)} + \ln f_{jj}(r_k, \phi_j) \right\},$$

where $\phi = [\phi_1, \phi_2, \ldots, \phi_M]^T$ and $\mathbf{s} = [s_1, s_2, \ldots, s_M]^T$. According to (8.3), this can be rewritten as

$$L(\phi; \mathbf{W}, \mathbf{X}) = -\frac{1}{2} \sum_{j=1}^{M} \sum_{k=0}^{T-1} \left\{ \frac{\mathbf{e}_j^T \mathbf{W} \tilde{\mathbf{f}}(r_k, \mathbf{X}) \mathbf{W}^T \mathbf{e}_j}{f_{jj}(r_k, \phi_j)} + \ln f_{jj}(r_k, \phi_j) \right\} + T \ln |\det(\mathbf{W})|, \tag{8.4}$$

where $\mathbf{e}_j \in \mathbb{R}^M$ with the jth element being 1 and others being 0.

Using prewhitened data, Lee et al. [11] proposed a Lagrange multiplier method by minimizing a constraint negative log-likelihood,

$$F^{PICS}(\phi; \mathbf{W}; \mathbf{X}) = -L(\phi; \mathbf{W}, \mathbf{X}) + \lambda^T \mathbf{C}, \tag{8.5}$$

where $\lambda \in \mathbb{R}^{M(M+1)/2}$ is the Lagrange multiplier, and $\mathbf{C} \in \mathbb{R}^{M(M+1)/2}$ with $\mathbf{C}_{(i-1)M+j} = (\mathbf{W}\mathbf{W}^T - \mathbf{I}_{M \times M})_{ij}$ for $i = 1, \ldots, M$ and $j = 1, \ldots, i$. The algorithm alternatively updates $(\mathbf{W}, \lambda)$ while holding the time series parameters ϕ fixed, then switch to update the latter by fixing the former. The iteration continues until the Amari distance falls below some threshold. Details can be found in Lee et al. [11].

8.3   GROUP PARAMETRIC INDEPENDENT COLORED SOURCES (GPICS)

Suppose there are $g = 1, 2, \ldots, G$ independent groups. Then (8.5) can be generalized to

$$F^{GPICS}(\phi; \mathbf{W}; \mathbf{X}) = \sum_g F^{PICS}(\phi^g; \mathbf{W}^g; \mathbf{X}^g), \tag{8.6}$$

where ϕ^g, $\mathbf{W}^g$ and $\mathbf{X}^g$ ($g = 1, \ldots, G$) are group-specific parameters and data, respectively. This will be useful for modeling group structures. For example, suppose there are two groups. Then the specification of $\phi^1 = \phi^2$ in the constraint optimization of (8.6) yields an estimation problem with two groups sharing the common time series features while the spatial features $\mathbf{W}^1$ and $\mathbf{W}^2$ can vary freely, so on and so forth.

The algorithm for this hypothesis-driven procedure can be viewed as a constraint optimization of several PICS problems.

Algorithm 1 GPICS algorithm

Initialize unmixing matrices $\mathbf{W}$ or $\mathbf{W}^g$ based on group structure hypothesis. Alternatively do the following two steps until the Amari distance of $\mathbf{W}$ or maximum distances of $\mathbf{W}^g$ between two iteration steps are below the threshold.

1. Recover the source signals by $\mathring{\mathbf{S}}^g = \mathring{\mathbf{W}}^g \mathbf{X}^g$ or $\mathring{\mathbf{S}}^g = \mathring{\mathbf{W}} \mathbf{X}^g$. Estimate time series parameters for the jth source by Yule–Walker method on the jth row of members of $\mathring{\mathbf{S}}^g$.
2. Update unmixing matrices and Lagrange multipliers via Equation (8.6).

8.4 SIMULATIONS

In this section, we examine GPICS performance through simulations and compare our approach with the widely used GIFT approach. The improvement of the former over the latter will be demonstrated as a result of incorporating the temporal correlation structure into constructing ICA.

8.4.1 BLIND SOURCE SEPARATION

In the first simulaton, we study GPICS performance of recovering blind sources. We test the procedure under three different cases of group structures.

1. Common space $\mathbf{W} = \mathbf{A}^{-1}$ and time $\mathbf{S}$.
2. Common space $\mathbf{W}$.
3. Common time $\mathbf{S}$.

For each case, we generate 2 groups of data with 5 subjects in each group. The ith subject data is generated from the model $\mathbf{X}^{gi} = \mathbf{A} \mathbf{S}^{gi}$ for Cases 1 and 2, and $\mathbf{X}^{gi} = \mathbf{A}^g \mathbf{S}^{gi}$ for Case 3, $g = 1, 2$ and $i = 1, 2, \ldots, 5$, where $\mathbf{A}$ and $\mathbf{A}^g$ are of size 4×4 and $\mathbf{S}^{gi}$ has dimension 4×512. Hence, the number of ICs is $M = 4$ and the number of time points is $T = 512$.

In each simulation run, we randomly generate one orthogonal mixing matrix for the first two cases and two for the last. Let ϕ^g_j, $j = 1, \ldots, 4$, denote the coefficients for the jth source of group g. Each row of $\mathbf{S}^{gi}$ is generated from an AR model with coefficients given in Table 8.1. In the first two cases, $\phi^1_j = \phi^2_j$, we take the set 1 coefficients for all subjects. In Case 3, we take the set 1 and set 2 coefficients for the subjects in the first and second group respectively. The noise of all sources are generated from uniform$[-\sqrt{3}, \sqrt{3}]$.

Table 8.1

AR coefficients used in the simulation experiment.

	ϕ_1	ϕ_2	ϕ_3	ϕ_4
set 1	[0.8]	[−0.6, −0.5]	[0.1, −0.8]	[−0.85, −0.7, 0.2]
set 2	[0.5, −0.12]	[0.6]	[−0.7, −0.3]	[0.3]

We estimate $\mathbf{A}$ or $\mathbf{A}^g$ and time series parameters by both GPICS and fastICA-based approaches. The GPICS approach was described in Section 8.3. In the first two cases, the fastICA-based approaches concatenate data matrices along temporal direction as $[\mathbf{X}^{11}, \mathbf{X}^{12}, \ldots, \mathbf{X}^{25}]$ and do an ICA decomposition via fastICA on this aggregation matrix to derive a common mixing matrix. In Case 3, we do fastICA separately on two aggregation matrices $[\mathbf{X}^{g1}, \mathbf{X}^{g2}, \ldots, \mathbf{X}^{g5}]$ for $g = 1, 2$ to obtain group-specific mixing matrices. The Amari distances between the estimates and the truth for both approaches are calculated. The temporal courses are recovered by $\hat{\mathbf{S}}^{gi} = \hat{\mathbf{W}} \mathbf{X}^{gi}$ or $\hat{\mathbf{S}}^{gi} = \hat{\mathbf{W}}^g \mathbf{X}^{gi}$. Then we calculate all pairs of correlation coefficients between the estimated temporal courses and the truth. We did 100 runs for each group structure. The results are shown in Figure 8.1 for Amari distances and Figure 8.2 for temporal correlations.

Figure 8.1 shows the boxplots of Amari distance between the estimation mixing matrices and the truth for 100 simulation runs. Each subplot corresponds to one group structure. We can see that our GPICS method performs consistently better than fastICA-based approaches in all group structures.

Figure 8.2 shows the boxplots of temporal correlations of estimated temporal courses and the truth. Each column corresponds to one group structure and each row represents one IC. In most cases, GPICS has correlations, of nearly 1 and they are higher than fastICA-based approaches in all cases.

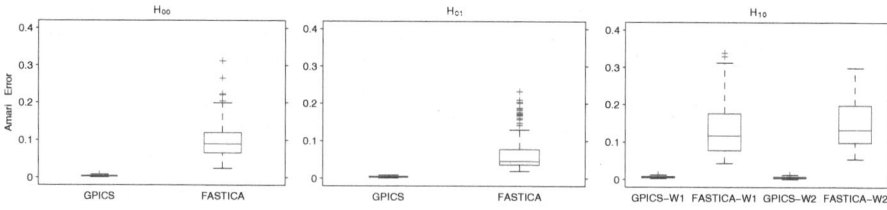

Figure 8.1: Boxplots of 100 simulation runs of Amari distances of estimated mixing matrices and the truth.

Our simulation results show that GPICS works better than fastICA-based approaches in blind source separation problems. This is expected in this experiment

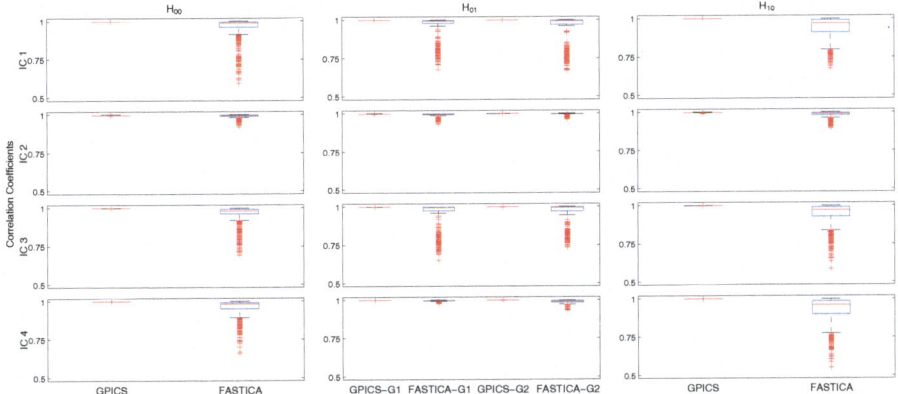

Figure 8.2: Boxplots of 100 simulation runs for temporal correlations of the estimation sources and the truth.

as temporal correlation will reveal more information than just the marginal distribution of the latent sources. Moreover, we have also found that GPICS detects more accurately the spatial features than its peers.

8.5   REAL DATA ANALYSIS

In this section, we apply our GPICS approach to one of the open-access *resting-state fMRI* (rs-fMRI) datasets, the ADHD-200 Sample [13]. The dataset contains a large collection of rs-fMRI and structural MRI scans of 491 typically developing controls (normal subjects) and 285 *attention deficit hyperactivity disorder* (ADHD) patients from 8 paticipating sites (for more details, see http://fcon_1000.projects. nitrc.org/indi/adhd200/).

For this study, we utilize the publicly available pre-processed data by the Athena pipeline (http://neurobureau.projects.nitrc.org/ADHD200/ Introduction.html). Briefly, the pipeline to fMRI data includes slice timing correction, deobliquing, motion correction, registration into MNI152 standard brain space at 4mm × 4mm × 4mm resolution, nuisance variance removal, and a 6-mm full width at half maximum (FWHM) Gaussian filter spatial smoothing. Detailed information regarding the pre-processing steps can be found at http://www.nitrc.org/plugins/mwiki/index.php/neurobureau: AthenaPipeline. Each processed fMRI image is of size $49 \times 58 \times 47$. We further apply a mask onto each image to extract voxels inside the brain and results in a vector of length $30,316$. The number of time points of each subject is 120. Hence, each fMRI data matrix, as the inputs of GPICS, is of size $30,316 \times 120$.

We choose 10 subjects from the control group of the site based in the Kennedy Krieger Institute, Johns Hopkins University. For this group, we assume that subjects are homogeneous in space but not so in time. This group structure assumption is consistent with the one commonly used in multisubject GICA approaches such as fastICA-based approaches. Since the number of voxels is relatively large, a Singular Value Decomposition (SVD)-based dimension-reduction method was applied to preprocess the data prior to optimizing (8.6) according to the constraint conditions imposed by Cases 1 and 2. In order to identify the active voxels, the z scores were calculated by standardizing each IC and are those voxels with $|z| > 2$ marked as active. Finally, the active voxels will be presented by overlaying their z scores onto an anatomical template.

The identified Default Mode Network (DMN), Visual Network (VN), and Auditory Network (AN) are provided in Figure 8.3. The networks identified by GPICS and fastICA-based approaches are shown on the left and the right side, respectively. Each row is a type of brain network. We can see that GPICS has more complete identifications on the DMN and Auditory network, especially the Auditory one in this case.

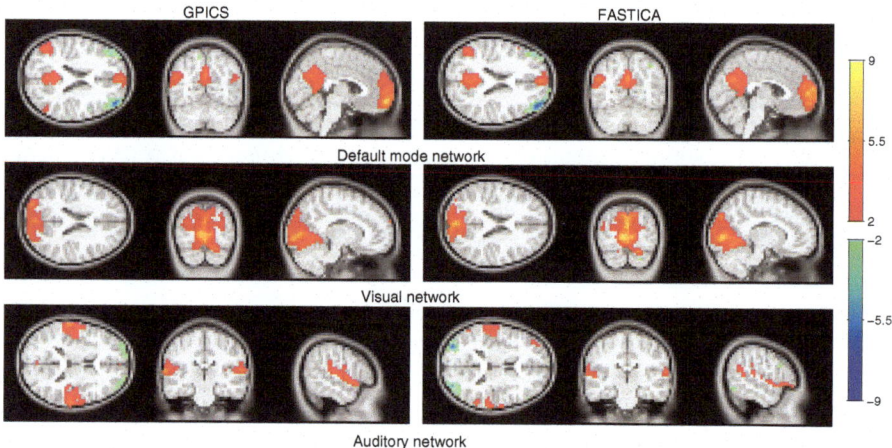

Figure 8.3: A comparison of brain networks selected by GPICS (left) and fastICA-based approaches (right). The analysis is based on 10 controls from the KKI site.

8.6 DISCUSSIONS AND CONCLUSIONS

From the simulated and real data analysis, it is observed that GPICS has performed as expected since data have shown some temporal correlations. This may cause some challenging problems for fastICA-based approaches as they totally ignored such features. However, fastICA-based approaches are very fast in computing as it is a much simpler procedure. It may work well when the shapes of the marginal distributions of the sources can be separated, and the ability for fastICA to detect the sources depends on how well the shapes are separated. It is worth knowing that GPICS offers more flexibility in exploring group structures in real data situations since it is desirable to examine the hypotheses of group commonality. It is also easier to interpret the result as many physiological components have a relatively well defined temporal correlation structure.

In fact, a typical way to present ICA results is depicted Figure 8.4. The wide and short panel on the top is the temporal component which may represent some physiological time series, the matrix of images right below it is the associated spatial brain maps. The GPICS approach offers several ways to study the group structure while fastICA-based approaches offer only the common spatial map approach. As a byproduct of GPICS, the time series model is estimated simultaneously with the brain map. This will be useful in some brain-related research where the correlation or spectral frequency may play a dominant role, which will be helpful for locating the desirable source. Approaches based on fastICA have to do this in two steps and this may pose some issues for testing hypotheses, which will be an important comparison to be carried out in our future research.

Another useful feature of GPICS is its ability to select the brain maps associated with some task-related time series, which will improve the efficiency in searching for the required brain network. In our simulated experiments, the whole 20 (the default number of ICs) sets of spatial maps were examined by visual inspection in order to match the well-defined ones. A more preferred way is by casting this procedure in the framework of statistical testing of a hypothesis, it is one of the important features to be implemented as an extension of GPICS.

In summary, this chapter presented a new approach to blind source separation problems based on group data. This is useful in many types of human brain research where group comparison is the main concern. The method was developed based on a very flexible way to model the temporal correlation feature, and the parameters were estimated using a robust likelihood approach. The numerical performance was carried out in a comparative manner using both simulated and real human brain data. The proposed method also offered many important extensions to address and improve practical issues in blind source detection problems.

8.7 M-FILES

The MATLAB functions used in the chapter are:

```
gcica_bss_dwst.m
gcica_bss_swdt.m
gcica_bss_swst.m
gcica_cica.m
```

The code of each function will be listed in the following subsections.

8.7.1 GCICA H_{10}

This is a MATLAB function for ICA involving a group of subjects who have similar or common spectral properties of the latent temporal sources with varying latent spatial sources.

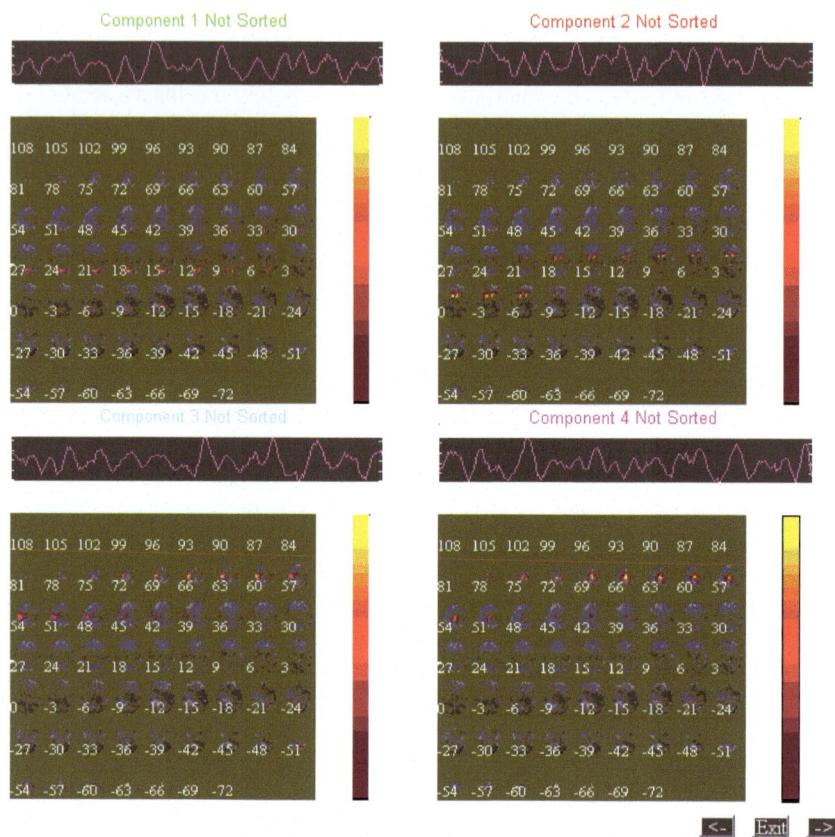

Figure 8.4: The first four brain maps and temporal components from a control.

```
% gcica_bss_dwst.m

function [W outoptarg] = gcica_bss_dwst(Xc,opt)
% ======================= PURPOSE ===============================
%
% Implement different W same T group-colorICA algorithm.
%
% Xc is a G by NG cell-array. G is the number of groups and NG
% is the number of subjects in group g. Each cell-matrix in Xc
% has dimension M by T where M is the number of mixtures and T
% is the number of time points. The (un)mixing matrix for each
% subject is an M by M invertible matrix.
%
```

```
% In this dwst algorithm, we assume that the mixing matrix and
% source time series models are the same within each group.
% Across groups, % mixing matrices are different but time series
% models are the same.
%
% ====================== INPUTS ============================
%
% Xc   G by NG cell-array contains observed        [cell-array]
%      data matrices. G is the number of groups and NG is the
%      number of subjects in each group. Each component of the
%      cell-array is a matrix with dimension M by T where M is
%      the number of mixtures and T is the number of time points.
%      Each data matrix has been pre-whitened and centered.
% opt  Optional input arguments                        [struct]
% .tol                                                 [scalar]
%          Tolerance for the convergence. [default: 1e-10]
% .verbose                                             [string]
%          'on' or 'off'.    [default: 'off']        [
% .centering                                           [string]
%        How to center the data matrix X
%        'row': extract row means from each row so that each row
%               sums to 0
%        'rowcol': Extract both column and row means so that each
%               row and column sum to 0
%        [default: 'row']
% .maxit                                               [scalar]
%          Maximum number of iterations. [default:200]
% .prewhite                                            [scalar]
%        Whether to prewhite the data.
%             1: prewhite the data. 0: no prewhite
%             [default: 1]
% .tsms_method                                         [string]
%        Source time series model select (tsms) method
%        'AIC' 'AICC' 'BIC'   [default: 'AIC']
% .initial_W                                           [matrix]
%        Initial of the unmixing matrix. Default is M by M
%        identity matrix
% .arma                                                [scalar]
%        AR model or ARMA model for sources.
%             0: AR model  1: ARMA model     [default: 0]
% .scovmatdnmntropt                                    [string]
%        Denominator used to calculate sample covariance matrix
%        'sz': using sample size T [default]
%        'szm1': sample size - 1 (T - 1)
```

```
%  .num_initials                                    [scalar]
%        Number of initial W tried. [default: 10]
%  .armsearlystopiter                               [scalar]
%        Under armsearlystopiter orders, time series model
%        selection will calculate the actual score. After this
%        number, if the algorithm finds a score that is smaller
%        than previous one, the algorithm stops and return the
%        current order.   [default: 20]
%  .logabsWterm                                     [string]
%        Whether to contain the log term (\ln \abs{det(W)}
%        in the likelihood function.
%        'logterm': contain log term.  [default]
%        'nologterm': do NOT contain log term.
%  .restartiter                                     [scalar]
%        At this iteration step, compare the distance with
%        opt.restartvalue. If the distance is larger than
%        opt.restartvalue, start the algorithm from another
%        initial guess.
%  .restartvalue                                    [scalar]
%        See opt.restartiter
%
%======================= OUTPUTS ============================
%
% W        M by M unmixing matrix                   [matrix]
% outoptarg  Optional output arguments              [struct]
%  .phi                                             [cellarray]
%        M by 1 cell array contains time series parameters
%        of each sources
%  .sigmasq                                         [vector]
%        M by 1 vector contains variances of each
%        time series model
%  .p                                               [vector]
%        M by 1 vector contains order of time series models
%  .iter                                            [scalar]
%        Number of iterations used to converge
%  .num_sg                                          [scalar]
%        Number of singular Hessian matrices found
%  .num_initials                                    [scalar]
%        Number of initial guess used
%
%======================= NOTES ============================
%
% DVW made minor modification of colorICA_pmm.m by Seonjoo Lee.
%
```

```
%========================= REFERENCES ==========================
%
% Seonjoo Lee, Haipeng Shen, Young Truong, Michelle Lewis, X.
% Huang (2011). Independent Component Analysis Involving
% Auto-correlated Sources with an Application to fMRI.
% Journal of the American Statistical Association,
% 106, 1009-1024.
%
%======== created by Seonjoo Lee, Jan 2009 ===================

%-------------------------------------------------------------
%--------- INPUT and OUTPUT arguments check -------------------
%-------------------------------------------------------------
error(nargchk(1,2,nargin,'struct'));
error(nargoutchk(0,2,nargout,'struct'));

%-------------------------------------------------------------
%-------------------- INPUT -----------------------------------
%-------------------------------------------------------------
if nargin < 2
  opt = struct('tol',1e-10,'verbose','off','centering','row',...
       'maxit',200,'prewhite',1,'tsms_method','AIC',...
       'initial_W',[],'arma',0,'scovmatdnmntropt','sz',...
       'num_initials',10,'armsearlystopiter',20,...
       'logabsWterm','logterm','restartiter',50,...
       'restartvalue',1e-2,'ts_est_method','cat');
end

if ~isstruct(opt)
  error('opt should be a struct array');
end

if (~isfield(opt,'tol') || isempty(opt.tol))
    opt.tol = 1e-10; end
if (~isfield(opt,'verbose') || isempty(opt.verbose))
  opt.verbose = 'off';
end
if (~isfield(opt,'centering') || isempty(opt.centering))
  opt.centering = 'row';
end
if (~isfield(opt,'maxit') || isempty(opt.maxit))
  opt.maxit = 200;
end
if (~isfield(opt,'prewhite') || isempty(opt.prewhite))
```

```
    opt.prewhite = 1;
end
if (~isfield(opt,'tsms_method') || isempty(opt.tsms_method))
    opt.tsms_method = 'AIC';
end
if (~isfield(opt,'initial_W') || isempty(opt.initial_W))
 opt.initial_W = repmat({eye(size(Xc{1}{1},1))},[1 length(Xc)]);
end
if (~isfield(opt,'arma') || isempty(opt.arma)) opt.arma = 0; end
if (~isfield(opt,'scovmatdnmntropt') || ...
    isempty(opt.scovmatdnmntropt))
    opt.scovmatdnmntropt = 'sz';
end
if (~isfield(opt,'num_initials') || ...
    isempty(opt.num_initials))
    opt.num_initials = 10;
end
if (~isfield(opt,'armsearlystopiter') || ...
    isempty(opt.armsearlystopiter))
    opt.armsearlystopiter = 20;
end
if (~isfield(opt,'logabsWterm') || ...
    isempty(opt.logabsWterm))
    opt.logabsWterm = 'logterm';
end
if (~isfield(opt,'restartiter') || ...
    isempty(opt.restartiter))
    opt.restartiter = 50;
end
if (~isfield(opt,'restartvalue') || ...
    isempty(opt.restartvalue))
    opt.restartvalue = 1e-2;
end
if (~isfield(opt,'ts_est_method') || ...
    isempty(opt.ts_est_method))
    opt.ts_est_method = 'cat';
end


%-----------------------------------------------------------------
%---------------- PREPROCESSING ----------------------------------
%-----------------------------------------------------------------
% M: Number of  mixtures (number of sources)
% T: number of time points
[M T] = size(Xc{1}{1});
```

```
% number of groups
num_group = length(Xc);

% number of subjects in each group
num_subj_group = zeros(1,num_group);

clear ik
for ik = 1:num_group
  num_subj_group(ik) = length(Xc{ik});
end

switch opt.scovmatdnmntropt
 case 'sz'
  opt.scovmatdnmntr = T;
 case 'szm1'
  opt.scovmatdnmntr = T - 1;
end

% prewhite
if (opt.prewhite == 1)
  for ik = 1:num_group
    for jk = 1:num_subj_group(ik)
      clear Xprew invsqrtcovmat
      [Xprew invsqrtcovmat] = mat_prewhitening(Xc{ik}{jk}, ...
       opt.scovmatdnmntr);
      Xc{ik}{jk} = Xprew;
      clear Xprew;
    end
  end
end

% Discrete Fourier Transformation
clear ik jk mk X_dftall
for ik = 1:num_group
  for jk = 1:num_subj_group(ik)
    X_dftall{ik}{jk} = zeros(M,T);
    clear mk
    for mk = 1:M
      X_dftall{ik}{jk}(mk,:) = ...
              fft(Xc{ik}{jk}(mk,:))/sqrt(2*pi*T);
    end
  end
end
```

```
% Frequencies
freq = (0:T-1)'/T*2*pi;


%------------------------------------------------------------------
%------ Initial W and source time series parameters ----------
%------------------------------------------------------------------
% initial W
W1 = opt.initial_W;
% initial source
clear ik jk
for ik = 1:num_group
  for jk = 1:num_subj_group(ik)
    wX{ik}{jk} = W1{ik} * Xc{ik}{jk};
  end
end


% g: f_jj spectral density
% f_jj(r) = \frac{\sigma^2_j}{2 \pi \abs{\Phi_j(\exp(-ir))}^2}

% Estimate time series order p and parameters phi
clear sourcetsik
sourcetsik = cell(1,M);

for mk = 1:M
  sourcetsik{mk} = zeros(sum(num_subj_group),T);
  lk = 0;
  for ik = 1:num_group
    for jk = 1:num_subj_group(ik)
      lk = lk + 1;
      sourcetsik{mk}(lk,:) = wX{ik}{jk}(mk,:);
    end
  end
end

for mk = 1:M
  clear iphi isigmasq ip
  if opt.arma == 1
    disp(['We only consider AR model for now!']);
    return;
  elseif opt.arma==0
    [iphi isigmasq ip] = ...
        gcica_ar_ms_mts(sourcetsik{mk},struct('ms_method', ...
  opt.tsms_method,'earlystopiter',opt.armsearlystopiter,...
```

```
    'est_method',opt.ts_est_method));

    if ip == 0
      g(mk,:) = isigmasq/(2*pi)*ones(1,T);
    else
      g(mk,:) = isigmasq/(2*pi)./(abs(1-(exp(-1i * ...
              freq * (1:ip) ) * iphi')).^2)';
    end
  else
    error('Wrong opt.arma value!');
  end
  p(mk) = ip;
  sigmasq(mk) = isigmasq;
  phi{mk} = iphi;
end


%------------------------------------------------------------------
%------------ iterative updating algorithm  --------------------
%------------------------------------------------------------------
% Distance between W1 and W2 (previous and current iteration)
wddist = 1;
% loop control
iter = 0;
% How many initial W tried
num_initials = 1;
% How many singular Hessian matrix found
num_sg = 0;
%
lambda = repmat({ones(M*(M+1)/2,1)},[1 num_group]);
% 111222333
index1 =  reshape(reshape(repmat(1:M,1,M),M,M)', 1,M^2);
% 123123123
index2 = repmat(1:M,1,M);
tempmx11 = reshape(index2,M,M);
tempmx22 = reshape(index1,M,M);
% row indices for Lower triangular elements
tempmx1 = tempmx11(index1<=index2);
% column indices for lower triangular elements
tempmx2 = tempmx22(index1<=index2);
clear tempmx11 tempmx22
%
indx_temp = [ eye(M*(M+1)/2), zeros(M*(M+1)/2,M*(M-1)/2)];

while (wddist > opt.tol && iter < opt.maxit + 1 && ...
```

```
              num_initials < opt.num_initials + 1 )
%------------------------------------------------------------------
%- Newton-Raphson method with Lagrange multiplier for updating W
%------------------------------------------------------------------
  iter = iter+1;

  clear X_wdft W1_inv
  for ik = 1:num_group
    for jk = 1:num_subj_group(ik)
      X_dft{ik}{jk} = X_dftall{ik}{jk};
      X_wdft{ik}{jk} = W1{ik} * X_dft{ik}{jk};
    end
    W1_inv{ik} = inv(W1{ik});
  end

  for ik = 1:num_group

    clear Score_l1_w Score_l2_w
    Score_l1_w = zeros(M^2,1);
    Score_l2_w = zeros(M^2,1);

    for jk = 1:num_subj_group(ik)
      Score_l1_w = Score_l1_w + ...
                sum(real(conj(X_wdft{ik}{jk}(index1,:)) ...
        .* X_dft{ik}{jk}(index2,:))./g(index1,:),2);
      Score_l2_w = Score_l2_w - ...
                T * reshape( W1_inv{ik}, M^2, 1);
    end

    % dCC / d(W_11, W_12, W_13,...,W_MM)
    clear J_e1 J_e2 Score_cc_w PP
    J_e1 = W1{ik}(tempmx1,index2).* indx_temp(tempmx2,index1);
    J_e2 = W1{ik}(tempmx2,index2).* indx_temp(tempmx1,index1);
    PP = (J_e1 + J_e2)';
    Score_cc_w = PP * lambda{ik};

    % dCC / d(lambda)
    clear c_e Score_cc_lambda
    c_e = W1{ik} * W1{ik}' - eye(M);
    Score_cc_lambda = (c_e(index1 <= index2))';

    % first derivative w.r.t. to
    % (W_11 W_12,...,W_MM,lambda_1,...lambda_{M(M+1)/2}
    clear score_nr
```

```
      switch opt.logabsWterm
       case 'logterm'
        score_nr = [Score_l1_w + ...
                    Score_l2_w + Score_cc_w; Score_cc_lambda];
       case 'nologterm'
        score_nr = [Score_l1_w + ...
                    Score_cc_w; Score_cc_lambda];
      end


      %==== Second derivativefor Hessian
      % d^2 CC / dW_jk dW_pq
      clear h_cc_ww ihl ihll
      h_cc_ww = zeros(M^2);
      ihl = zeros(M);
      ihl(index1 <= index2) = lambda{ik};
      ihll = ihl + ihl';
      h_cc_ww = kron(ihll,eye(M));


      % d^2 (L1) / dW_jk dW_pq
      h_l1_ww_all = zeros(M^2);
      for jk = 1:num_subj_group(ik)
        clear h_l1_ww ihl1ww ihl1ww2
        h_l1_ww = zeros(M^2);
        ihl1ww = real(conj(X_dft{ik}{jk}(index1,:)) .* ...
                    X_dft{ik}{jk}(index2,:));
        for mk = 1:M
clear ihl1ww2
ihl1ww2 = sum(ihl1ww./repmat(g(mk,:),[M^2 1]),2);
h_l1_ww(((mk-1)*M + 1):(mk*M),((mk-1)*M+1):(mk*M)) = ...
    (reshape(ihl1ww2,[M M]))';
        end
        h_l1_ww_all = h_l1_ww_all + h_l1_ww;
      end
      clear h_l1_ww
      h_l1_ww = h_l1_ww_all;
      clear h_l1_ww_all


      % d^2 (L2) / dW_jk dW_pq
      clear h_l2_ww
      h_l2_ww = num_subj_group(ik) * T * ...
        (W1_inv{ik}(index2,index1).* W1_inv{ik}(index2,index1)');


      % d^2 CC/ dW_jk d(lambda_r)
      clear h_cc_wlambda
```

```
    h_cc_wlambda = PP;

    % d^2 CC / d(lambda_r) dW_jk
    clear h_cc_lambdaw
    h_cc_lambdaw = PP';

    % d^2 CC / d(lambda_r) d(lambda_s) = 0
    clear h_cc_lambdalambda
    h_cc_lambdalambda = zeros(length(lambda{ik}));

    % Hessian matrix w.r.t (W_11
    % W_12,...,W_MM,lambda_1,...lambda_{M(M+1)/2}
    clear h_nr

    switch opt.logabsWterm
     case 'logterm'
      h_nr = [h_cc_ww + h_11_ww + h_12_ww h_cc_wlambda; ...
      h_cc_lambdaw h_cc_lambdalambda];
     case 'nologterm'
      h_nr = [h_cc_ww + h_11_ww  h_cc_wlambda; ...
      h_cc_lambdaw h_cc_lambdalambda];
    end

    % Check singularity of Hessian matrix h_nr
    rcondh = rcond(h_nr);

    if  rcondh>= 10^(-15) && all(abs(isigmasq) < Inf)
       clear iW W2
       iW = vertcat(reshape(W1{ik}', M^2,1),lambda{ik}) - ...
                h_nr\score_nr;
       W2 = reshape(iW(1:M^2),M,M)';
       lambda{ik} = iW((M^2+1) : (M*(3*M+1)/2));
       wddistall(ik) = gcica_amaridist(W2,W1{ik});
       W1{ik} = NaN;
       W1{ik} = W2;

    if strcmp(opt.verbose,'on')
       fprintf('%d %0.5g\n',iter,wddist);
    end

    end

    if  rcondh< 10^(-15) || any( abs(isigmasq) == Inf) || ...
  (iter == opt.maxit) || ...
```

```
(iter == opt.restartiter && wddist > opt.restartvalue)
    W2 = rand_orth(M);
    lambda{ik} = ones(M*(M+1)/2,1);
    iter = 0;
    num_initials = num_initials +1;
    W1{ik} = W2;

    if num_initials == opt.num_initials + 1
W1 = [];
phi = [];
p = [];
sigmasq = [];
disp('No convergence!');
break
    end

    if strcmp(opt.verbose,'on')
if rcondh< 10^(-15)
  disp('Hessian matrix is close to singular!');
  num_sg = num_sg + 1;
end
disp('Start from another initial guess!')
    end
  end
end

wddist = max(wddistall);

if isempty(W1)
  break
end

%----------------------------------------------------------------
%------ Updating source time series parameters  --------------
%----------------------------------------------------------------
clear ik jk
for ik = 1:num_group
  for jk = 1:num_subj_group(ik)
    wX{ik}{jk} = W1{ik} * Xc{ik}{jk};
  end
end

% g: f_jj spectral density
% f_jj(r) = \frac{\sigma^2_j}{2 \pi \abs{\Phi_j(\exp(-ir))}^2}
```

```
% Estimate time series order p and parameters phi
clear sourcetsik
sourcetsik = cell(1,M);

for mk = 1:M
  sourcetsik{mk} = zeros(sum(num_subj_group),T);
  lk = 0;
  for ik = 1:num_group
    for jk = 1:num_subj_group(ik)
lk = lk + 1;
sourcetsik{mk}(lk,:) = wX{ik}{jk}(mk,:);
    end
  end
end

for mk = 1:M
  clear iphi isigmasq ip
  if opt.arma == 1
    disp(['We only consider AR model for now!']);
    return;
  elseif opt.arma==0
    [iphi isigmasq ip] = ...
    gcica_ar_ms_mts(sourcetsik{mk},struct('ms_method', ...
opt.tsms_method,'earlystopiter',opt.armsearlystopiter,...
'est_method',opt.ts_est_method));

    if ip == 0
g(mk,:) = isigmasq/(2*pi)*ones(1,T);
    else
g(mk,:) =isigmasq/(2*pi)./(abs(1-(exp(-1i*freq*(1:ip)) ...
          * iphi')).^2)';
    end
  else
    error('Wrong opt.arma value!');
  end

  p(mk) = ip;
  sigmasq(mk) = isigmasq;
  phi{mk} = iphi;
  end
end

W = W1;
```

```
if nargout > 1
  outoptarg = struct('p',p,'sigmasq',sigmasq, 'num_iter',...
              iter, 'num_initials',num_initials,'num_sg',...
              num_sg,'phi',phi);
end
```

8.7.2   GCICA H_{01}

This is a MATLAB function for ICA involving a group of subjects who have similar or common latent spatial sources with varying latent temporal sources.

```
% gcica_bss_swdt.m

function [W outoptarg] = gcica_bss_swdt(Xc,opt)
%========================= PURPOSE ===========================
%
% Implement same W different T group-colorICA algorithm.
%
% Xc is a G by NG cell-array. G is the number of groups and NG
% is the number of subjects in group g. Each cell-matrix in Xc
% has dimension M by T where M is the number of mixtures and T
% is the number of time points. The (un)mixing matrix for each
% subject is an M by M invertible matrix.
%
% In this swdt algorithm, we assume that the mixing matrix and
% source time series models are the same within each group.
% Across groups, mixing matrices are the same but time series
% models differ.
%
% ====================== INPUTS =============================
%
% Xc   G by NG cell-array contains observed         [cell-array]
%      data matrices. G is the number of groups and NG is the
%      number of subjects in each group. Each component of the
%      cell-array is a matrix with dimension M by T where M is
%      the number of mixtures and T is the number of time points.
%      Each data matrix has been pre-whitened and centered.
% opt  Optional input arguments                      [struct]
% .tol                                               [scalar]
%        Tolerance for the convergence. [default: 1e-10]
% .verbose                                           [string]
```

```
%           'on' or 'off'.    [default: 'off']        [
% .centering                                        [string]
%       How to center the data matrix X
%       'row': extract row means from each row so that each row
%               sums to 0
%       'rowcol': Extract both column and row means so that each
%               row and column sum to 0
%       [default: 'row']
% .maxit                                            [scalar]
%       Maximum number of iterations. [default:200]
% .prewhite                                         [scalar]
%       Whether to prewhite the data.
%           1: prewhite the data. 0: no prewhite
%       [default: 1]
% .tsms_method                                      [string]
%       Source time series model select (tsms) method
%       'AIC' 'AICC' 'BIC'  [default: 'AIC']
% .initial_W                                        [matrix]
%       Initial of the unmixing matrix. Default is M by M
%       identity matrix
% .arma                                             [scalar]
%       AR model or ARMA model for sources.
%           0: AR model  1: ARMA model    [default: 0]
% .scovmatdnmntropt                                 [string]
%       Denominator used to calculate sample covariance matrix
%       'sz': using sample size T [default]
%       'szm1': sample size - 1 (T - 1)
% .num_initials                                     [scalar]
%       Number of initial W tried. [default: 10]
% .armsearlystopiter                                [scalar]
%       Under armsearlystopiter orders, time series model
%       selection will calculate the actual score. After this
%       number, if the algorithm finds a score that is smaller
%       than previous one, the algorithm stops and return the
%       current order.    [default: 20]
% .logabsWterm                                      [string]
%       Whether to contain the log term (\ln \abs{det(W)}
%       in the likelihood function.
%       'logterm': contain log term.   [default]
%       'nologterm': do NOT contain log term.
% .restartiter                                      [scalar]
%       At this iteration step, compare the distance with
%       opt.restartvalue. If the distance is larger than
%       opt.restartvalue, start the algorithm from another
```

```
%          initial guess.
% .restartvalue                                   [scalar]
%          See opt.restartiter
%
%======================= OUTPUTS ===========================
%
% W         M by M unmixing matrix                [matrix]
% outoptarg Optional output arguments             [struct]
% .phi                                            [cellarray]
%          M by 1 cell array contains time series parameters
%          of each sources
% .sigmasq                                        [vector]
%          M by 1 vector contains variances of each
%          time series model
% .p                                              [vector]
%          M by 1 vector contains order of time series models
% .iter                                           [scalar]
%          Number of iterations used to converge
% .num_sg                                         [scalar]
%          Number of singular Hessian matrices found
% .num_initials                                   [scalar]
%          Number of initial guess used
%
%======================= NOTES ============================
%
% DVW made minor modification of colorICA_pmm.m by Seonjoo Lee.
%
%======================= REFERENCES ========================
%
% Seonjoo Lee, Haipeng Shen, Young Truong, Michelle Lewis, X.
% Huang (2011). Independent Component Analysis Involving
% Auto-correlated Sources with an Application to fMRI.
% Journal of the American Statistical Association,
% 106, 1009-1024.
%
%======== created by Seonjoo Lee, Jan 2009 ==================
%= Last modified by dvwang-at-alumni.princeton.edu, 11/04/2013 =

%----------------------------------------------------------
%--------- INPUT and OUTPUT arguments check --------------
%----------------------------------------------------------
error(nargchk(1,2,nargin,'struct'));
error(nargoutchk(0,2,nargout,'struct'));
```

```
%------------------------------------------------------------------
%------------------- INPUT ----------------------------------------
%------------------------------------------------------------------
if nargin < 2
  opt = struct('tol',1e-10,'verbose','off','centering','row',...
       'maxit',200,'prewhite',1,'tsms_method','AIC',...
       'initial_W',[],'arma',0,'scovmatdnmntropt','sz',...
       'num_initials',10,'armsearlystopiter',20,...
       'logabsWterm','logterm','restartiter',50,...
       'restartvalue',1e-2,'ts_est_method','cat');
end

if ~isstruct(opt)
  error('opt should be a struct array');
end

if (~isfield(opt,'tol') || isempty(opt.tol))
       opt.tol = 1e-10; end
if (~isfield(opt,'verbose') || isempty(opt.verbose))
  opt.verbose = 'off';
end
if (~isfield(opt,'centering') || isempty(opt.centering))
  opt.centering = 'row';
end
if (~isfield(opt,'maxit') || isempty(opt.maxit))
  opt.maxit = 200;
end
if (~isfield(opt,'prewhite') || isempty(opt.prewhite))
  opt.prewhite = 1;
end
if (~isfield(opt,'tsms_method') || isempty(opt.tsms_method))
  opt.tsms_method = 'AIC';
end
if (~isfield(opt,'initial_W') || isempty(opt.initial_W))
  opt.initial_W = eye(size(Xc{1}{1},1));
end
if (~isfield(opt,'arma') || isempty(opt.arma)) opt.arma = 0; end
if (~isfield(opt,'scovmatdnmntropt') || ...
    isempty(opt.scovmatdnmntropt))
  opt.scovmatdnmntropt = 'sz';
end
if (~isfield(opt,'num_initials') || ...
    isempty(opt.num_initials))
  opt.num_initials = 10;
```

```
end
if (~isfield(opt,'armsearlystopiter') || ...
    isempty(opt.armsearlystopiter))
  opt.armsearlystopiter = 20;
end
if (~isfield(opt,'logabsWterm') || ...
    isempty(opt.logabsWterm))
  opt.logabsWterm = 'logterm';
end
if (~isfield(opt,'restartiter') || ...
    isempty(opt.restartiter))
  opt.restartiter = 50;
end
if (~isfield(opt,'restartvalue') || ...
    isempty(opt.restartvalue))
  opt.restartvalue = 1e-2;
end
if (~isfield(opt,'ts_est_method') || ...
    isempty(opt.ts_est_method))
  opt.ts_est_method = 'cat';
end


%----------------------------------------------------------------
%---------------- PREPROCESSING ---------------------------------
%----------------------------------------------------------------
% M: Number of  mixutures (number of sources)
% T: number of time points
[M T] = size(Xc{1}{1});

% number of groups
num_group = length(Xc);

% number of subjects in each group
num_subj_group = zeros(1,num_group);

clear ik
for ik = 1:num_group
  num_subj_group(ik) = length(Xc{ik});
end

switch opt.scovmatdnmntropt
  case 'sz'
  opt.scovmatdnmntr = T;
  case 'szm1'
```

```
    opt.scovmatdnmntr = T - 1;
end


% prewhite
if (opt.prewhite == 1)
  for ik = 1:num_group
    for jk = 1:num_subj_group(ik)
      clear Xprew invsqrtcovmat
      [Xprew invsqrtcovmat] = mat_prewhitening(Xc{ik}{jk}, ...
       opt.scovmatdnmntr);
      Xc{ik}{jk} = Xprew;
      clear Xprew;
    end
  end
end


% Discrete Fourier Transformation
clear ik jk mk X_dftall
for ik = 1:num_group
  for jk = 1:num_subj_group(ik)
    X_dftall{ik}{jk} = zeros(M,T);
    clear mk
    for mk = 1:M
      X_dftall{ik}{jk}(mk,:) = ...
            fft(Xc{ik}{jk}(mk,:))/sqrt(2*pi*T);
    end
  end
end


% Frequencies
freq = (0:T-1)'/T*2*pi;

%------------------------------------------------------------------
%------ Initial W and source time series parameters  ----------
%------------------------------------------------------------------
% initial W
W1 = opt.initial_W;
% initial source
clear ik jk
for ik = 1:num_group
  for jk = 1:num_subj_group(ik)
    wX{ik}{jk} = W1 * Xc{ik}{jk};
  end
end
```

```
% g: f_jj spectral density
% f_jj(r) = \frac{\sigma^2_j}{2 \pi \abs{\Phi_j(\exp(-ir))}^2}

% Estimate time series order p and parameters phi
for ik = 1:num_group

  clear sourcetsik
  sourcetsik = cell(1,M);

  for mk = 1:M
    sourcetsik{mk} = zeros(num_subj_group(ik),T);
    for jk = 1:num_subj_group(ik)
      sourcetsik{mk}(jk,:) = wX{ik}{jk}(mk,:);
    end
  end

  for mk = 1:M
    clear iphi isigmasq ip
    if opt.arma == 1
      disp(['We only consider AR model for now!']);
      return;
    elseif opt.arma==0
      [iphi isigmasq ip] = ...
        gcica_ar_ms_mts(sourcetsik{mk},struct('ms_method', ...
opt.tsms_method,'earlystopiter',opt.armsearlystopiter,...
'est_method',opt.ts_est_method));

      if ip == 0
g{ik}(mk,:) = isigmasq/(2*pi)*ones(1,T);
      else
g{ik}(mk,:) =isigmasq/(2*pi)./(abs(1-(exp(-1i*freq*(1:ip)) ...
                * iphi')).^2)';
      end
    else
      error('Wrong opt.arma value!');
    end

    p{ik}(mk) = ip;
    sigmasq{ik}(mk) = isigmasq;
    phi{ik}{mk} = iphi;
  end
end
```

```
%------------------------------------------------------------------
%------------ iterative updating algorithm  -------------------
%------------------------------------------------------------------
% Distance between W1 and W2 (previous and current iteration)
wddist = 1;
% loop control
iter = 0;
% How many initial W tried
num_initials = 1;
% How many singular Hessian matrix found
num_sg = 0;
%
lambda = ones(M*(M+1)/2,1);
index1 =  reshape(reshape(repmat(1:M,1,M),M,M)', 1,M^2);
index2 = repmat(1:M,1,M);
tempmx11 = reshape(index2,M,M);
tempmx22 = reshape(index1,M,M);
% row indices for Lower triangular elements
tempmx1 = tempmx11(index1<=index2);
% column indices for lower triangular elements
tempmx2 = tempmx22(index1<=index2);
clear tempmx11 tempmx22
%
indx_temp = [ eye(M*(M+1)/2), zeros(M*(M+1)/2,M*(M-1)/2)];

while (wddist > opt.tol && iter < opt.maxit + 1 && ...
           num_initials < opt.num_initials + 1 )
%------------------------------------------------------------------
%- Newton-Raphson method with Lagrange multiplier for updating W
%------------------------------------------------------------------
  iter = iter+1;

  clear X_wdft W1_inv
  for ik = 1:num_group
    for jk = 1:num_subj_group(ik)
      X_dft{ik}{jk} = X_dftall{ik}{jk};
      X_wdft{ik}{jk} = W1 * X_dft{ik}{jk};
    end
  end

  W1_inv = inv(W1);

  clear Score_l1_w Score_l2_w
  Score_l1_w = zeros(M^2,1);
```

```
Score_12_w = zeros(M^2,1);
for ik = 1:num_group
  for jk = 1:num_subj_group(ik)
    Score_11_w = ...
      Score_11_w + sum(real(conj(X_wdft{ik}{jk}(index1,:)) ...
       .* X_dft{ik}{jk}(index2,:))./g{ik}(index1,:),2);
    Score_12_w = Score_12_w - T * reshape( W1_inv, M^2, 1);
  end
end

% dCC / d(W_11, W_12, W_13,...,W_MM)
clear J_e1 J_e2 Score_cc_w PP
J_e1 = W1(tempmx1,index2).* indx_temp(tempmx2,index1);
J_e2 = W1(tempmx2,index2).* indx_temp(tempmx1,index1);
PP = (J_e1 + J_e2)';
Score_cc_w = PP * lambda;

% dCC / d(lambda)
clear c_e Score_cc_lambda
c_e = W1 * W1' - eye(M);
Score_cc_lambda = (c_e(index1 <= index2))';

% first derivative w.r.t. to
% (W_11 W_12,...,W_MM,lambda_1,...lambda_{M(M+1)/2}
clear score_nr
switch opt.logabsWterm
  case 'logterm'
    score_nr = [Score_11_w + Score_12_w + ...
                Score_cc_w; Score_cc_lambda];
  case 'nologterm'
    score_nr = [Score_11_w + Score_cc_w; Score_cc_lambda];
end

%==== Second derivativefor Hessian
% d^2 CC / dW_jk dW_pq
clear h_cc_ww ihl ihll
h_cc_ww = zeros(M^2);
ihl = zeros(M);
ihl(index1 <= index2) = lambda;
ihll = ihl + ihl';
h_cc_ww = kron(ihll,eye(M));

% d^2 (L1) / dW_jk dW_pq
h_11_ww_all = zeros(M^2);
```

```
    for ik = 1:num_group
      for jk = 1:num_subj_group(ik)
        clear h_l1_ww ihl1ww ihl1ww2
        h_l1_ww = zeros(M^2);
        ihl1ww = real(conj(X_dft{ik}{jk}(index1,:)) .* ...
                 X_dft{ik}{jk}(index2,:));
        for mk = 1:M
clear ihl1ww2
ihl1ww2 = sum(ihl1ww./repmat(g{ik}(mk,:),[M^2 1]),2);
h_l1_ww(((mk-1)*M + 1):(mk*M),((mk-1)*M+1):(mk*M)) = ...
    (reshape(ihl1ww2,[M M]))';
        end
        h_l1_ww_all = h_l1_ww_all + h_l1_ww;
      end
    end
    clear h_l1_ww
    h_l1_ww = h_l1_ww_all;
    clear h_l1_ww_all

    % d^2 (L2) / dW_jk dW_pq
    clear h_l2_ww
    h_l2_ww = sum(num_subj_group) * ...
        T * (W1_inv(index2,index1).* W1_inv(index2,index1)');

    % d^2 CC/ dW_jk d(lambda_r)
    clear h_cc_wlambda
    h_cc_wlambda = PP;

    % d^2 CC / d(lambda_r) dW_jk
    clear h_cc_lambdaw
    h_cc_lambdaw = PP';

    % d^2 CC / d(lambda_r) d(lambda_s) = 0
    clear h_cc_lambdalambda
    h_cc_lambdalambda = zeros(length(lambda));

    % Hessian matrix w.r.t (W_11
    % W_12,...,W_MM,lambda_1,...lambda_{M(M+1)/2}
    clear h_nr

    switch opt.logabsWterm
     case 'logterm'
      h_nr = [h_cc_ww + h_l1_ww + h_l2_ww h_cc_wlambda; ...
      h_cc_lambdaw h_cc_lambdalambda];
```

```
  case 'nologterm'
    h_nr = [h_cc_ww + h_ll_ww  h_cc_wlambda; ...
    h_cc_lambdaw h_cc_lambdalambda];
  end

  % Check singularity of Hessian matrix h_nr
  rcondh = rcond(h_nr);

  if  rcondh>= 10^(-15) && all(abs(isigmasq) < Inf)
    clear iW W2
    iW = vertcat(reshape(W1', M^2,1),lambda) - h_nr\score_nr;
    W2 = reshape(iW(1:M^2),M,M)';
    lambda = iW((M^2+1) : (M*(3*M+1)/2));
    wddist = gcica_amaridist(W2,W1);
    clear W1
    W1 = W2;

    if strcmp(opt.verbose,'on')
      fprintf('%d %0.5g\n',iter,wddist);
    end

  end

  if  rcondh< 10^(-15) || any( abs(isigmasq) == Inf) || ...
(iter == opt.maxit) || ...
(iter == opt.restartiter && wddist > opt.restartvalue)
    W2 = rand_orth(M);
    lambda = ones(M*(M+1)/2,1);
    iter = 0;
    num_initials = num_initials +1;
    W1 = W2;

    if num_initials == opt.num_initials + 1
      W1 = [];
      phi = [];
      p = [];
      sigmasq = [];
      disp('No convergence!');
      break
    end

    if strcmp(opt.verbose,'on')
      if rcondh< 10^(-15)
disp('Hessian matrix is close to singular!');
```

```
num_sg = num_sg + 1;
      end
      disp('Start from another initial guess!')
    end
  end


%------------------------------------------------------------
%------ Updating source time series parameters  --------------
%------------------------------------------------------------
  clear ik jk wX
  for ik = 1:num_group
    for jk = 1:num_subj_group(ik)
      wX{ik}{jk} = W1 * Xc{ik}{jk};
    end
  end

  % g: f_jj spectral density
  % f_jj(r) = \frac{\sigma^2_j}{2 \pi \abs{\Phi_j(\exp(-ir))}^2}
  clear g

  % Estimate time series order p and parameters phi
  for ik = 1:num_group

    clear sourcetsik
    sourcetsik = cell(1,M);

    for mk = 1:M
      sourcetsik{mk} = zeros(num_subj_group(ik),T);
      for jk = 1:num_subj_group(ik)
sourcetsik{mk}(jk,:) = wX{ik}{jk}(mk,:);
      end
    end

    for mk = 1:M
      clear iphi isigmasq ip
      if opt.arma == 1
disp(['We only consider AR model for now!']);
return;
      elseif opt.arma==0
[iphi isigmasq ip] = ...
        gcica_ar_ms_mts(sourcetsik{mk},struct('ms_method', ...
  opt.tsms_method,'earlystopiter',opt.armsearlystopiter,...
  'est_method',opt.ts_est_method));
```

```
if ip == 0
  g{ik}(mk,:) = isigmasq/(2*pi)*ones(1,T);
else
  g{ik}(mk,:) =isigmasq/(2*pi)./(abs(1-(exp(-1i * ...
                freq * (1:ip) ) * iphi')).^2)';
end
      else
error('Wrong opt.arma value!');
      end
      p{ik}(mk) = ip;
      sigmasq{ik}(mk) = isigmasq;
      phi{ik}{mk} = iphi;
    end
  end
end


W = W1;

if nargout > 1
  outoptarg = struct('p',p,'sigmasq',sigmasq, 'num_iter',...
    iter, 'num_initials',num_initials,'num_sg',num_sg,'phi',phi);
end
```

8.7.3   GCICA H_{00}

This is a MATLAB function for ICA involving a group of subjects who have similar or common latent spatial and temporal sources.

```
% gcica_bss_swst.m

function [W outoptarg] = gcica_bss_swst(Xc,opt)
% ====================== PURPOSE =========================
%
% Implement same W same T group-colorICA algorithm.
%
% Xc is a G by NG cell-array. G is the number of groups and NG
% is the number of subjects in group g. Each cell-matrix in Xc
% has dimension M by T where M is the number of mixtures and T
% is the number of time points. The (un)mixing matrix for each
% subject is an M by M invertible matrix.
%
% In the swst algorithm, we assume that the mixing matrix and
```

```
% source time series models are the same within and across each
% group.
%
% ======================= INPUTS =============================
%
% Xc   G by NG cell-array contains observed         [cell-array]
%       data matrices. G is the number of groups and NG is the
%       number of subjects in each group. Each component of the
%       cell-array is a matrix with dimension M by T where M is
%       the number of mixtures and T is the number of time points.
%       Each data matrix has been pre-whitened and centered.
% opt   Optional input arguments                        [struct]
% .tol                                                  [scalar]
%           Tolerance for the convergence. [default: 1e-10]
% .verbose                                              [string]
%           'on' or 'off'.    [default: 'off']        [
% .centering                                            [string]
%       How to center the data matrix X
%       'row': extract row means from each row so that each row
%               sums to 0
%       'rowcol': Extract both column and row means so that each
%               row and column sum to 0
%           [default: 'row']
% .maxit                                                [scalar]
%           Maximum number of iterations. [default:200]
% .prewhite                                             [scalar]
%           Whether to prewhite the data.
%               1: prewhite the data. 0: no prewhite
%               [default: 1]
% .tsms_method                                          [string]
%           Source time series model select (tsms) method
%           'AIC' 'AICC' 'BIC'  [default: 'AIC']
% .initial_W                                            [matrix]
%           Initial of the unmixing matrix. Default is M by M
%           identity matrix
% .arma                                                 [scalar]
%           AR model or ARMA model for sources.
%               0: AR model  1: ARMA model    [default: 0]
% .scovmatdnmntropt                                     [string]
%           Denominator used to calculate sample covariance matrix
%           'sz': using sample size T [default]
%           'szm1': sample size - 1 (T - 1)
% .num_initials                                         [scalar]
%           Number of initial W tried. [default: 10]
```

```
%   .armsearlystopiter                               [scalar]
%        Under armsearlystopiter orders, time series model
%        selection will calculate the actual score. After this
%        number, if the algorithm finds a score that is smaller
%        than previous one, the algorithm stops and return the
%        current order.    [default: 20]
%   .logabsWterm                                     [string]
%        Whether to contain the log term (\ln \abs{det(W)}
%        in the likelihood function.
%        'logterm': contain log term.    [default]
%        'nologterm': do NOT contain log term.
%   .restartiter                                     [scalar]
%        At this iteration step, compare the distance with
%        opt.restartvalue. If the distance is larger than
%        opt.restartvalue, start the algorithm from another
%        initial guess.
%   .restartvalue                                    [scalar]
%        See opt.restartiter
%
%======================= OUTPUTS ============================
%
% W        M by M unmixing matrix                    [matrix]
% outoptarg  Optional output arguments               [struct]
%   .phi                                           [cellarray]
%        M by 1 cell array contains time series parameters
%        of each sources
%   .sigmasq                                         [vector]
%        M by 1 vector contains variances of each
%        time series model
%   .p                                               [vector]
%        M by 1 vector contains order of time series models
%   .iter                                            [scalar]
%        Number of iterations used to converge
%   .num_sg                                          [scalar]
%        Number of singular Hessian matrices found
%   .num_initials                                    [scalar]
%        Number of initial guess used
%
%======================= NOTES ============================
%
% DVW made minor modification of colorICA_pmm.m by Seonjoo Lee.
%
%======================= REFERENCES ========================
%
```

```
% Seonjoo Lee, Haipeng Shen, Young Truong, Michelle Lewis, X.
% Huang (2011). Independent Component Analysis Involving
% Auto-correlated Sources with an Application to fMRI.
% Journal of the American Statistical Association,
% 106, 1009-1024.
%
%======== created by Seonjoo Lee, Jan 2009 ====================
%== Last modified by dvwang-at-alumni.princeton.edu, 07/01/2013

%-----------------------------------------------------------
%--------- INPUT and OUTPUT arguments check --------------------
%-----------------------------------------------------------
error(nargchk(1,2,nargin,'struct'));
error(nargoutchk(0,2,nargout,'struct'));


%-----------------------------------------------------------
%------------------- INPUT -----------------------------------
%-----------------------------------------------------------
if nargin < 2
  opt = struct('tol',1e-10,'verbose','off','centering','row',...
        'maxit',200,'prewhite',1,'tsms_method','AIC',...
        'initial_W',[],'arma',0,'scovmatdnmntropt','sz',...
        'num_initials',10,'armsearlystopiter',20,...
        'logabsWterm','logterm','restartiter',50,...
        'restartvalue',1e-2,'ts_est_method','cat');
end

if ~isstruct(opt)
  error('opt should be a struct array');
end

if (~isfield(opt,'tol') || isempty(opt.tol))
        opt.tol = 1e-10; end
if (~isfield(opt,'verbose') || isempty(opt.verbose))
  opt.verbose = 'off';
end
if (~isfield(opt,'centering') || isempty(opt.centering))
  opt.centering = 'row';
end
if (~isfield(opt,'maxit') || isempty(opt.maxit))
  opt.maxit = 200;
end
if (~isfield(opt,'prewhite') || isempty(opt.prewhite))
  opt.prewhite = 1;
```

```
end
if (~isfield(opt,'tsms_method') || isempty(opt.tsms_method))
  opt.tsms_method = 'AIC';
end
if (~isfield(opt,'initial_W') || isempty(opt.initial_W))
  opt.initial_W = eye(size(Xc{1}{1},1));
end
if (~isfield(opt,'arma') || isempty(opt.arma)) opt.arma = 0; end
if (~isfield(opt,'scovmatdnmntropt') || ...
    isempty(opt.scovmatdnmntropt))
  opt.scovmatdnmntropt = 'sz';
end
if (~isfield(opt,'num_initials') || ...
    isempty(opt.num_initials))
  opt.num_initials = 10;
end
if (~isfield(opt,'armsearlystopiter') || ...
    isempty(opt.armsearlystopiter))
  opt.armsearlystopiter = 20;
end
if (~isfield(opt,'logabsWterm') || ...
    isempty(opt.logabsWterm))
  opt.logabsWterm = 'logterm';
end
if (~isfield(opt,'restartiter') || ...
    isempty(opt.restartiter))
  opt.restartiter = 50;
end
if (~isfield(opt,'restartvalue') || ...
    isempty(opt.restartvalue))
  opt.restartvalue = 1e-2;
end
if (~isfield(opt,'ts_est_method') || ...
    isempty(opt.ts_est_method))
  opt.ts_est_method = 'cat';
end


%------------------------------------------------------------------
%---------------- PREPROCESSING ----------------------------------
%------------------------------------------------------------------
% number of groups
num_group = length(Xc);
```

```
% number of subjects in each group
for ik = 1:num_group
  num_subj_group(ik) = length(Xc{ik});
end

lk = 0;
for ik = 1:num_group
  for jk = 1:num_subj_group(ik)
    lk = lk + 1;
    Xnew{1}{lk} = Xc{ik}{jk};
  end
end

[W outoptarg] = gcica_bss_swdt(Xnew,opt);
```

8.7.4   GCICA H_{11}

This is a MATLAB function for ICA involving a group of subjects.

```
% gcica_cica.m

function [W outoptarg] = gcica_cica(X,opt)
%========================== PURPOSE ============================
%
% To extract source signals from a mixture X. X is an M by T
% matrix where M index the voxles and T index time points.
%
% We assume that the unmixing matrix W has dimension M by M
% and is invertible.
%
%======================== INPUTS ===============================
%
% X        M by T matrix. M is the number of observed    [matrix]
%          mixtures (also the number of sources) and T is the
%          number of time series points. X will be centered
%          within the algorithm.
% opt      Optional input arguments                      [struct]
%  .tol                                                  [scalar]
%          Tolerance for the convergence. [default: 1e-10]
%  .verbose                                              [string]
%          'on' or 'off'.   [default: 'off']
%  .centering                                            [string]
%          How to center the data matrix X
%          'row': extract row means from each row so that each row
```

```
%                    sums to 0
%          'rowcol': Extract both column and row means so that each
%                    row and column sum to 0
%          [default: 'row']
%    .maxit                                         [scalar]
%          Maximum number of iterations. [default:200]
%    .prewhite                                      [scalar]
%          Whether to prewhite the data.
%             1: prewhite the data. 0: no prewhite
%          [default: 1]
%    .tsms_method                                   [string]
%          Source time series mdoel select (tsms) method
%          'AIC' 'AICC' 'BIC'  [default: 'AIC']
%    .initial_W                                     [matrix]
%          Initial of the unmixing matrix. Default is M by M
%          identity matrix
%    .arma                                          [scalar]
%          AR model or ARMA model for sources.
%             0: AR model  1: ARMA model    [default: 0]
%    .scovmatdnmntropt                              [string]
%          Denominator used to calculate sample covariance matrix
%          'sz': using sample size T [default]
%          'szm1': sample size - 1 (T - 1)
%    .num_initials                                  [scalar]
%          Number of initial W tried. [default: 10]
%    .armsearlystopiter                             [scalar]
%          Under armsearlystopiter orders, time series model
%          selction will calculate the actual score. After this
%          number, if the algorithm finds a score that is smaller
%          than previous one, the algorithm stops and return the
%          current order.    [default: 20]
%    .logabsWterm                                   [string]
%          Whether to contain the log term (\ln \abs{det(W)} in the
%          likelihood function.
%          'logterm': contain log term.    [default]
%          'nologterm': do NOT contain log term.
%    .restartiter                                   [scalar]
%          At this iteration step, compare the distance with
%          opt.restartvalue. If the distance is larger than
%          opt.restartvalue, start the algorithm from another
%          initial guess.
%    .restartvalue                                  [scalar]
%          See opt.restartiter
%
```

```
%========================= OUTPUTS ============================
%
% W        M by M unmixitng matrix                     [matrix]
% outoptarg  Optional output arguments                 [struct]
% .phi                                               [cellarray]
%        M by 1 cell array contains time series parameters of
%             each sources
% .sigmasq                                              [vector]
%             M by 1 vector contains variances of each time series
%             model
% .p                                                    [vector]
%             M by 1 vector contains order of time series models
% .iter                                                 [scalar]
%             Number of iterations used to converge
% .num_sg                                               [scalar]
%             Number of singular Hessian matrices found
% .num_initials                                         [scalar]
%             Number of initial guess used
%
%========================= NOTES ============================
%
% DVW made minor modification of colorICA_pmm.m by Seonjoo Lee.
%
%========================= REFERENCES ========================
%
% Seonjoo Lee, Haipeng Shen, Young Truong, Michelle Lewis, X.
% Huang (2011). Independent Component Analysis Involving
% Auto-correlated Sources with an Application to fMRI.
% Journal of the American Statistical Association,
% 106, 1009-1024.
%
%======== created by Seonjoo Lee, Jan 2009 ==================
%== Last modified by dvwang-at-alumni.princeton.edu, 10/07/2013


%-----------------------------------------------------------------
%--------- INPUT and OUTPUT arguments check --------------------
%-----------------------------------------------------------------
error(nargchk(1,2,nargin,'struct'));
error(nargoutchk(0,2,nargout,'struct'));


%-----------------------------------------------------------------
%-------------------- INPUT ---------------------------------------
%-----------------------------------------------------------------
if nargin < 2
```

```
    opt = struct('tol',1e-10,'verbose','off','centering','row',...
        'maxit',200,'prewhite',1,'tsms_method','AIC',...
        'initial_W',[],'arma',0,'scovmatdnmntropt','sz',...
        'num_initials',10,'armsearlystopiter',20,...
        'logabsWterm','logterm','restartiter',50,...
        'restartvalue',1e-2);
end

if ~isstruct(opt)
  error('opt should be a struct array');
end

if (~isfield(opt,'tol') || isempty(opt.tol))
          opt.tol = 1e-10; end
if (~isfield(opt,'verbose') || isempty(opt.verbose))
  opt.verbose = 'off';
end
if (~isfield(opt,'centering') || isempty(opt.centering))
  opt.centering = 'row';
end
if (~isfield(opt,'maxit') || isempty(opt.maxit))
  opt.maxit = 200;
end
if (~isfield(opt,'prewhite') || isempty(opt.prewhite))
  opt.prewhite = 1;
end
if (~isfield(opt,'tsms_method') || isempty(opt.tsms_method))
  opt.tsms_method = 'AIC';
end
if (~isfield(opt,'initial_W') || isempty(opt.initial_W))
  opt.initial_W = eye(size(X,1));
end
if (~isfield(opt,'arma') || isempty(opt.arma))
          opt.arma = 0; end
if (~isfield(opt,'scovmatdnmntropt') || ...
    isempty(opt.scovmatdnmntropt))
  opt.scovmatdnmntropt = 'sz';
end
if (~isfield(opt,'num_initials') || ...
    isempty(opt.num_initials))
  opt.num_initials = 10;
end
if (~isfield(opt,'armsearlystopiter') || ...
    isempty(opt.armsearlystopiter))
```

```
    opt.armsearlystopiter = 20;
end
if (~isfield(opt,'logabsWterm') || ...
    isempty(opt.logabsWterm))
  opt.logabsWterm = 'logterm';
end
if (~isfield(opt,'restartiter') || ...
    isempty(opt.restartiter))
  opt.restartiter = 50;
end
if (~isfield(opt,'restartvalue') || ...
    isempty(opt.restartvalue))
  opt.restartvalue = 1e-2;
end


%-------------------------------------------------------------------
%---------------- PREPROCESSING ------------------------------------
%-------------------------------------------------------------------
% centering
Xc = matcenter(X,opt.centering);
clear X

% M: Number of  mixutures (number of sources)
% T: number of time points
[M T] = size(Xc);

switch opt.scovmatdnmntropt
 case 'sz'
  opt.scovmatdnmntr = T;
 case 'szm1'
  opt.scovmatdnmntr = T - 1;
end

% prewhite
if (opt.prewhite == 1)
 [Xprew invsqrtcovmat] = mat_prewhitening(Xc,opt.scovmatdnmntr);
  clear Xc
  Xc = Xprew;
  clear Xprew;
end

% Discrete Fourier Transformation
X_dft = zeros(M,T);
clear mk
```

```
for mk = 1:M
  X_dft(mk,:) = fft(Xc(mk,:))/sqrt(2*pi*T);
end

% Frequencies
freq = (0:T-1)'/T*2*pi;

%----------------------------------------------------------------
%------ Initial W and source time series parameters  ----------
%----------------------------------------------------------------
% initial W
W1 = opt.initial_W;
% initial source
wX = W1 * Xc;

% g: f_jj spectral density
% f_jj(r) = \frac{\sigma^2_j}{2 \pi \abs{\Phi_j(\exp(-ir))}^2}
g = zeros(M,T);

% Estimate time series order p and parameters phi
for  mk = 1:M
  clear iphi isigmasq ip
  if opt.arma == 1
    disp(['We only consider AR model for now!']);
    return;
  elseif opt.arma==0

    [iphi isigmasq ip] = ...
        gcica_ar_ms_sts(wX(mk,:),struct('method', ...
opt.tsms_method,'earlystopiter',opt.armsearlystopiter));

    if ip == 0
      g(mk,:) = isigmasq/(2*pi)*ones(1,T);
    else
      g(mk,:) =isigmasq/(2*pi)./(abs(1-(exp(-1i * ...
              freq * (1:ip) ) * iphi')).^2)';
    end
  else
    error('Wrong opt.arma value!');
  end
  p(mk) = ip;
  sigmasq(mk) = isigmasq;
  phi{mk} = iphi;
end
```

```
%----------------------------------------------------------------
%----------- iterative updating algorithm  -------------------
%----------------------------------------------------------------
% Distance between W1 and W2 (previous and current iteration)
wddist = 1;
% loop control
iter = 0;
% How many initial W tried
num_initials = 1;
% How many singular Hessian matrix found
num_sg = 0;
%
lambda = ones(M*(M+1)/2,1);
index1 =  reshape(reshape(repmat(1:M,1,M),M,M)', 1,M^2);
index2 = repmat(1:M,1,M);
tempmx11 = reshape(index2,M,M);
tempmx22 = reshape(index1,M,M);
% row indices for Lower triangular elements
tempmx1 = tempmx11(index1<=index2);
% column indices for lower triangular elements
tempmx2 = tempmx22(index1<=index2);
clear tempmx11 tempmx22
indx_temp = [ eye(M*(M+1)/2), zeros(M*(M+1)/2,M*(M-1)/2)];

while (wddist > opt.tol && iter < opt.maxit + 1 && ...
           num_initials < opt.num_initials + 1 )
%----------------------------------------------------------------
%- Newton-Raphson method with Lagrange multiplier for updating W
%----------------------------------------------------------------
  iter = iter+1;

  clear X_wdft W1_inv
  X_wdft = W1 * X_dft;
  W1_inv = inv(W1);

  clear Score_l1_w Score_l2_w
  Score_l1_w = sum(real(conj(X_wdft(index1,:)) .* ...
    X_dft(index2,:))./ g(index1,:),2);
  Score_l2_w = - T * reshape( W1_inv, M^2, 1);

  % dCC / d(W_11, W_12, W_13,...,W_MM)
  clear J_e1 J_e2 Score_cc_w PP
  J_e1 = W1(tempmx1,index2).* indx_temp(tempmx2,index1);
```

```
J_e2 = W1(tempmx2,index2).* indx_temp(tempmx1,index1);
PP = (J_e1 + J_e2)';
Score_cc_w = PP * lambda;

% dCC / d(lambda)
clear c_e Score_cc_lambda
c_e = W1 * W1' - eye(M);
Score_cc_lambda = (c_e(index1 <= index2))';

% first derivative w.r.t. to
% (W_11 W_12,...,W_MM,lambda_1,...lambda_{M(M+1)/2}
clear score_nr
switch opt.logabsWterm
  case 'logterm'
   score_nr = [Score_11_w + ...
           Score_12_w + Score_cc_w; Score_cc_lambda];
  case 'nologterm'
   score_nr = [Score_11_w + Score_cc_w; Score_cc_lambda];
end

%==== Second derivativefor Hessian
% d^2 CC / dW_jk dW_pq
clear h_cc_ww ihl ihll
h_cc_ww = zeros(M^2);
ihl = zeros(M);
ihl(index1 <= index2) = lambda;
ihll = ihl + ihl';
h_cc_ww = kron(ihll,eye(M));

% d^2 (L1) / dW_jk dW_pq
clear h_11_ww ihl1ww ihl1ww2
h_11_ww = zeros(M^2);
ihl1ww = real(conj(X_dft(index1,:)) .* X_dft(index2,:));
for mk = 1:M
  clear ihl1ww2
  ihl1ww2 = sum(ihl1ww./repmat(g(mk,:),[M^2 1]),2);
  h_11_ww(((mk-1)*M + 1):(mk*M),((mk-1)*M+1):(mk*M)) = ...
   (reshape(ihl1ww2,[M M]))';
end

% d^2 (L2) / dW_jk dW_pq
clear h_12_ww
h_12_ww = T * (W1_inv(index2,index1).* ...
W1_inv(index2,index1)');
```

```
% d^2 CC/ dW_jk d(lambda_r)
clear h_cc_wlambda
h_cc_wlambda = PP;

% d^2 CC / d(lambda_r) dW_jk
clear h_cc_lambdaw
h_cc_lambdaw = PP';

% d^2 CC / d(lambda_r) d(lambda_s) = 0
clear h_cc_lambdalambda
h_cc_lambdalambda = zeros(length(lambda));

% Hessian matrix w.r.t (W_11
% W_12,...,W_MM,lambda_1,...lambda_{M(M+1)/2}
clear h_nr

switch opt.logabsWterm
 case 'logterm'
  h_nr = [h_cc_ww + h_11_ww + h_12_ww h_cc_wlambda; ...
  h_cc_lambdaw h_cc_lambdalambda];
 case 'nologterm'
  h_nr = [h_cc_ww + h_11_ww  h_cc_wlambda; ...
  h_cc_lambdaw h_cc_lambdalambda];
end

% Check singularity of Hessian matrix h_nr
rcondh = rcond(h_nr);

if  rcondh>= 10^(-15) && all(abs(isigmasq) < Inf)
  clear iW W2
  iW = vertcat(reshape(W1', M^2,1),lambda) - h_nr\score_nr;
  W2 = reshape(iW(1:M^2),M,M)';
  lambda = iW((M^2+1) : (M*(3*M+1)/2));
  wddist = gcica_amaridist(W2,W1);
  clear W1
  W1 = W2;

  if strcmp(opt.verbose,'on')
    fprintf('%d %0.5g\n',iter,wddist);
  end

end
```

```
 if  rcondh< 10^(-15) || any( abs(isigmasq) == Inf) || ...
(iter == opt.maxit) || ...
(iter == opt.restartiter && wddist > opt.restartvalue)
    W2 = rand_orth(M);
    lambda = ones(M*(M+1)/2,1);
    iter = 0;
    num_initials = num_initials +1;
    W1 = W2;

    if num_initials == opt.num_initials + 1
      W1 = [];
      phi = [];
      p = [];
      sigmasq = [];
      disp('No convergence!');
      break
    end

    if strcmp(opt.verbose,'on')
      if rcondh< 10^(-15)
disp('Hessian matrix is close to singular!');
num_sg = num_sg + 1;
      end
      disp('Start from another initial guess!')
    end
  end

  %----------------------------------------------------------------
  %------ Updating source time series parameters  --------------
  %----------------------------------------------------------------
  wX = W2 * Xc;

  clear g phi sigmasq p
  g = zeros(M,T);

  % Estimate time series order p and parameters phi
  clear p sigmasq phi
  for  mk = 1:M
    clear iphi isigmasq ip
    if opt.arma == 1
      disp(['We only consider AR model for now!']);
      return;
    elseif opt.arma==0
      [iphi isigmasq ip] = gcica_ar_ms_sts(wX(mk,:), ...
```

```
struct('method',opt.tsms_method, ...
'earlystopiter',opt.armsearlystopiter));

    if ip == 0
g(mk,:) = isigmasq/(2*pi)*ones(1,T);
    else
g(mk,:) =isigmasq/(2*pi)./(abs(1-(exp(-1i * ...
            freq * (1:ip) ) * iphi')).^2)';
    end
  else
    error('Wrong opt.arma value!');
  end
  p(mk) = ip;
  sigmasq(mk) = isigmasq;
  phi{mk} = iphi;
end

end


W = W1;

if nargout > 1
  outoptarg = ...
    struct('p',p,'sigmasq',sigmasq, 'num_iter',iter, ...
'num_initials',num_initials,'num_sg',num_sg,'phi',phi);
end
```

Bibliography

1. F. R. Bach and M. I. Jordan. Kernel independent component analysis. *The Journal of Machine Learning Research*, 3:1–48, 2002.
2. A. J. Bell and T. J. Sejnowski. An information-maximization approach to blind separation and blind deconvolution. *Neural Computation*, 7(6):1129–1159, 1995.
3. D. R. Brillinger. *Time Series: Data Analysis and Theory*, volume 36. SIAM, 2001.
4. P. J. Brockwell and R. A. Davis. *Time Series: Theory and Methods*. Springer, 2009.
5. V. D. Calhoun and T. Adali. Multisubject independent component analysis of fMRI: A decade of intrinsic networks, default mode, and neurodiagnostic discovery. *IEEE Reviews in Biomedical Engineering*, 5:60–73, 2012.

6. V. D. Calhoun, T. Adali, G. D. Pearlson, and J. J. Pekar. A method for making group inferences from functional MRI data using independent component analysis. *Human Brain Mapping*, 14:140–151, 2001.

7. V. D. Calhoun, J. Liu, and T. Adali. A review of group ICA for fMRI data and ICA for joint inference of imaging, genetic, and ERP data. *NeuroImage*, 45: S163–S172, 2009.

8. T. Hastie and R. Tibshirani. Independent components analysis through product density estimation. In *Advances in Neural Information Processing Systems*, pages 649–656, 2002.

9. A. Hyvärinen. Independent component analysis: Recent advances. *Philosophical Transactions of the Royal Society A: Mathematical, Physical and Engineering Sciences*, 371(1984), 2013.

10. A. Hyvärinen, J. Karhunen, and E. Oja. *Independent Component Analysis*. Wiley, New York, 2001.

11. S. Lee, H. Shen, Y. Truong, M. Lewis, and X. Huang. Independent component analysis involving autocorrelated sources with an application to functional magnetic resonance imaging. *Journal of the American Statistical Association*, 106(495):1009–1024, 2011.

12. M. J. McKeown, S. Makeig, G. G. Brown, T. Jung, S. S. Kindermann, A. J. Bell, and T. J. Sejnowski. Analysis of fMRI data by blind separation into independent spatial components. *Human Brain Mapping*, 6(3):160–188, 1998.

13. M. P. Milham, D. Fair, M. Mennes, and S. H. Mostofsky. The ADHD-200 consortium: A model to advance the translational potential of neuroimaging in clinical neuroscience. *Frontiers in Systems Neuroscience*, 6(62), 2012.

14. J. V. Stone. *Independent Component Analysis: A Tutorial Introduction*. MIT Press, Cambridge, MA, 2004.

15. P. Whittle. Some results in time series analysis. *Skandivanisk Aktuarietidskrift*, 35:48–60, 1952.

16. K. J. Worsley, C. H. Liao, J. Aston, V. Petre, G. H. Duncan, F. Morales, and A. C. Evans. A general statistical analysis for fMRI data. *NeuroImage*, 15(1): 1–15, 2002.

9 Diagnostic Probability Modeling for Longitudinal Structural Brain MRI Data Analysis

Atsushi Kawaguchi
Kyoto University Graduate School of Medicine

CONTENTS

This chapter describes a procedure for estimating the diagnostic probability of the onset of dementia from longitudinal structural brain magnetic resonance imaging (MRI) data. The method was developed and tested by using MRI data of elderly persons, some of whom had been diagnosed with dementia. The image intensity of MRI data is represented by millions of voxels that form a 3D array, and a longitudinal study yields multiple images per subject. Processing such a large amount of data remains a computational challenge. In contrast to the approach followed by existing methods, this work involved the development of a diagnostic method for Alzheimer's disease that uses longitudinal whole-brain MRI data based on a spatiotemporal functional logistic model. The performance of the new method was tested by applying it to real data obtained from the Open Access Series of Imaging Studies. It was shown that the proposed method offers feasible classification functions and a reasonable prediction accuracy based on a receiver operating characteristic analysis.

9.1 INTRODUCTION

Magnetic resonance imaging (MRI), a noninvasive neuroimaging technique that provides detailed images of the brain structure, has been emerging as an important diagnostic tool for Alzheimer's disease (AD) [15, 12, 5, 26, 1]. The collection of structural brain MRI data that involves repeated observations of a person over a number of years, i.e., longitudinally, serves as an objective and quantitative way to investigate both normal aging and the progression of brain diseases as a function of time. The brain MRI data consist of image intensities of millions of voxels that collectively form a 3D array. As one voxel corresponds to one variable in a statistical term, the data are extremely high-dimensional. Thus, the large number of voxels that would be generated by including the whole brain in a longitudinal MRI study would be computationally prohibitive. This is the main reason why some existing neuroimaging methods [11, 18] limit the amount of data by only considering pre-specified brain regions. As shown by [29] and [23], there are two distinct existing models for analyzing an aging brain structure. The first is a generative model [13], in which the brain image is considered the response and age is the predictor that is analyzed by the parametric or nonparametric regression model. The second is the recognition model [6, 14] in which the brain image is considered the predictor and age the response. This model can be applied in high-dimensional procedures, such as machine learning. In addition, the use of the brain structure to predict disease has also been studied [8, 9, 16].

In contrast to these methods, this chapter proposes a diagnostic method for predicting AD that involves using longitudinal whole-brain MRI data based on a spatiotemporal functional logistic model. It emphasizes that development and aging can be studied by taking three aspects into account, each of which is incorporated in the model as a separate phase. The first phase consists of addressing the high-dimensionality of the data by including two functional representations in the model that are motivated by a reduction in the spatial dimensions and the evaluation of age-related temporal changes. The reduction of the spatial dimension can be achieved by functionalization through the supervised composite basis expansion, which is the linear combination of simple basis functions for the purpose of data-driven selection of the related brain region. The combination components are determined through the use of regularized singular value decomposition supervised by the outcome information for the purpose of improving prediction accuracy. The second phase takes into account that different individuals display different aging patterns, and obtains the functional age-related changes in the brain for each individual using a mixed effect model, with each resulting composite basis function obtained as a function of age with its correlated errors. The third phase combines the outcomes of the first two phases by applying a functional logistic model that uses the age-related changes as predictors to obtain the diagnosis probability as a function of the brain aging trajectory. The penalized likelihood with shrinkage regularization is used to estimate the parameters in the model and to select predictive age changes. The results of the proposed method can be viewed in a reconstructed brain image and placed into the original space that shows the age effect.

This chapter is organized as follows. Section 9.2 describes the proposed statistical methods that were used to implement each of the three phases mentioned aboce. In Section 9.3, the proposed method is applied to real data from the Open Access Series of Imaging Studies (OASIS). Section 9.4 presents a discussion of the results of the receiver operating characteristic (ROC) analysis. Section 9.5 summarizes and concludes the chapter.

9.2 METHODS

9.2.1 NOTATION AND CONCEPTUAL MODEL

Consider n patients with the data notation:

$$\{(y_\alpha, x_\alpha, s_\alpha(t_{\alpha\tau}), t_{\alpha\tau}); \quad \tau = 1, \ldots, T_\alpha; \quad \alpha = 1, \ldots, n\}$$

where y is a binary outcome with a converter ($y = 1$) or a non-converter $y = 0$ to dementia, and x is the p-dimensional potential covariate vector, $s_\alpha(t) = (s_\alpha(w_1, t), \ldots, s_\alpha(w_N, t))^\top$ is the vectorized image data at age t where $w_i \in \mathbb{Z}^3 (i = 1, \ldots, N)$ is the voxel location, and N is the number of voxels. The functional form for $s(w, t)$ is unobserved and has to be estimated, and $t_{\alpha\tau}$ is the τ-th measurement time for the patient α. The number of measurements can differ across subjects. The spatial-temporal functional logistic model is

$$\text{logitPr}(y_\alpha = 1 | x_\alpha, s_\alpha(t), T_{max}, T_{min}) = x_\alpha^\top \beta_1 + \int_{T_{min}}^{T_{max}} s_\alpha(t)^\top f(t) dt \qquad (9.1)$$

where $f(t) = (f(w_1, t), \ldots, f(w_N, t))^\top$ are the N-dimensional time-varying coefficients for images, β_1 is a p-dimensional coefficient vector for covariates, and T_{min} and T_{max} are the minimum and maximum measurement times for all patients, respectively. The integral in Equation (9.1) will be approximated to the summation later.

9.2.2 SPATIAL MODELING

In the first phase, the number of dimensions is reduced by applying the composite basis expansion to the coefficient for voxel w and time t

$$f(w, t) = \sum_{m=1}^{M} \psi_m(w) \beta_{2m}(t) \qquad (9.2)$$

where $\beta_{2m}(t)$ is the time-dependent coefficient and the m-th component basis function is $\psi_m(w) = \sum_{k=1}^{K} \phi_k(w) v_{km}$, which is the linear combination of the radial B-spline function [24]

$$\phi_k(w) = \frac{1}{4h^2} \begin{cases} h^3 + 3h^2(h - d_k(w)) \\ \quad + 3h(h - d_k(w))^2 - 3(h - d_k(w))^3 & (d_k(w) \le h) \\[2ex] (2h - d_k(w))^3 & (h < d_k(w) \le 2h) \\[2ex] 0 & (d_k(w) > 2h) \end{cases}$$

where $d_k(w) = \|w - r_k\|$ and $r_k \in \mathbb{Z}^3 (k = 1, \cdots, K)$ represent location of knots on voxels which is equally spaces on three-dimensional space. $h > 0$ is the distance between adjacent knots. The number of the radial B-spline functions K is determined by the number of voxels and the distance between pre-specified knots. We refer $\phi_k(w)$ to the solo basis function. The basis function reduces the number of parameters from N to M, i.e., by converting each voxel to a component of the function. It should be noted that the use of a basis expansion of this nature as a preprocessing procedure for reducing the number of dimensions was useful for neuroimaging analysis as in, for example, [22], [3], and [28]. As shown in Figure 9.1, the solo basis function $\phi_k(w)$ has a spherical shape. On the other hand, it would be possible for the composite basis $\psi_k(w)$ to represent a flexible shape.

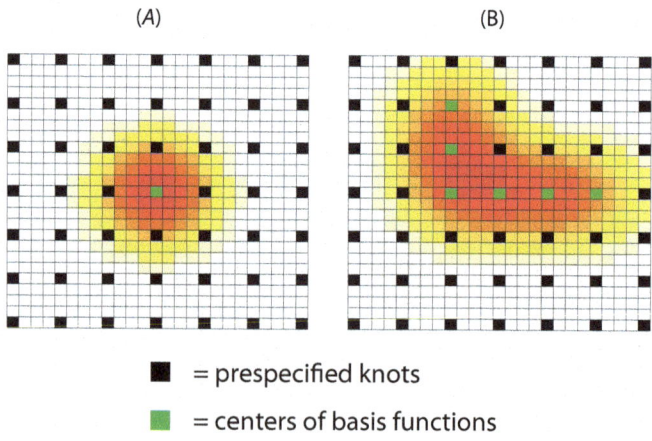

(A)   (B)

■ = prespecified knots

■ = centers of basis functions

Figure 9.1: Example of a 2D image for the solo (A) and composite (B) basis functions.

The sparse method is applied for the selection of the solo basis functions to construct the composite basis function. The weights $v_m = (v_{1m}, v_{2m}, \ldots, v_{Km})^\top$ in the composite basis function were estimated based on the regularized supervised singular value decomposition (SVD). Let $u_m = Av_m$ be a score vector where $A = SB$ is the $T \times K$ matrix with the $T \times N$ matrix $S = \{s_\alpha(t_{\alpha\tau})\}$, $T = \sum_{\alpha=1}^{n} T_\alpha$, and $B = \{\phi_k(w_j)\}$ is the $N \times K$ matrix. When matrix A is normalized by its column, v_m is obtained by maximizing the following equation,

$$L(v_m) = \mu u_m^\top u_m + (1 - \mu) u_m y - \lambda_1 \|v_m\|_1 \tag{9.3}$$

where $0 \le \mu \le 1$ is the proportion of the supervision and $\mu = 0.5$ (as used in this work). Note that when $\mu = 0$, the regularized SVD is induced [25, 27]. λ_1 is the regularized parameter that is used to control the sparsity. The maximization of (9.3) is implemented by the algorithm presented in Table 9.1.

Table 9.1

Algorithm for the regularized supervised singular value decomposition.

Set $m = 1$ and repeat below until $m = 1, \dots, M$.

Step 1. Initialize u_m.
Step 2. Repeat the following until convergence.
   (a) For fixed u_m, $v_m = h_{\lambda_1}(A^\top \{\mu u_m + (1 - \mu)y\})$ where $h_\lambda(sign(y)(|y| > \lambda)_+)$
     and normalize $v_m \leftarrow v_m / \|v_m\|$.
   (b) $u_m = Av_m$.
Step 3. $A \leftarrow A - u_m v_m^\top$ and $m \leftarrow m + 1$.

The larger value of the regularization parameter λ_1 has many $\phi_k(w_j)$ values from which its optimal value is selected by minimizing the Bayesian information criterion (BIC).

$$BIC_1(\lambda_1) = \log \left(\frac{\|A - UV^\top\|}{TK} \right) + \frac{\log(TK)}{TK} df_1$$

where $U = (u_1, \dots, u_M)$, $V = (v_1, \dots, v_M)$, and df_1 is the number of effective parameters, which depend on the value of λ_1. This type of BIC has been used in the framework of matrix decomposition [17, 2]. Thus, the image data for which the number of dimensions has been reduced is obtained as

$$\tilde{s}_{\alpha m}(t) = s_\alpha(t)^\top \Psi_m \tag{9.4}$$

where $\Psi_m = Bv_m$ and $m = 1, \dots, M$.

9.2.3 TEMPORAL MODELING

In the second phase, a mixed-effect linear regression model (with the intercept being the random effect and the slope being the age of the subjects) is used to estimate the age trajectory from the dimension-reduced image data for each component. For each patient, let $\tilde{s}_{\alpha m} = (\tilde{s}_{\alpha m}(t_{\alpha 1}), \ldots, \tilde{s}_{\alpha m}(t_{\alpha T_\alpha}))^\top$ be the individual dimension-reduced image; then the model is given by

$$\tilde{s}_{\alpha m} = X_\alpha \gamma_m + Z_\alpha b_{\alpha m} + \varepsilon_{\alpha m} \quad (m = 1, \ldots, M)$$

where the design matrices X_α and Z_α are given by

$$X_\alpha = Z_\alpha = \begin{bmatrix} 1 & \cdots & 1 \\ t_{\alpha 1} & \cdots & t_{\alpha T_\alpha} \end{bmatrix}^\top$$

and $\gamma_m = (\gamma_{m0}, \gamma_{m1})^\top$ is the coefficient vector, $b_{\alpha m} = (b_{\alpha m0}, b_{\alpha m1})^\top$ is the random effect vector, $b_{\alpha m} \sim N(0, D_m)$ with $D_m = diag(D_{m1}, D_{m2})^\top$ are variances of the random effects, $\varepsilon_{\alpha m}$ is the residual error vector following the multivariate normal distribution with mean vector 0, and the $T_\alpha \times T_\alpha$ variance–covariance matrix $\Sigma_{\alpha m}$ of which the (t_1, t_2)−element is $\sigma_{m t_1 t_2} = \sigma_m \rho_m^{|t_1 - t_2|}$, that is, it has an AR(1) structure. Note that it would be possible for the variance–covariance matrix to have other structures, but these structures are beyond the scope of this chapter.

Let $\tilde{s}_m = (\tilde{s}_{1m}^\top, \ldots, \tilde{s}_{nm}^\top)^\top$ be the vector consisting of dimension-reduced images for time points for all patients, and let $b_m = (b_{1m}^\top, \ldots, b_{nm}^\top)^\top$ and $\varepsilon_m = (\varepsilon_{1m}^\top, \ldots, \varepsilon_{nm}^\top)^\top$ similarly be the random effects and the residual error vectors, respectively. The design matrixes are then determined as follows:

$$X = [X_1^\top, \ldots, X_n^\top]^\top, \quad Z = \begin{bmatrix} Z_1 & 0 & \cdots & 0 \\ 0 & \ddots & \ddots & \vdots \\ \vdots & \ddots & \ddots & 0 \\ 0 & \cdots & 0 & Z_n \end{bmatrix}.$$

Then, from the patient-wise model, the following model is obtained:

$$\tilde{s}_m = X\gamma_m + Zb_m + \varepsilon_m \quad (m = 1, \ldots, M)$$

with $b_m \sim N(0, D)$ and $\varepsilon_m \sim N(0, \Sigma_m)$ where

$$D = I_n \otimes D_m, \quad \Sigma_m = \begin{bmatrix} \Sigma_{1m} & 0 & \cdots & 0 \\ 0 & \ddots & \ddots & 0 \\ \vdots & \ddots & \ddots & 0 \\ 0 & \cdots & 0 & \Sigma_{nm} \end{bmatrix}$$

where I_n is the $n \times n$ identity matrix and $\otimes$ denotes the (right) Kronecker product. Thus, since the general form of the linear mixed-effects model was obtained, the parameters can be estimated by using the restricted maximum likelihood (REML), which is a standard theory for the details; see for example, [10] and [7].

$$\hat{\gamma}_m = \left(\sum_{\alpha=1}^n X_\alpha^\top \Sigma_{\alpha m}^{-1} X_\alpha \right)^{-1} \left(\sum_{\alpha=1}^n X_\alpha^\top \Sigma_{\alpha m}^{-1} \tilde{s}_{\alpha m} \right), \quad \hat{\sigma}_m = \frac{1}{n-2} \sum_{\alpha=1}^n r_\alpha^\top \Sigma_{\alpha m}^{-1} r_\alpha$$

where $r_\alpha = \tilde{s}_{\alpha m} - X_\alpha \hat{\gamma}_m$. The remaining parameter ρ_m in $\Sigma_{\alpha m}$ is estimated by maximizing the log-profile likelihood function and is plugged into $\Sigma_{\alpha m}$ to obtain the estimator $\hat{\Sigma}_{\alpha m}$. The estimator for $b_{\alpha m}$ is based on the best linear unbiased prediction (BLUP):

$$\hat{b}_{\alpha m} = \hat{D}_m Z_\alpha^\top \hat{\Sigma}_{\alpha m}^{-1} (\tilde{s}_{\alpha m} - X_\alpha \hat{\gamma}_m).$$

Thus, the estimator for the age trajectory is obtained by

$$\hat{s}_{\alpha m}(t) = X(t)\hat{\gamma}_m + Z(t)\hat{b}_{\alpha m} \quad (m = 1,\ldots,M) \tag{9.5}$$

where $X(t) = Z(t) = (1,t)$. Note that this model will be useful in applications where subjects may have varying number of observations.

9.2.4 FINAL MODEL

The estimated age trajectory is included in the spatial-temporal functional logistic model, where the integral part in (9.1) is approximated by the summation and is summarized as follows.

$$\int_{T_{min}}^{T_{max}} \tilde{s}_\alpha(t)^\top \beta_2(t) dt \approx \sum_{t=T_{min}}^{T_{max}} \tilde{s}_\alpha(t)^\top \beta_2(t)$$

$$= \sum_{t=T_{min}}^{T_{max}} \sum_{m=1}^{M} \tilde{s}_{\alpha m}(t)^\top \beta_{2m}(t)$$

$$= \sum_{m=1}^{M} \left\{ \sum_{t=T_{min}}^{T_{max}} \tilde{s}_{\alpha m}(t)^\top \beta_{2m}(t) \right\}.$$

Replacing $\tilde{s}_{\alpha m}(t)$ with the estimated age trajectory $\hat{s}_{\alpha m}(t)$ in (9.5), the spatial-temporal functional logistic model with the approximated integral is given by

$$\text{logitPr}(y_\alpha = 1 | x_\alpha, s_\alpha(t), T_{max}, T_{min}) = x_\alpha^\top \beta_1 + \sum_{m=1}^{M} \hat{S}_{\alpha m}^\top \beta_{2m},$$

where $\hat{S}_{\alpha m} = (\hat{s}_{\alpha m}(T_{min}), \hat{s}_{\alpha m}(T_{min} + 1), \ldots, \hat{s}_{\alpha m}(T_{max}))^\top$ and $\beta_{2m} = (\beta_{2m}(T_{min}), \beta_{2m}(T_{min} + 1), \ldots, \beta_{2m}(T_{max}))^\top$. In this model, our parameter estimation is based on the likelihood function and the selection of variables is conducted componentwise. This is achieved by using the regularized likelihood function to restrict the age trajectory to a specific age group, that is, the group lasso method is used [19]:

$$(\hat{\beta}_1, \hat{\beta}_2) = \text{argmax} \left\{ \ell(\beta_1, \beta_2) - n\lambda_2 \left(\|\hat{\beta}_1\|_1 + \sum_{m=1}^{M} \|\hat{\beta}_{2m}\|_2 \right) \right\}$$

where $\ell(\beta_1, \beta_2) = \sum_{\alpha=1}^{n} \{ y_\alpha A(\beta_1, \beta_2) - \log(1 + \exp A(\beta_1, \beta_2)) \}$, $A(\beta_1, \beta_2) = x_\alpha^\top \beta_1 + \sum_{m=1}^{M} \hat{S}_{\alpha m}^\top \beta_{2m}$, and $\beta_2 = (\beta_{21}, \ldots, \beta_{2M})^\top$. The result is interpreted by using the estimated Ψ_m in (9.4) and β_{2m} for the spatial and temporal parts, respectively.

Note that this results in a reconstructed brain image in the original space and an age effect. The small λ_2 has many non-zero values for β_{2m}, that is, there are many selected trajectories. The regularization parameter is selected by minimizing the BIC.

$$BIC_2(\lambda_2) = 2\ell(\hat{\beta}_1, \hat{\beta}_2) + \log(n)df_2,$$

where df_2 is the effective number of model parameters.

9.3 APPLICATION

This study used the OASIS longitudinal data from a project by Washington University that is aimed at making MRI data sets of the brain freely available to the scientific community. The study investigated the relationship between the results of brain MRI studies and the occurrence of AD. The data includes three or four T_1-weighted images in single scan sessions and a total of 14 AD converters from a study population of 86 participants aged 60–96 years. The brain of each participant was scanned during two or more visits, with the separation between visits being at least one year. All participants in the study were characterized as healthy at the time of their initial visit. The 14 AD subjects were considered to be those who were subsequently characterized as demented during a later visit. The controls were characterized as healthy throughout the course of the study. Our study considered gender, total brain volume, and educational history as potential covariates, all of which are candidates for selection as variables. The preprocessing of the images was conducted using the VBM8 toolbox (http://dbm.neuro.uni-jena.de/vbm/).

Our model was tested by applying it to the MRI data of all 86 participants described above. First, the voxels that represent the brain regions were extracted from among the total number of voxels, namely 2,122,945 ($121 \times 145 \times 121$), that were obtained for one subject, a process that resulted in $N = 839,089$ voxels. The dimension of the basis function is $K = 7,176$ because of the 4-voxel ($h_0 = 4$); therefore, $h = \sqrt{3 \times 4^2} = 6.93$) equal spacing knots. As a result, to enable trajectory selection by using the group lasso method, only one component ($M=1$) and no covariates were selected. The estimated temporal parameter β_{21} is shown in Figure 9.2, where it can be seen that the effect displayed by this parameter is almost linear and that the probability of being diagnosed with AD is highest for persons aged about 60. The corresponding region of the brain Ψ_1 in (9.4) is shown in Figure 9.3. Our model indicated that the cerebellum was the most important region in the AD diagnosis. Although the hippocampus, which is the region that primarily is affected by AD, also was selected, it did not show as high an association with AD as the cerebellum.

9.4 ROC ANALYSIS

The estimated diagnostic probability was evaluated by the receiver operating characteristic (ROC) analysis based on five repeated 10-fold cross-validated estimators for the diagnostic probability. Figure 9.4 shows the ROC curve. The proposed method, represented by the red curve, had an area under the curve (AUC) value of 0.812.

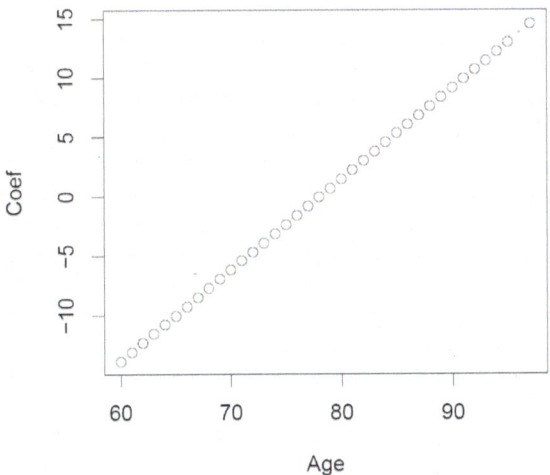

Figure 9.2: Estimated temporal coefficients.

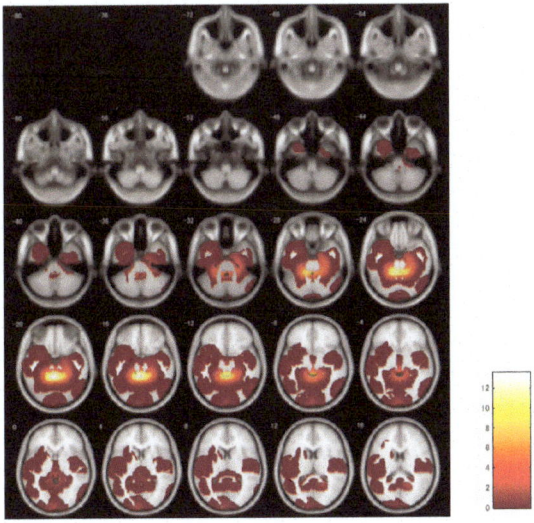

Figure 9.3: Spatial imaging region related to diagnostic probability.

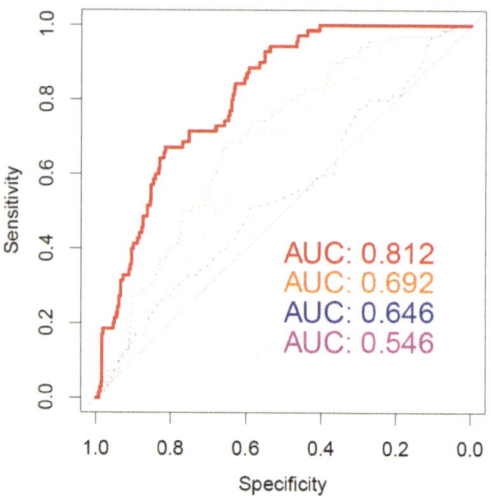

Figure 9.4: ROC curve for the cross-validated estimator for the diagnostic probability. The red line represents the result of the proposed method. The orange, blue, and purple lines represent the results of the unsupervised, the time-independent coefficient, and the solo basis methods, respectively.

For comparison purposes, the results for other methods are also shown. The other methods that were applied to the real data were the unsupervised method, which is the case where $\mu = 0$ in (9.3), the time-independent coefficient method, which is the case where $\beta_{2m}(t) = \beta_{2m}$ for any t in (9.2), and the solo basis method, which is the case where $\Psi_m = B$ in (9.4). The AUC values that were obtained using these methods were 0.692, 0.646, and 0.546, respectively. The figure shows that the proposed method is the most effective for making predictions. Additionally, lowering the AUC value obtained with the proposed method would impart important information regarding a model feature. In this respect, the composition of the basis function is the most important feature of the model, followed by the varying coefficient feature.

9.5   SUMMARY AND CONCLUSION

In summary, a method was developed to predict the possibility of a person developing dementia. The method takes, as its input, images of the brain that were obtained as a function of age. The method has three key features: (1) dimension reduction by using a composite basis function, (2) varying coefficients to estimate the diagnostic probability, and (3) using the supervised version of the regularized SVD, in the order of importance. The dimension reduction method described in this chapter is based on the basis expansion approach and it was not only effective from the point of view of performance, but also from that of computational cost. This was confirmed by the fact that implementing the method without the dimension reduction was found to be much more time consuming. It would be possible to extend the proposed method by, for example, considering a nonlinear approach in the case of more measurements per subject, in which case dimension reduction also would be desirable. There are other studies generating longitudinal data that would also benefit from using this approach to modeling [20].

Although the method is robust and generated good predictive results, there are some limitations. It may be necessary to fit the mixed-effects model to the entire sample again to estimate the random effect representing the individual profile. The accuracy of this process would be improved by modeling other kinds of data. Additionally, the time-dependent outcome could be incorporated into the generalized linear mixed-effects model. Our dimension reduction method would enable us to model complex structures, because of the reduction in the number of parameters.

The MRI brain region that was selected for analysis should be considered more carefully in the future. The particular area of the brain that was selected for this work was chosen to improve the utility of the supervised method proposed in this article. The results showed that the temporal lobe around the hippocampus, which is known to be the area where AD is the most apparent, did not produce the best result in the prediction. The successful application of our method would depend on the use of more sophisticated preprocessing, such as that used by [4]. This would also form part of future work. The potential advantages of using our method for longitudinal studies are that it has reasonable predictive performance, computational cost, and broad utility. In conclusion, our method will enable the construction of a diagnostic procedure for dementia from a longitudinal study.

9.6 SOFTWARE IMPLEMENTATION

The methods in this chapter were implemented by the R programming environment [21], which required three packages, msma, mgcv, and grpreg. The mgcv and grpreg packages are available from the Comprehensive R Archive Network (CRAN). The msma package is under development and will be uploaded.

The methods for the spatial modeling described in Section 2.2 were implemented with the functions in the msma package. First, the regularization parameter was selected by minimizing BIC with the code:

```
regpara1 = regparasearch(X = X, Z = Z, comp = 10, muX = 0.5,
method = "BIC")
```

where X is the dimension-reduced matrix A, and Z is the outcome variable y in the text. The number of components was fixed as 10 in this step. Next, the number of components was selected by minimizing BIC with the code:

```
params = ncompsearch(X = X, Z = Z, comps = min(dim(X)),
lambdaX = regpara1["optlambdaX"], muX = 0.5, regpara = FALSE,
method = "BIC")
```

where this statement used the output of the function `regparasearch`. Finally, the weight vector v_m was obtained by maximizing Equation (9.3) with the following code:

$$\mathtt{fit1} = \mathtt{msma}(X = X, Z = Z, \mathtt{comp} = \mathtt{params}[\text{"optncomp"}],$$
$$\mathtt{lambdaX} = \mathtt{params}[\text{"optlambdaX"}], \mathtt{muX} = 0.5)$$

where this statement used the output of the functions `regparasearch` and `ncompsearch`.

The methods for the temporal modeling described in Section 2.3 were implemented with the function gamm in the mgcv package:

$$\mathtt{gamm.fit} = \mathtt{gamm}(\mathtt{Vage}, \mathtt{data} = \mathtt{dataset00}, \mathtt{random} = \mathtt{list}(\mathtt{id} = \mathtt{age}),$$
$$\mathtt{correlation} = \mathtt{corAR1}(\mathtt{form} = 1|\mathtt{id}))$$

where V is the individual dimension-reduced image.

The methods for the final model described in Section 9.2.4 were implemented with the function grpreg in the grpreg package:

$$\mathtt{grpreg.fit} = \mathtt{grpreg}(\mathtt{funcdata}, Z, \mathtt{group0}, \mathtt{family} = \text{"binomial"},$$
$$\mathtt{penalty} = \text{"gLasso"})$$

where `funcdata` represents the predicted values from the output of the function gamm. The optimal regularization parameter was selected with the function `select` in the grpreg package:

$$\mathtt{finalmodel} = \mathtt{select}(\mathtt{grpreg.fit}, \mathtt{criterion} = \text{"BIC"},$$
$$\mathtt{df.method} = \text{"default"})$$

This final output was used the result display and the evaluation.

ACKNOWLEDGMENT

This research was supported in part by Grants-in-Aid from the Ministry of Education, Culture, Sport, Science and Technology of Japan (24700286). For this research work, we used the supercomputer of ACCMS, Kyoto University.

Bibliography

1. Paul A Adlard, Bob A Tran, David I Finkelstein, Patricia M Desmond, Leigh A Johnston, Ashley I Bush, and Gary F Egan. A review of β-amyloid neuroimaging in Alzheimer's disease. *Frontiers in Neuroscience*, 8, 2014.

2. Genevera I Allen, Logan Grosenick, and Jonathan Taylor. A generalized least-square matrix decomposition. *Journal of the American Statistical Association*, 109(505):145–159, 2014.

3. Yuko Araki, Atsushi Kawaguchi, and Fumio Yamashita. Regularized logistic discrimination with basis expansions for the early detection of Alzheimer's disease based on three-dimensional MRI data. *Advances in Data Analysis and Classification*, 7(1):109–119, 2013.

4. John Ashburner and Gerard R Ridgway. Symmetric diffeomorphic modeling of longitudinal structural MRI. *Frontiers in Neuroscience*, 6, 2012.

5. Thomas G Beach, Sarah E Monsell, Leslie E Phillips, and Walter Kukull. Accuracy of the Clinical Diagnosis of Alzheimer Disease at National Institute on Aging Alzheimer's Disease Centers, 2005–2010. *Journal of Neuropathology and Experimental Neurology*, 71(4):266, 2012.

6. Signe Bray, Catie Chang, and Fumiko Hoeft. Applications of multivariate pattern classification analyses in developmental neuroimaging of healthy and clinical populations. *Frontiers in Human Neuroscience*, 3, 2009.

7. Tomasz Burzykowski et al. *Linear mixed-effects models using R: A step-by-step approach*. Springer Science & Business Media, 2013.

8. Ramon Casanova, Christopher T Whitlow, Benjamin Wagner, Jeff Williamson, Sally A Shumaker, Joseph A Maldjian, and Mark A Espeland. High dimensional classification of structural MRI Alzheimer's disease data based on large scale regularization. *Frontiers in Neuroinformatics*, 5, 2011.

9. Rémi Cuingnet, Emilie Gerardin, Jérôme Tessieras, Guillaume Auzias, Stéphane Lehéricy, Marie-Odile Habert, Marie Chupin, Habib Benali, Olivier Colliot, Alzheimer's Disease Neuroimaging Initiative, et al. Automatic classification of patients with Alzheimer's disease from structural MRI: A comparison of ten methods using the ADNI database. *NeuroImage*, 56(2):766–781, 2011.

10. Eugene Demidenko. *Mixed Models: Theory and Applications with R*. John Wiley & Sons, 2013.

11. Yong Fan, Dinggang Shen, Ruben C Gur, Raquel E Gur, and Christos Davatzikos. Compare: classification of morphological patterns using adaptive regional elements. *Medical Imaging, IEEE Transactions on*, 26(1):93–105, 2007.

12. Luiz Kobuti Ferreira and Geraldo F Busatto. Neuroimaging in Alzheimer's disease: Current role in clinical practice and potential future applications. *Clinics*, 66:19–24, 2011.

13. Anders M Fjell, Kristine B Walhovd, Lars T Westlye, Ylva Østby, Christian K Tamnes, Terry L Jernigan, Anthony Gamst, and Anders M Dale. When does brain aging accelerate? Dangers of quadratic fits in cross-sectional studies. *NeuroImage*, 50(4):1376–1383, 2010.

14. Katja Franke, Eileen Luders, Arne May, Marko Wilke, and Christian Gaser. Brain maturation: Predicting individual BrainAGE in children and adolescents using structural MRI. *NeuroImage*, 63(3):1305–1312, 2012.

15. Giovanni B Frisoni, Nick C Fox, Clifford R Jack, Philip Scheltens, and Paul M Thompson. The clinical use of structural MRI in Alzheimer disease. *Nature Reviews Neurology*, 6(2):67–77, 2010.

16. Logan Grosenick, Brad Klingenberg, Kiefer Katovich, Brian Knutson, and Jonathan E Taylor. Interpretable whole-brain prediction analysis with GraphNet. *NeuroImage*, 72:304–321, 2013.

17. Mihee Lee, Haipeng Shen, Jianhua Z Huang, and JS Marron. Biclustering via sparse singular value decomposition. *Biometrics*, 66(4):1087–1095, 2010.

18. Benoît Magnin, Lilia Mesrob, Serge Kinkingnéhun, Mélanie Pélégrini-Issac,

Olivier Colliot, Marie Sarazin, Bruno Dubois, Stéphane Lehéricy, and Habib Benali. Support vector machine-based classification of Alzheimer's disease from whole-brain anatomical MRI. *Neuroradiology*, 51(2):73–83, 2009.

19. Lukas Meier, Sara Van De Geer, and Peter Bühlmann. The group lasso for logistic regression. *Journal of the Royal Statistical Society: Series B (Statistical Methodology)*, 70(1):53–71, 2008.

20. Edward Ofori, Guangwei Du, Debra Babcock, Xuemei Huang, and David E Vaillancourt. Parkinson's disease biomarkers program brain imaging repository. *NeuroImage*, 2015.

21. R Core Team. *R: A Language and Environment for Statistical Computing*. R Foundation for Statistical Computing, Vienna, Austria, 2015. URL `http://www.R-project.org/`.

22. Philip T Reiss and R Todd Ogden. Functional generalized linear models with images as predictors. *Biometrics*, 66(1):61–69, 2010.

23. Timothy A Salthouse. Neuroanatomical substrates of age-related cognitive decline. *Psychological Bulletin*, 137(5):753, 2011.

24. Afsar Saranli and Buyurman Baykal. Complexity reduction in radial basis function (RBF) networks by using radial B-spline functions. *Neurocomputing*, 18(1):183–194, 1998.

25. Haipeng Shen and Jianhua Z Huang. Sparse principal component analysis via regularized low rank matrix approximation. *Journal of Multivariate Analysis*, 99(6):1015–1034, 2008.

26. Michael W Weiner, Dallas P Veitch, Paul S Aisen, Laurel A Beckett, Nigel J Cairns, Robert C Green, Danielle Harvey, Clifford R Jack, William Jagust, Enchi Liu, et al. The Alzheimer's disease neuroimaging initiative: A review of papers published since its inception. *Alzheimer's & Dementia*, 8(1):S1–S68, 2012.

27. Daniela M Witten, Robert Tibshirani, and Trevor Hastie. A penalized matrix decomposition, with applications to sparse principal components and canonical correlation analysis. *Biostatistics*, 10(3):515–534, 2009.

28. Hisako Yoshida, Atsushi Kawaguchi, and Kazuhiko Tsuruya. Radial basis function-sparse partial least squares for application to brain imaging data. *Computational and Mathematical Methods in Medicine*, 2013, 2013.

29. Gabriel Ziegler, Robert Dahnke, Christian Gaser, Alzheimer's Disease Neuroimaging Initiative, et al. Models of the aging brain structure and individual decline. *Frontiers in Neuroinformatics*, 6, 2012.

10 Supervised SVD of fMRI Data with Time-Varying Frequency Components

Avner Halevy
University of North Carolina at Chapel Hill

Young K. Truong
University of North Carolina at Chapel Hill

CONTENTS

10.1 INTRODUCTION

Functional magnetic resonance imaging (fMRI) is a procedure for studying the brain in action. Providing an unprecedented ability to safely and noninvasively image brain activity, since the early 1990s, it has become widely popular among scientists in both the clinical and especially cognitive research settings. In the clinical setting it can be used to diagnose brain disorders like autism, schizophrenia, and Alzheimer's disease. In the cognitive research setting it is used to study how we remember, learn, and manipulate information and to elucidate the neurological underpinnings of diseases like Parkinson's. It provides very good spatial resolution and relatively good temporal resolution relative to previous methods like positron emission tomography (PET) (whose use was also more limited due to safety concerns about radiation) [5].

fMRI exploits the fact that when neurons in the brain become active, increased blood flow to the active area leads to a relative surplus of local blood oxygen. The signal measured in fMRI depends on this change in oxygenation and is known as the blood-oxygenation-level-dependent (BOLD) signal. A single image usually consists of a number of slices where each slice consists of individual cuboid elements called *voxels*. This three-dimensional matrix of voxels is sampled at regular time intervals. The resulting data set is usually represented as a spatio-temporal matrix $\mathbf{X}$ of dimension $M \times N$, where M represents the total number of voxels and N represents the total number of time points. In most experiments the number of time points is much smaller than the number of voxels ($N \ll M$) [2].

Statistical methods for analyzing fMRI data can be categorized as model-driven or data-driven. A leading data-driven approach has been Independent Component Analysis (ICA) [4]. ICA is capable of extracting from an fMRI signal the individual signals that correspond to different sources, such as the experimental stimulus and cardiac and respiratory effects. However, one weakness of ICA algorithms is sensitivity to outliers. An algorithm known as Supervised Singular Value Decomposition (SSVD) has been proposed as a preprocessing step to make ICA algorithms more robust against outliers [2].

fMRI data are usually of high dimension. For example, a typical matrix $\mathbf{X}$ might be $153,594 \times 200$. To make ICA computationally feasible, dimension reduction is performed in a preprocessing stage using techniques like singular value decomposition (SVD), which is unsupervised and purely data-driven. Unfortunately, SVD is highly susceptible to outliers. SSVD uses information about the experimental design, particularly the frequencies of the stimulus, in order to introduce supervision into the optimization problem. This results in a decomposition that is less sensitive to outliers.

An example of a popular experiment where frequency information is readily available is that of the right/left hand finger tapping exercise that has been used to study motor dysfunction in patients with Parkinson's disease. The algorithm set forth in [2] was shown to perform well given this information. However, until now, SSVD has been limited to signals with fixed frequency. In the present chapter, we aim to extend this algorithm in order to deal with signals whose frequency may vary with time. At the heart of this extension lies the use of wavelets, which are well suited for detecting (temporally) local structure.

10.2   INDEPENDENT COMPONENT ANALYSIS (ICA)

10.2.1   OVERVIEW OF ICA

The basic idea behind ICA comes from a problem in signal processing known as *blind source separation*. Suppose a number of signals are emitted by some physical sources and a number of sensors are recording different mixtures of these signals. For example, the sources could be people at a party emitting speech signals, and the sensors could be microphones at different locations in the room. The goal of blind source separation is to find the original signals from the mixtures. ICA solves this problem by attempting to transform the mixtures into signals that are statistically independent, called *independent components*.

Mathematically, let $\mathbf{y}$ and $\mathbf{s}$ be r-dimensional random vectors representing the observed mixtures and the unknown sources, respectively. We assume the sources are independent and that the two vectors are related by

$$\mathbf{y} = \mathbf{Ms}, \tag{10.1}$$

where $\mathbf{M}$ is an $r \times r$ unknown mixing matrix. The goal of ICA is to find a linear transformation represented by a matrix $\mathbf{W}$ so that $\mathbf{s} = \mathbf{Wy}$. Assuming $\mathbf{M}$ is invertible and the sources are non-Gaussian, the solution is well-defined up to the variances, signs, and order of the independent components.

10.2.2   ICA FOR FMRI

ICA can be applied to fMRI data in two different ways [1]. One way is to extract temporal signals that are statistically independent, as in the problem of blind source separation described above, in which case the procedure is known as *temporal ICA*. In this case the weights in the mixing matrix carry spatial information. The other way is to extract spatial maps that are statistically independent, in which case the procedure is known as spatial ICA. In this case, the weights in the mixing matrix carry temporal information. Spatial ICA is more common and shall be our focus. The spatial maps we seek are believed to correspond to functionally independent neural networks. The temporal weights tell us how to combine the activity in these during each scan in order to produce the activity detected in each voxel.

Mathematically, denote the kth independent component (spatial map) identified by ICA by

$$\mathbf{c}_k = \begin{bmatrix} c_{k1} \\ c_{k2} \\ \vdots \\ c_{kM} \end{bmatrix},$$

where c_{kj} is the activation of voxel j in component k. Spatial ICA assumes that the observed response at each scan is a weighted sum of these components. Denote the weight on component k at scan i by m_{ik} and the response in all M voxels at scan i by

$$\mathbf{x}_i = \begin{bmatrix} x_{i1} \\ x_{i2} \\ \vdots \\ x_{iM} \end{bmatrix}.$$

ICA extracts as many components as there are time scans. Then with N scans we have

$$\mathbf{x}_i = m_{i1}\mathbf{c}_1 + m_{i2}\mathbf{c}_2 + \ldots + m_{iN}\mathbf{c}_N.$$

Note that we know nothing about the right side of this equation. The weights and the components must be estimated from the data under the assumption that the components are statistically independent and non-normal. To remove further ambiguity, they are also assumed to have unit variance. Putting together the information from all the voxels and all the scans leads to the matrix equation

$$\mathbf{X}^T = \mathbf{M}\mathbf{C}^T, \tag{10.2}$$

where the spatio-temporal data matrix $\mathbf{X}$ is $M \times N$ (as before), $\mathbf{M}$ is $N \times N$, and $\mathbf{C}$ is $M \times N$.

10.2.3 DIMENSION REDUCTION

As mentioned above, fMRI data are usually of high dimension, especially in the spatial domain. Hence, in order to reduce the burden of computing ICA, the dimension of the data is typically reduced in a preprocessing step using SVD. Precisely, suppose rank$(\mathbf{X}) = r \leq \min(M, N)$. Then we can use SVD to decompose $\mathbf{X}$ as

$$\mathbf{X} = \mathbf{U}\mathbf{D}\mathbf{V}^T, \tag{10.3}$$

where $\mathbf{U}$ is an $M \times r$ matrix of orthonormal left singular vectors, $\mathbf{V}$ is an $N \times r$ matrix of orthonormal right singular vectors, and $\mathbf{D}$ is an $r \times r$ diagonal matrix of positive singular values.

If $\mathbf{X}$ is a spatio-temporal matrix representing an fMRI data set, then we can think of $\mathbf{U}$ and $\mathbf{V}$ as basis vectors spanning the spatial patterns and temporal sequences, respectively. Furthermore, by letting $\mathbf{A} = \mathbf{U}\mathbf{D}$ and $\mathbf{S} = \mathbf{V}^T$, we can decompose $\mathbf{X}$ as

$$\mathbf{X} = \mathbf{A}\mathbf{S}, \tag{10.4}$$

where $\mathbf{A}$ is $M \times r$ and $\mathbf{S}$ is $r \times N$. Recall that this matrix product can be computed as the sum of r outer products between the columns of $\mathbf{A}$ and the rows of $\mathbf{S}$. Precisely, if we denote the columns of $\mathbf{A}$ by $\mathbf{a}_1, \ldots, \mathbf{a}_r$ and the rows of $\mathbf{S}$ by $\mathbf{s}_1, \ldots, \mathbf{s}_r$, then

$$\mathbf{X} = \sum_{i=1}^{r} \mathbf{a}_i \mathbf{s}_i^T. \tag{10.5}$$

Therefore we can think of $\mathbf{X}$ as the sum of products of spatial maps (columns of $\mathbf{A}$) and associated time courses (rows of $\mathbf{S}$). However, this decomposition does not guarantee that the spatial maps are statistically independent. This will be achieved by applying ICA to $\mathbf{A}$ (as is done in spatial ICA). By applying ICA to $\mathbf{A}$, which has only r columns, instead of $\mathbf{X}$, which has N, we can compute ICA more efficiently.

10.3   SUPERVISED SVD

SVD is sensitive to spikes that are inevitable in fMRI data. In [2] it was proposed to replace the conventional SVD used in the data reduction step described above with a modified version that uses information about the experimental design in order to avoid the traps introduced by these spikes. This makes the subsequent ICA more robust because it focuses on the correct subspace.

10.3.1   LOW RANK APPROXIMATION

Recall that SVD provides an optimal low-rank approximation. Precisely, in the SVD of $\mathbf{X}$ in (10.3), let $\mathbf{U} = [\mathbf{u}_1,\ldots,\mathbf{u}_r]$, $\mathbf{V} = [\mathbf{v}_1,\ldots,\mathbf{v}_r]$, and $\mathbf{D} = \mathrm{diag}\{d_1,\ldots,d_r\}$. Recall the Frobenius norm, which is defined as $\|\mathbf{X}\|_F^2 = \mathrm{tr}\{\mathbf{X}^T\mathbf{X}\}$. For an integer $l \leq r$, let $\mathbf{X}^{(l)} = \sum_{k=1}^{l} d_k \mathbf{u}_k \mathbf{v}_k^T$. Then $\mathbf{X}^{(l)} = \mathrm{argmin}\|\mathbf{X} - \mathbf{X}^*\|_F$ where the minimum is taken over all matrices $\mathbf{X}^*$ of rank no greater than l. In particular, the best rank 1 approximation of $\mathbf{X}$ is given by $d_1\mathbf{u}_1\mathbf{v}_1^T$. Subsequent singular triples $(\mathbf{u}_k, d_k, \mathbf{v}_k)$ provide the best rank 1 approximation of the residual matrices. For example, $d_2\mathbf{u}_2\mathbf{v}_2^T$ is the best rank 1 approximation of $\mathbf{X} - d_1\mathbf{u}_1\mathbf{v}_1^T$.

10.3.2   SUPERVISED SVD

In [2] it was recognized that when an experimental task results in temporal data with a significant periodic component, this information can be used effectively to supervise the SVD. This is accomplished by constraining the singular vector $\mathbf{v}$ to lie in the span of a basis of sinusoidal functions with a frequency known from the design (and otherwise estimated from the data). Precisely, if $\mathbf{B}$ represents the basis, the optimization problem becomes

$$\min_{\mathbf{u},\psi} \|\mathbf{X} - \mathbf{u}\mathbf{v}^T\|_F \quad \text{subject to} \quad \mathbf{v} = \mathbf{B}\psi. \tag{10.6}$$

To make the model identifiable, $\mathbf{u}$ and $\mathbf{v}$ are standardized and a slope parameter d is introduced. This leads to

$$\min_{d,\mathbf{u},\psi} \|\mathbf{X} - d\mathbf{u}\mathbf{v}^T\|_F \quad \text{subject to} \quad \mathbf{v} = \mathbf{B}\psi, \quad \mathbf{u}^T\mathbf{u} = 1, \quad \mathbf{v}^T\mathbf{v} = 1. \tag{10.7}$$

This problem was reduced to the solution of two generalized eigenvalue problems. The resulting algorithm was implemented and shown to perform effectively using both simulated and real data.

10.4   EXTENSION TO TIME-VARYING FREQUENCY

The basis **B** used in the previous section consisted of sinusoidal functions with a fixed frequency. However, the solution of the corresponding optimization problem holds for any basis. Therefore, the singular vector **v** can be constrained to lie in any subspace for which we have a basis. If the signal of interest has a time-varying frequency, we should look for a basis that is flexible enough to represent all of the relevant frequencies. We consider two such approaches. First, we use short-time (that is, local as opposed to global) sinusoids. Second, we consider wavelets, which offer a wealth of options for bases that can adapt to the (possibly varying) local structure of the signal. By using one of these approaches, we can thus expand the applicability of Supervised SVD, as we show below.

10.4.1   SUPERVISED SVD WITH LOCALIZED SINUSOIDS

Suppose the signal is known to contain several frequencies. Assuming we know what these are and when changes in frequency occur, we can construct short-time sinusoids to adapt to the local character of the signal. For example, in Figure 10.1 we can represent the signal in the top left, which contains three different frequencies, with a linear combination of the other three signals, which contain local sinusoids. If we don't know what the frequencies are or when they change, we can estimate this information by performing spectral analysis on the temporal components extracted from a conventional SVD.

10.4.2   WAVELETS

Just like the sinusoidal functions of Fourier analysis, wavelets are used to approximate arbitrary functions with simple ones. However, where the classical Fourier bases are global, wavelets are local in nature and therefore well suited for the analysis of signals whose character changes with time. A wavelet basis consists of shifted and scaled versions of a prototype function, often called a *mother wavelet*. Although there are many types of wavelets, all of them offer a trade-off between frequency resolution and time resolution.

10.4.3   SUPERVISED SVD WITH WAVELETS

Given the frequencies contained in the signal and the points in time at which they change, we can construct an approximation of the signal of interest that is correct up to phase shifts. We can then use a predefined dictionary of wavelets to construct a greedy approximation of the signal using orthogonal pursuit. An example of typical wavelets used in this study is shown in Figure 10.2. We then let the basis **B** consist of the chosen wavelets. From that point, the algorithm for supervised SVD is identical to the one presented in [2].

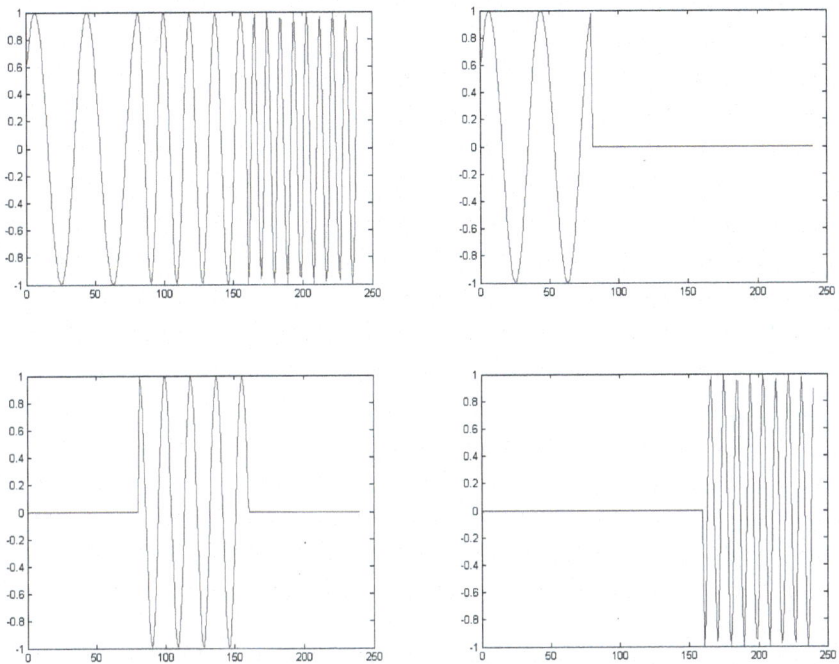

Figure 10.1: In the upper left panel is a signal with a time-varying frequency. We can use the short-time sinusoids in the remaining panels to recover it.

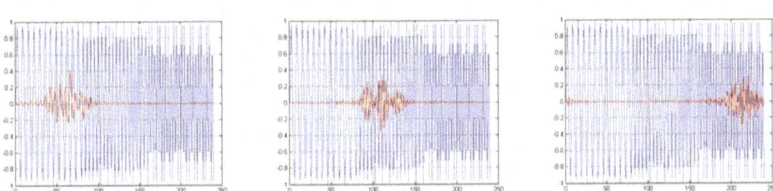

Figure 10.2: Wavelets superimposed on a signal with a time-varying frequency

10.5 SIMULATION STUDIES

We now use simulated data in order to compare the performance of SSVD-ICA with conventional SVD-ICA. The data is constructed to contain at least one signal with a time-varying frequency. It is modeled after real fMRI data. Evaluation of the method using real data is left for future work.

10.5.1 DATA DESCRIPTION

To simulate an $N \times M$ data matrix $\mathbf{X}$ with r components, we simulate an $N \times r$ spatial matrix $\mathbf{A}$ and an $r \times N$ temporal matrix $\mathbf{S}$ separately and then multiply them according to (10.4). In this simulation, we set $r = 4, M = 30 \times 30 \times 10$, and $N = 240$ (see Figure 10.3). Each column of $\mathbf{A}$ is a spatial map containing voxel information for 10 slices, each of which is 30×30. Each row of $\mathbf{S}$ is a time series of length 240 corresponding to one of these maps. One of the time series contains a time-varying frequency. To each of the components we add noise that is uniformly distributed on $[-0.05, 0.05]$. The amplitudes of the four components are 0.5, 0.7, 0.35, and 0.45, respectively. The frequencies of the three fixed components are 0.06 Hz, 0.3 Hz, and 0.7 Hz. The test signal contains three different frequencies (0.5 Hz, 0.75 Hz, and 1 Hz) during three equal-length intervals. To simulate the spikes (outliers) we randomly select 10% of the elements and replace them with noise uniformly distributed on $[-8, -2]$ and [2,8].

10.5.2 ANALYSIS

Following standard practice, we first normalize the data matrix $\mathbf{X}$ by centering the columns and standardizing the rows. Then, we preprocess the data in four different ways and subsequently apply ICA. First, we preprocess with conventional SVD. Second, momentarily focusing only on the signal containing a time-varying frequency, we preprocess using the SSVD proposed in [2] by using the average frequency. Third, we preprocess with SSVD using local sinusoids. Finally, we preprocess with SSVD using wavelets. In all cases, ICA is computed using the FastICA algorithm (see [3]). To display the activated voxels in the spatial maps effectively, values are standardized to z-scores and those with $|z| \geq 1$ are identified as active.

10.5.3 RESULTS

First, Figure 10.4 shows our motivation for extending SSVD: Using the average frequency with the original SSVD is not sufficient to recover a signal containing a time-varying frequency. Next, Figure 10.5 compares the results obtained using SSVD with localized sinusoids with those obtained using conventional SVD, both followed by ICA. In the former, all four components are recovered reasonably well and the signal containing a time-varying frequency is recovered perfectly. In the latter, only one component is recovered, but the signal of interest has all but lost its essential characteristics. A similar comparison, this time using wavelets, is shown in Figure 10.6.

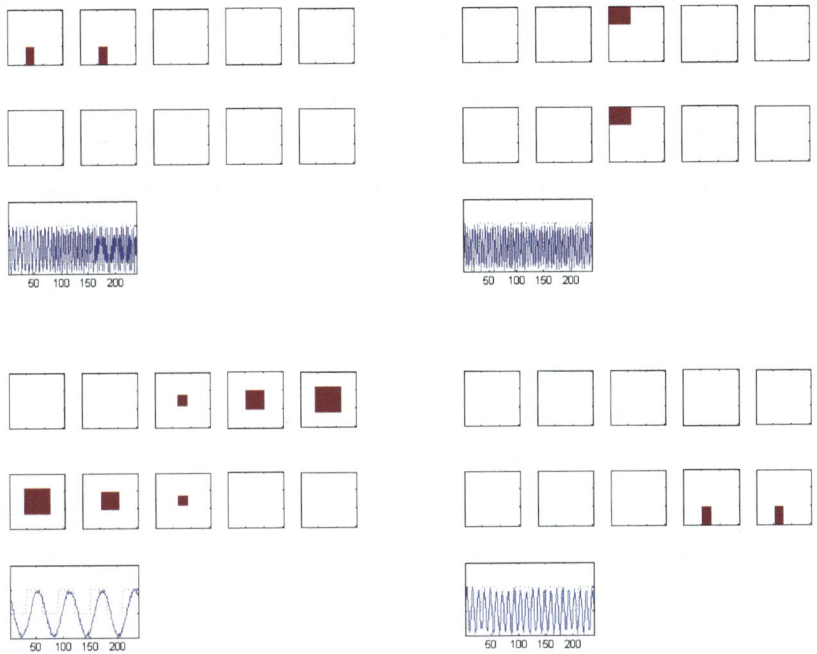

Figure 10.3: The four components used in the simulation. In each panel, the first 10 images are the spatial component maps (one column of **A**), with the red areas denoting activated voxels. The solid line in the subsequent panel is the corresponding time series (one row of **S**). The dotted line denotes the rest-activation block design. The upper left panel contains a signal with a time-varying frequency.

The results are similar, although the signal of interest is not recovered quite as precisely as before.

10.6 CONCLUSION

In this chapter, we extended the technique of supervised SVD in order to allow for the detection of signals with a time-varying frequency. This was accomplished by using a basis of signals that are local rather than global in time. Our simulation shows that this makes the extracted components less sensitive to outliers and noise. While we assumed that the different frequencies as well as the times at which they change are known, these can be adequately estimated from the data when they are not.

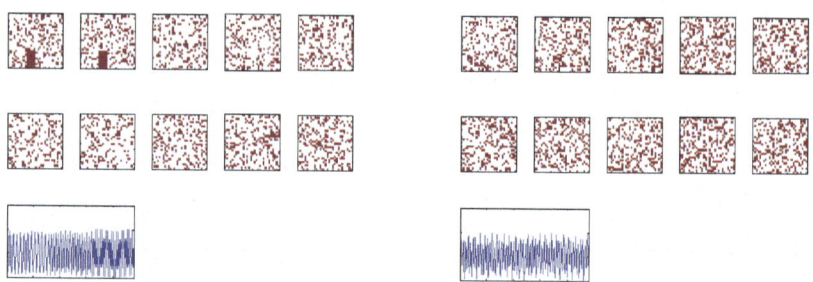

Figure 10.4: The original SSVD (right) fails to recover a signal with a time-varying frequency, but the extended SSVD (left) succeeds.

10.7 M-FILES

Three MATLAB functions used in the chapter are:

```
EXT_base_rSVD.m
EXT_plotcomponents.m
EXT_SSVD_ICA.m
```

```
% EXT_base_rSVD.m
%% SSVD: base_rSVD
%%
%% Authors: Ping Bai, Seonjoo Lee, Haipeng Shen and Young Truong
%%
%% MATLAB functions and scripts to support the paper:
%% P. Bai, H. Shen, X. Huang and Y. Truong (2008).
%% A Supervised Singular Value Decomposition for
%% Independent Component Analysis of fMRI.
%% Statistica Sinica, 18, 1233--1252.
%%
%% Updated on 4/22/15 by Avner Halevy to allow for
%% time-varying frequency
%%
%% omega is an array containing as many structures as there
%% are components. The fields of each structure are:
%% m = method (0 = fixed frequency,
%%              1 = local sines and cosines,
%%              2 = wavelets)
%% f = one or more frequencies in the signal
%% t = time points at which each frequency ends
```

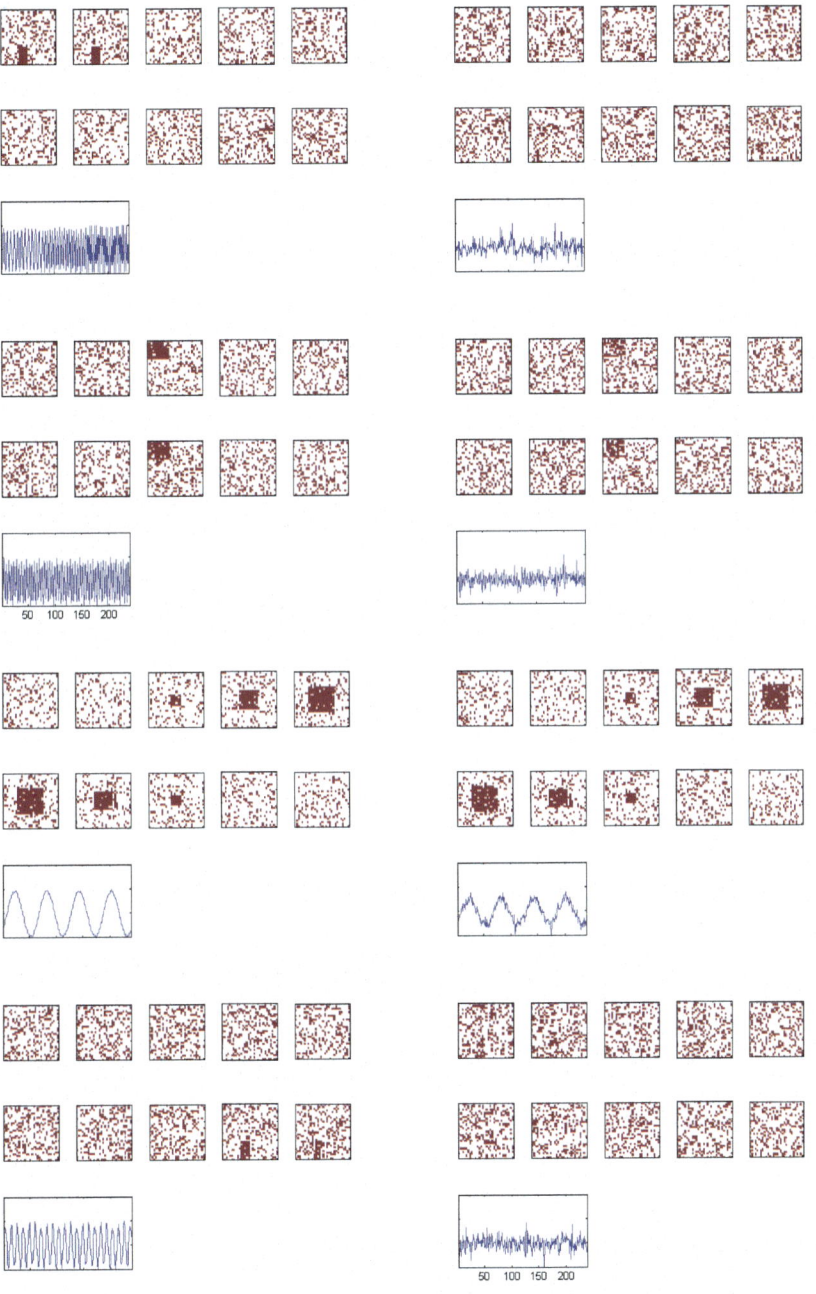

Figure 10.5: Comparison of the results obtained using localized sinusoids (left column) and the conventional approach (right column). The top two panels show the superior results in recovering the signal containing a time-varying frequency.

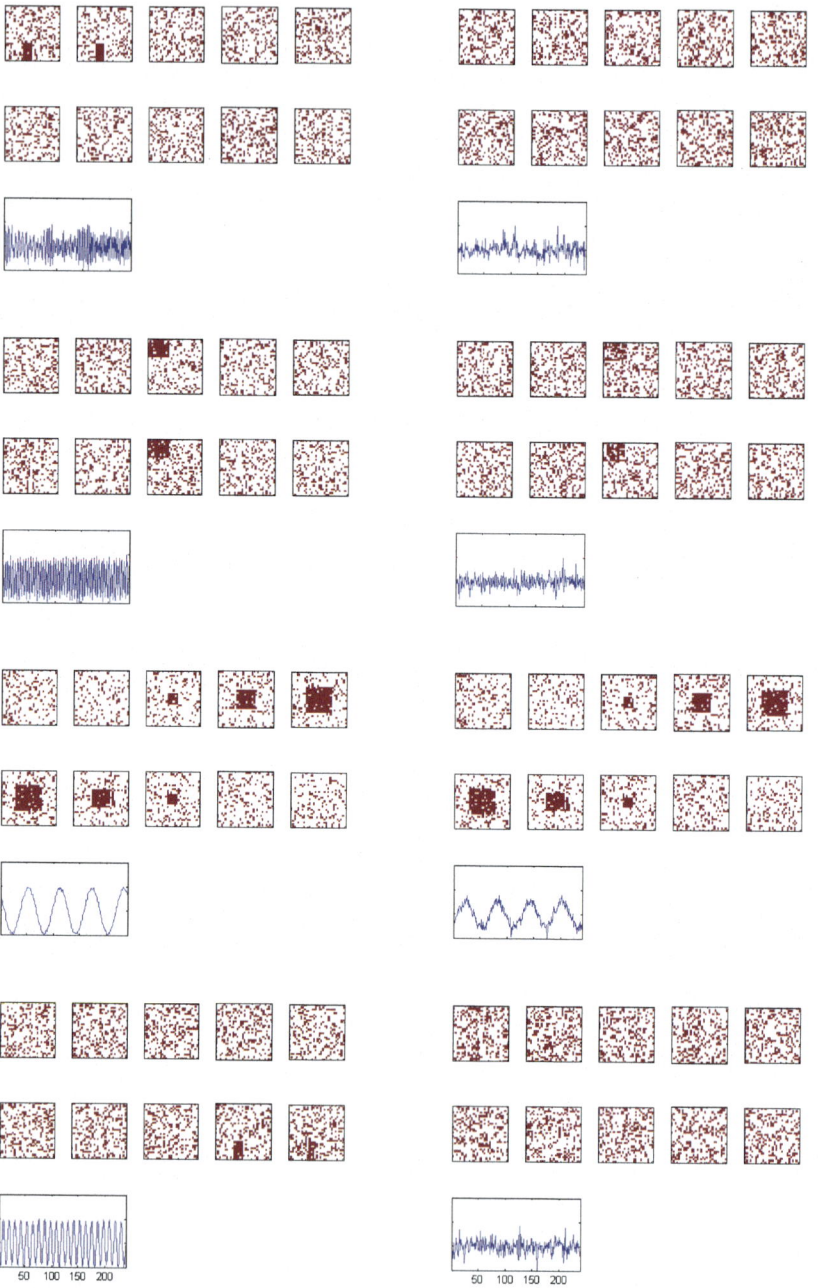

Figure 10.6: Comparison of the results obtained using wavelets (left column) and the conventional approach (right column). The top two panels show the superior results in recovering the signal containing a time-varying frequency.

```
%%

function [U, D, V] = EXT_base_rSVD(X, omega, TR)

[m n] = size(X);
ncomp = length(omega)

U = zeros(m, ncomp);
D = zeros(ncomp,1);
V = zeros(n,ncomp);

T = TR * (n - 1);  % time period 0 ~ T
ind = (0 : TR : T)';

X1 = X;
for i = 1:ncomp
  if omega(i).m == 0
    V1 = [sin(2*pi*omega(i).f*ind), cos(2*pi*omega(i).f*ind)];
  else

  num_freq = length(omega(i).f);

  B = zeros(n,2*num_freq);
  start = 1;
  for j=1:num_freq
    B(start:omega(i).t(j),2*j-1) = ...
        sin(2*pi*omega(i).f(j)*ind(start:omega(i).t(j)));
    B(start:omega(i).t(j),2*j) = ...
        cos(2*pi*omega(i).f(j)*ind(start:omega(i).t(j)));
    start = omega(i).t(j) + 1;
  end

  if omega(i).m == 1

  V1 = B;

  else

    heart = zeros(n,1);
    for j=1:num_freq
      heart = heart + B(:,2*j-1);
    end

    [mpdict,~,~,longs] = ...
```

```
        wmpdictionary(240,'1stcpt',{{'sym4',5},{'wpsym4',5}});
      [YFIT1,R1,COEFF1,IOPT1,QUAL1] = ...
        wmpalg('BMP',heart,mpdict,'itermax',30);
      V1 = mpdict(:,IOPT1);

  end
end

      [U1, D1, Vi] = svd(X1,0);
      UtU = U1' * U1;

      R1 = chol(UtU);

      VtV = V1' * V1;
      R2 = chol(VtV);

      [tu, d, tv] = svd(inv(R1)' * U1' * X1 * V1 * inv(R2));
      u1 = U1 * inv(R1) * tu(:,1);
      v1 = V1 * inv(R2) * tv(:,1);
      s1 = v1' * X1' * u1;

      U(:, i) = u1;
      D(i) = s1;
      V(:,i) = v1;

      X1 = X1 - s1 * u1 * v1';
end;

% EXT_plotcomponents.m
%% SSVD: Plot IC
%%
%% Authors: Ping Bai, Seonjoo Lee, Haipeng Shen and Young Truong
%%
%% MATLAB functions and scripts to support the paper:
%% P. Bai, H. Shen, X. Huang and Y. Truong (2008).
%% A Supervised Singular Value Decomposition for
%% Independent Component Analysis of fMRI.
%% Statistica Sinica, 18, 1233--1252.

% plot the components (as appeared in the paper)
nvox = 30*30; % number of voxels on one slice
ns = 10; % number of slice
ntp = 240; % number of time points
ncomp = 4; % change this to 10 for SVD analysis
```

```
method % 1 -- SSVD-ICA(specify frequencies);
       % 2 -- SSVD-ICA(estimate freq);
       % 3 -- SVD-ICA

if method == 1
    TCtitle = 'TC_ssvd_boxcar.txt';
    ICtitle = 'IC_ssvd_boxcar.txt';
    plottitle = 'SSVD-ICA';
elseif method == 2
    TCtitle = 'TC_ssvd_est_boxcar.txt';
    ICtitle = 'IC_ssvd_est_boxcar.txt';
    plottitle = 'SSVD-EST-ICA';
elseif method == 3
    TCtitle = 'TC_svd_boxcar.txt';
    ICtitle = 'IC_svd_boxcar.txt';
    plottitle = 'SVD-ICA';
end;

ptreshold = 0.09 * ones(1, ncomp);
t = [1.1 1.1 1.1 1.1];
TR = 0.3;
dim = [30 30];
block = zeros(ntp,1);
for i = 1:4
    block(30*(2*i-1)+1 : 30*2*i) = ones(30,1);
end;

P = block;


%fid = fopen('TC_ssvd_boxcar.txt');
fid = fopen(TCtitle);
TC = fscanf(fid, '%f', [ncomp, ntp]);
TC = TC';
S = TC;
fclose(fid);

%fid = fopen('IC_ssvd_boxcar.txt');
fid = fopen(ICtitle);
IC = fscanf(fid, '%f', [ncomp, nvox*ns]);
IC = IC';
fclose(fid);
```

```
for i = 1 : ncomp
    figure;
    % image
    ICi = IC(:,i);
  m = mean(ICi);
 s=std(ICi);
    stICi = abs((ICi-m)/s);
    ICi(find(stICi>=t(i))) = 1;
    ICi(find(stICi<t(i))) = 0;

    for j = 1 : ns
      subplot(3, 5, j);

      Aj=reshape(ICi(1+(j-1)*900:900+(j-1)*900), dim(1), dim(2));

      % rescale
      Aj = Aj * 32 + 32;

      mapout = Aj;

      co = white;
      co(64,:)=[0.5625,0,0]; % dark red

      %figure(1);
      image(mapout);

      colormap(co);

      set(gca, 'YTicklabel', ' ');
      set(gca, 'XTicklabel', ' ');
      axis square;

      if j == 2
        %title(str(i));
        %str = sprintf('SVD-ICA; component %d',i);
        %str = sprintf('SSVD-ICA; component %d',i);
        %str = strcat(plottitle, sprintf('; component %d', i));
        str = plottitle;
        %title(str);
      end;
    end;

    % time
```

```
    subplot(3, 5, 11:12);
    Ti = S(:,i)/max(abs(S(:,i)));
    plot(1:length(block), Ti, '-');
    axis([1 length(block) -1 2]);

    set(gca, 'YTicklabel', ' ');
    if i ~= ncomp
        set(gca, 'XTicklabel', ' ');
    end;
end;

% EXT_SSVD_ICA.m
%% SSVD: Run the simulation study
%%
%% Authors: Ping Bai, Seonjoo Lee, Haipeng Shen and Young Truong
%%
%% MATLAB functions and scripts to support the paper:
%% P. Bai, H. Shen, X. Huang and Y. Truong (2008).
%% A Supervised Singular Value Decomposition for
%% Independent Component Analysis of fMRI.
%% Statistica Sinica, 18, 1233--1252.
%%
%% Run generatedata.m before this.
%% Calls fastica.
%% FastICA_2.5.tar.gz http://research.ics.aalto.fi/ica/fastica/
%% Input:
%%          X_3D_Outliers10_boxcar.txt
%% Output:
%%          TC_ssvd_boxcar.txt
%%          IC_ssvd_boxcar.txt
%%
%% Updated on 4/22/15 by Avner Halevy to allow for
%% time-varying frequency
%%
%% omega is an array containing as many structures as there
%% are components. The fields of each structure are:
%% m = method (0 = fixed frequency,
%%             1 = local sines and cosines,
%%             2 = wavelets)
%% f = one or more frequencies in the signal
%% t = time points at which each frequency ends
%%

nvox = 9000;    % number of voxels
```

```
ntp = 240;         % number of time points
ncomp = 4;         % number of components

method % 1 -- SSVD-ICA(specify frequencies);
       % 2 -- SSVD-ICA(estimate freq);
       % 3 -- SVD-ICA

% load data: run generatedata.m
fid = fopen('X_3D_Outliers10.txt');
%fid = fopen('X_3D_NoOutliers_boxcar.txt');
X = fscanf(fid, '%f', [ntp, nvox]);
X = X';   % nvox by ntp
fclose(fid);

omega(1).m = 0;
omega(1).f = 1/18;
omega(1).t = 0;

omega(2).m = 1;
omega(2).f = [0.5 0.75 1];
omega(2).t = [80 160 ntp];

omega(3).m = 0;
omega(3).f = 0.3;
omega(3).t = 0;

omega(4).m = 0;
omega(4).f = 0.7;
omega(4).t = 0;

TR = 0.3;
block = zeros(ntp,1);
for i = 1:4
    block(30*(2*i-1)+1 : 30*2*i) = ones(30,1);
end;

%% standardize
% column center
X = X - repmat(mean(X), size(X,1),1);
% row normalization
for i = 1 : nvox
    tem = X(i,:);
    tem1 = tem - mean(tem);
    tem2 = tem1/std(tem(:));
```

```
      X(i,:) = tem2;
end;

%% Run SSVD
if method == 1
    % supervised SVD by specifying the frequencies
    [U, D, V] = EXT_base_rSVD(X, omega, TR);
    TCtitle = 'TC_ssvd_boxcar.txt';
    ICtitle = 'IC_ssvd_boxcar.txt';
elseif method == 2
    % supervised SVD by estimating the frequencies
    [U, D, V] = base_rSVD_est(X, ncomp, TR);
    TCtitle = 'TC_ssvd_est_boxcar.txt';
    ICtitle = 'IC_ssvd_est_boxcar.txt';
elseif method == 3
    % conventional SVD
    [U, D, V] = svd(X,0);
    D = diag(D);
    TCtitle = 'TC_svd_boxcar.txt';
    ICtitle = 'IC_svd_boxcar.txt';
end;

%%
ni = sum(D > 10e-4);
    % the # of significant components for the ith slice
ni = min(ni, ncomp);
%U = U(:, 1:ni);
%D = D(1:ni);
%V = V(:, 1:ni);

D = diag(D);

% FastICA_2.5.tar.gz http://research.ics.aalto.fi/ica/fastica/
cd FastICA_25

% get reduced data Y
Y = (U(:, 1:ni) * D(1:ni, 1:ni))'; % ncomp by nvox;  spatial
% Y = D(1:ni,1:ni) * V(:,1:ni)'; % temporal

[icasig, A, W] = fastica (Y, 'numOfIC', ni);
size(A)

T = V(:, 1:ni) * A;
S = icasig;
```

```
% fastICA algorithm does the data reduction automatically
% when the number of components required is larger than
% the singularity of covariance matrix
ni = min(ni, size(S,1))
cd ..

form = [];
for j = 1:ni
    form = [form, '%f\t'];
end;
form = [form, '\n'];

fid = fopen(TCtitle, 'w');
for j = 1 : ntp
    fprintf(fid, form, T(j,:));
end;
fclose(fid);

fid = fopen(ICtitle,'w');
S = S'; % nvox by ni
for j = 1 : nvox
    fprintf(fid, form, S(j,:));
end;
fclose(fid);
```

Bibliography

1. F. G. Ashby. *Statistical analysis of fMRI data.* MIT press, 2011.
2. P. Bai, H. Shen, X. Huang, and Y. Truong. A Supervised Singular Value Decomposition for Independent Component Analysis of fMRI. *Statistica Sinica,* 18:1233–1252, 2008.
3. A. Hyvärinen, J. Karhunen, and E. Oja. *Independent component analysis,* volume 46. John Wiley & Sons, 2004.
4. M.J. McKeown, S. Makeig, G.G. Brown, T.P. Jung, S.S. Kindermann, A.J. Bell, and T.J. Sejnowski. Analysis of fMRI Data by Blind Separation into Independent Spatial Components. *Human Brain Mapping,* 6(3):160–188, 1998.
5. R. A. Poldrack, J. A. Mumford, and T. E. Nichols. *Handbook of functional MRI data analysis.* Cambridge University Press, 2011.

A Discrete Fourier Transform

CONTENTS

The definitions of *Discrete Fourier Transform* (DFT), multivariate normal and its complex extension will be given in this appendix. See also Chapter 3 and books on time series analysis [1, 2].

A.1 DISCRETE FOURIER TRANSFORM (DFT)

The *Discrete Fourier Transform* (DFT) of a vector-valued time series $\mathbf{Y}_t$, $t = 0, 1, \ldots, T - 1$, is given by

$$\varphi_Y(k) = \sum_{t=0}^{T-1} \mathbf{Y}_t \exp(-i2\pi kt/T), \quad k = 0, \pm 1, \pm 2, \ldots, \pm[(T-1)/2]. \qquad (A.1)$$

Under reasonable assumptions, it can be shown that $\varphi_Y(k)$, $k = 0, \pm 1, \ldots$ are complex normal [1].

In numerical computational problems involving R or MATLAB, the `fft` function computes the DFT of a time series using a fast Fourier transform (FFT) algorithm.

A.2 MULTIVARIATE NORMAL DISTRIBUTION

In probability theory and statistics, the multivariate normal distribution or multivariate Gaussian distribution, is a generalization of the one-dimensional (univariate) normal distribution to higher dimensions. Given a d-dimensional mean vector

$$\mu = (\mu_1, \mu_2, \ldots, \mu_d)$$

and $d \times d$ (symmetric, positive-definite) covariance matrix

$$\Sigma = \begin{pmatrix} \sigma_{11} & \sigma_{12} & \cdots & \sigma_{1d} \\ \sigma_{21} & \sigma_{22} & \cdots & \sigma_{2d} \\ \vdots & \vdots & \ddots & \vdots \\ \sigma_{d1} & \sigma_{d2} & \cdots & \sigma_{dd} \end{pmatrix}, \qquad \sigma_{ij} \in \mathbb{R}, \quad i, j = 1, 2, \ldots, d,$$

a d-dimensional random vector $\mathbf{X} = (x_1, x_2, \ldots, x_d)$ is said to have a multivariate normal distribution, abbreviated by $\mathbf{X} \sim N_d(\mu, \Sigma)$, iff $Y = \mathbf{a}'\mathbf{X} = a_1 X_1 + \cdots + a_d X_d$ is normal with mean $\mathbf{a}'\mu$ and variance $\mathbf{a}'\Sigma\mathbf{a}$.

It can be shown that the probability density function of $\mathbf{X}$ is given by

$$f(\mathbf{x}) = \frac{1}{\sqrt{(2\pi)^d |\Sigma|}} \exp\left(-\frac{1}{2}(\mathbf{x} - \mu)^{\mathrm{T}} \Sigma^{-1} (\mathbf{x} - \mu)\right), \quad \mathbf{x} \in \mathbb{R}^d.$$

A.3   COMPLEX NORMAL DISTRIBUTION

Definition 2 *The Complex Multivariate Normal Distribution. If $\Sigma = \Sigma_1 + i\Sigma_2$ is a complex-valued $m \times m$ matrix such that $\Sigma = \Sigma^\tau$ and $\mathbf{a}^\tau \Sigma a \geq 0$ for all $\mathbf{a} \in \mathbb{C}^m$, then we say that $\mathbf{Y} = \mathbf{Y}_1 + i\mathbf{Y}_2$ is a complex-valued multivariate normal random vector with mean $\mu = \mu_1 + i\mu_2$ and covariance matrix Σ if*

$$\begin{bmatrix} \mathbf{Y}_1 \\ \mathbf{Y}_2 \end{bmatrix} \sim N\left(\begin{bmatrix} \mu_1 \\ \mu_2 \end{bmatrix}, \frac{1}{2}\begin{bmatrix} \Sigma_1 & -\Sigma_2 \\ \Sigma_2 & \Sigma_1 \end{bmatrix}\right). \tag{A.2}$$

We then write $\mathbf{Y} \sim N_m^C(\mu, \Sigma)$.

Bibliography

1. D.R. Brillinger. *Time Series*. Holden-Day, 1981.
2. P.J. Brockwell and R.A. Davis. *Time Series: Theory and Methods*. Springer, second edition, 1991.

B The R Software Package

CONTENTS

B.1 SOFTWARE INFORMATION

The main tools we introduce in this appendix are the R language [4] and the `knitr` package [7, 8]. These are excellent tools for reproducible research and both packages are available on *Comprehensive R Archive Network* (CRAN) as free and open-source software. You may download them from the CRAN website:

$$\texttt{https://cran.r-project.org/}$$

or any of its mirrors, such as

$$\texttt{http://cran.rstudio.com}$$

It is also useful to consider using *RStudio* [5]:

$$\texttt{https://www.rstudio.com/}$$

B.2 INSTALLATION

Instructions for installation in `Windows`, `Mac OSX` and `Linux` are available in the above websites, and they are very straightforward to install.

B.2.1 WINDOWS INSTALLATION

For Windows, follow these steps:

1. Navigate to `https://cran.r-project.org/`
2. Click on `Download R for Windows`.
3. Execute the downloaded `.exe` to install R.

B.2.2 LINUX-UBUNTU INSTALLATION

For Linux-Ubuntu, follow these steps:

1. Navigate to `https://cran.r-project.org/`
2. Click on `Download R for Linux` and select `ubuntu`.
3. In the `/etc/apt/sources.list` file, add the CRAN mirror entry.
4. Download and update the package lists from the repositories using the `sudo apt-get update` command.
5. Install the R system using the `sudo apt-get install r-base` command.

B.2.3 LINUX-RHEL/CENTOS INSTALLATION

For Linux-RHEL/CentOS, follow these steps:

1. Navigate to `https://cran.r-project.org/`
2. Click on `Download R for Linux` and select `Red Hat OS`.
3. Download the `R-*core-*.rpm` file.
4. Install the `.rpm` package using the `rpm -ivh R-*core-*.rpm` command.
5. Install the R system using `sudo yum install R`.

B.2.4 MAC OS X INSTALLATION

For Mac, follow these steps:

1. Navigate to `https://cran.r-project.org/`
2. Click on `Download R for (Mac) OS X`.
3. Download the `R-*.pkg` file.
4. Click on `tools` or go to `https://cran.r-project.org/bin/macosx/tools`, download `gfortran-*.dmg` and `tcltk-*.dmg`.
5. Install the `R-*.pkg` file.
6. Install the `gfortran-*.dmg` and `tcltk-*.dmg` files.

Under Mac OS X, some R packages do not come with binaries, and so `gfortran` would come in handy for compiling from the sources.

After installing the `base R` package, it is advisable to install *RStudio*, which is a powerful and intuitive Integrated Development Environment (IDE) for R.

B.2.5 INSTALLING RSTUDIO

To install *RStudio*, follow the steps below:

1. Navigate to https://www.rstudio.com/products/rstudio/download/
2. Download the latest version of *RStudio* for your operating system.
3. Execute the installer file and install *RStudio*.

B.2.6 INSTALLING R PACKAGES, TASK VIEWS

There are several thousand R packages, consisting of collections of R functions, data, and compiled code. Browse for packages by name, author, title, and published date or find packages organized by task at:

https://cran.r-project.org/web/packages/

There are base packages (which come with R automatically), and contributed packages (which must be downloaded for installation).

There are instructions for installing these packages. For starters, we recommend using *RStudio*.

It is also very useful to check out Task Views:

https://mran.revolutionanalytics.com/taskview/

These are guides on CRAN that group sets of R packages and functions by type of analysis, fields, or methodologies. Instructions for installing Task Views are available at:

https://mran.revolutionanalytics.com/rpackages/

B.3 DOCUMENTATION

There is a section for R Manuals; browse them at:

https://cran.r-project.org/manuals.html

These documents are available in pdf, html, or epub. We recommend *An Introduction to R*

https://cran.r-project.org/doc/manuals/r-release/R-intro.pdf

to get started. Most of the examples described there can be run and reproduced using *RStudio*.

Some useful free books are available at

https://cran.r-project.org/other-docs.html

For starters, R can be approached as a programming language [3], or as a general statistical tool for applications [2]. The latter has a very informative website:

http://www.statmethods.net/

B.4   TUTORIALS

R by itself has the capabilities necessary to gather data, analyze it, and, with a little help from *knitr/rmarkdown* and markup languages, present results in a way that is highly reproducible. *RStudio* will do all of these things, simplify many of them, and navigate through them more easily. It also is a happy medium between R's text-based interface and a pure GUI.

To run the examples in this book, it is necessary to treat each chapter as a project as there are R functions written specifically for that chapter. This is another reason for using *RStudio* as it has an excellent project management feature. Moreover, *RStudio* is a highly advanced text editor for R, and has R's help system, version control, packages management, and many other useful features to expedite the program development process in a single application. *RStudio* does not perform any computation or statistical analysis; it only makes it easier for you to perform such tasks with R. Most importantly, *RStudio* offers many facilities that make working reproducibly a lot easier. A good place to learn more about it is [6].

We usually carry out the statistical analysis using R by grouping files and data into a so-called working directory, and files containing R functions will have an extension R. For example, collect all the R functions in Chapter 3 and put them in a folder or directory called ch03. The file BIV.R contains one or more R functions written by us. Now start *RStudio* and select New Project under File, then select Existing Directory in the pop-up panel Create project from:. Now navigate to the directory ch03 and click Create Project. After this, *RStudio* will then manage the files and data created in the project.

When *RStudio* is first started, there will be three panels: The long one on the left is the Console where R commands can be entered to conduct statistical operations. On the right, the top panel records the Workspace and History, and the bottom panel has several tabs. Click on File to display the contents of the working or project directory. The Plot is for graphical display of R's many excellent graphical procedures for data analysis. The Packages tab shows what R packages have been installed in your system. Here you can easily install or update packages. The next tab is Help, where all the R help files are kept. You can search them by entering the command name or some key-words.

To call the functions in BIV.R into R, issue the R command below in the *RStudio* Console:

```
> source("BIV.R")
```

Instead of typing the commands directly into the Console window, you can use *RStudio* to create a file and enter all the commands there. An example is given in the file `prep.R`. Each line or group of lines in this file can be run in R by highlighting them, followed with a click on the Run icon at the top of that window. The results should be reproduced through R.

R comes with base packages installed and they are loaded when R is launched. For example, the default base packages loaded at startup are

```
"datasets"  "utils"  "grDevices" "graphics" "stats"   "methods"
```

The command to get the base packages is:

```
> getOption("defaultPackages")
```

Projects such as those in Chapters 2, 6, or 7 will require additional R packages to be installed. These are called *contributed packages*. For example, Chapter 2 will call upon the R package `polspline`. This can be easily installed by clicking the Packages tab in the lower right panel of *RStudio*, and selecting Install. Enter `polspline` into the pop-up panel, then wait for *RStudio* to complete the installation. Alternatively, it can be installed by issuing the following command

```
> install.packages("polspline", depends = TRUE)
```

into the RStudio Console. After the package has been installed, load it into R by checking the box next to the package name, `polspline`, listed in the Packages panel, or enter the following command into the console:

```
> library(polspline)
```

To learn more about the package, click the name in the Packages panel. Before using any package, it may be a good idea to learn more about it by visiting its website. The package `polspline` is located at:

https://cran.r-project.org/web/packages/polspline/index.html

Next, repeat the steps described above to install the package `coloredICA`:

https://cran.r-project.org/web/packages/coloredICA/index.html

These packages will be stored in the memory as long as *RStudio* (or R) is up and running. To free up the memory space, uncheck that package if it is no longer needed in the statistical analysis. If it is needed again in the analysis, just check it. Also, all the loaded packages will be removed from the memory once *RStudio* has ended. After re-starting *RStudio*, the installed packages can be re-loaded into the memory by checking the boxes next to the package names. A list of installed packages is available for viewing in the Packages tab in *RStudio*.

After a project is completed, the next step is to copy and paste the results into a document in various formats (LATEX to pdf, Markdown to HTML). The R package *knitr* is ideal for this purpose [8, 7]. Install it via *RStudio* Packages/Install or using the Console with

```
> install.packages("knitr")
```

Together with a typesetting program such as TEX/L^ATEX, R, *RStudio* and *knitr* form a very powerful tool for reproducible research. Many excellent examples are located at:

http://yihui.name/knitr/

and

https://github.com/christophergandrud/Rep-Res-Book

The latter has provided the source to reproduce the whole book [1].

Bibliography

1. Christopher Gandrud. *Reproducible Research with R and RStudio*. CRC Press, 2013.
2. Robert Kabacoff. *R in Action*. Manning Publications Co., 2014.
3. Norman Matloff. *The Art of R Programming: A Tour of Statistical Software Design*. No Starch Press, 2011.
4. R Core Team. *R: A Language and Environment for Statistical Computing*. R Foundation for Statistical Computing, Vienna, Austria, 2013. URL http://www.R-project.org/.
5. RStudio Team. *RStudio: Integrated Development Environment for R*. RStudio, Inc., Boston, MA, 2015. URL http://www.rstudio.com/.
6. Mark PJ Van der Loo. *Learning RStudio for R Statistical Computing*. Packt Publishing Ltd, 2012.
7. Yihui Xie. *Dynamic Documents with R and knitr*. Chapman and Hall/CRC, Boca Raton, Florida, 2nd edition, 2015. URL http://yihui.name/knitr/. ISBN 978-1498716963.
8. Yihui Xie. *knitr: A General-Purpose Package for Dynamic Report Generation in R*, 2015. URL http://yihui.name/knitr/. R package.

C Matrix Computation

CONTENTS

C.1 GETTING STARTED

MATLAB (MATrix LABoratory) is a powerful tool for neurocomputing. The best way to get started is to launch the software and familiarize yourself with it by browsing the tutorials, which also contain a few short videos. After launching MATLAB, it waits for your instructions, which can be given next to >> with a blinking bar | in the (center) panel called the Command Window. Start your tutorials by clicking Getting Started in:

New to MATLAB? Watch this Video, see Examples, or read ...

Alternatively, after launching MATLAB, select

    HOME/Help/Examples/Getting Started.

Next, learn more about the MATLAB Desktop by watching the video:

    Working in The Development Environment.

After these videos, get a more comprehensive introduction to MATLAB by tabbing Help/Documentation and browse the topics there.

By now there are thousands of books on MATLAB. We specifically mention three: The short *Learning MATLAB* [1], a mathematically oriented text *MATLAB Guide* [2], and an introductory text in numerical methods, Matlab, and technical computing [3], which is also available at

    https://www.mathworks.com/moler/chapters.html

Experiments with MATLAB is a collection of mathematical and computational projects:

    http://www.mathworks.com/moler/exm

Finally, worth mentioning is a book on MATLAB for neuroscientists [4], which can be served as an extension of [3].

A few important things about MATLAB:

1. Use the up- or down-arrow keys to cycle through commands that have been issued.
2. Type edit in the Command Window to launch the MATLAB Editor for writing code.
3. Run your code step by step to debug.
4. Use the Window in the Tool Bar to locate windows that are related to the current project. This is also useful for showing the window of interest.
5. Save graphs from the Figure Window to the format of your choice.
6. Download and install MATLAB project manager from

> http://www.mathworks.com/matlabcentral/fileexchange/
> 41585-matlab-project-manager

C.2   TUTORIALS

Download Experiments with MATLAB (http://www.mathworks.com/moler/exm). Create a directory called exm and unpack the downloaded file into it.

> http://www.mathworks.com/moler/exm

Launch MATLAB by entering the command pathtool

```
>> help pathtool
>> pathtool
```

and navigate to the directory exm created above.

Assume that projects was properly installed and is in the MATLAB search path. Check it out by entering the following command:

```
>> help projects
... ...
Examples:
    projects list
    projects save myProject
    projects close
    projects load default
    projects rename myProject myLibrary
```

All projects are stored in the %userpath%/projects.mat. This file with empty "default" project is created at the first run of the script. If %userpath% is empty, the script will execute

```
userpath('reset').
```

First project always has name "default"

If it is not installed correctly or if `projects` is not in the MATLAB path, an error message will be shown. Suppose the above message is on the screen (i.e., MATLAB Command Window); enter the following:

```
>> projects
List of available projects:
-> 1: default
```

Since there are no named projects other than the current one, it is called the default. Now enter

```
>> projects save exm
Project "exm" was saved

>> projects
List of available projects:
   1: default
-> 2: exm
```

You just added a new project to MATLAB. To create another one, repeat the above steps after downloading

```
                https://www.mathworks.com/moler/ncm.zip
```

Now MATLAB should have your two projects listed:

```
>> projects
List of available projects:
   1: default
-> 2: exm
   3: ncm
```

Using this function to manage your projects is essential, though it may seem unnecessary at the beginning. Soon you will appreciate the project manager after checking through the MATLAB functions listed at the end of each chapter in this book. By creating a project for each chapter and examining the steps in each function, perhaps two or three files were opened in the editor in the project `chapter2`. If MATLAB is needed for something else outside the book chapters, close the project:

```
>> project close
```

so that the other task can be accomplished. After this, you can return to the previous project via

```
>> project load chapter2
```

The manager will remember where the exact settings of project A and the files opened in the MATLAB editor are located.

Bibliography

1. Tobin A Driscoll. *Learning MATLAB*. SIAM, 2009.
2. Nicholas J Higham. *The matrix computation toolbox for MATLAB (version 1.0)*. Manchester Centre for Computational Mathematics, 2002.
3. Cleve B Moler. *Numerical computing with MATLAB: revised reprint*. SIAM, 2008.
4. Pascal Wallisch, Michael E Lusignan, Marc D Benayoun, Tanya I Baker, Adam Seth Dickey, and Nicholas G Hatsopoulos. *MATLAB for neuroscientists: An introduction to scientific computing in MATLAB*. Academic Press, 2014.

D Singular Value Decomposition

CONTENTS

In matrix computation, *singular value decomposition* (SVD) is a factorization of a matrix into a product of special matrices. There are many important applications in numerical analysis, signal processing, and statistics [1]. It is an essential tool for dimension reduction in the methodology for ICA.

D.1 SINGULAR VALUE DECOMPOSITION (SVD)

Suppose $\mathbf{Y} \in \mathbb{R}^{m \times n}$ is an $m \times n$ matrix whose entries are real numbers. Then there exist orthogonal matrices

$$\mathbf{U} \in \mathbb{R}^{m \times m}, \qquad \mathbf{V} \in \mathbb{R}^{n \times n}$$

such that

$$\mathbf{U}'\mathbf{Y}\mathbf{V} = \mathrm{diag}(\sigma_1, \ldots, \sigma_p), \qquad p = \min\{m, n\}, \tag{D.1}$$

where $\mathrm{diag}(\sigma_1, \ldots, \sigma_p)$ is a diagonal matrix of non-negative *singular values* and

$$\sigma_1 \geq \sigma_2 \geq \cdots \geq \sigma_p \geq 0.$$

Rewrite (D.1) so that

$$\mathbf{Y} = \mathbf{U}\,\mathrm{diag}(\sigma_1, \ldots, \sigma_p)\,\mathbf{V}'.$$

This is a well-known factorization and is referred to as a *singular value decomposition* (SVD) of $\mathbf{Y}$ [1]. The SVD also holds for matrices with complex entries. In this case, $\mathbf{U}$ and $\mathbf{V}$ are unitary matrices.

D.2   COMPUTING SVD

In MATLAB the SVD of of a matrix Y is achieved by issuing the command:

```
>> [U,S,V] = svd(Y);
```

Below is an example of the SVD of a small, square matrix and is provided by one of the test matrices from the MATLAB gallery.

```
>> Y = gallery(3)

Y =

    -149    -50    -154
     537    180     546
     -27     -9     -25

>> [U,S,V] = svd(Y);
>> U

U =

    -0.2691    -0.6798     0.6822
     0.9620    -0.1557     0.2243
    -0.0463     0.7167     0.6959

>> S

S =

   817.7597          0          0
          0     2.4750          0
          0          0     0.0030

>> V

V =

     0.6823    -0.6671     0.2990
     0.2287    -0.1937    -0.9540
     0.6944     0.7193     0.0204
```

The expression U*S*V' generates the original matrix to within round-off error.

```
>> U*S*V'

ans =

 -149.0000   -50.0000  -154.0000
  537.0000   180.0000   546.0000
  -27.0000    -9.0000   -25.0000
```

D.3   SVD IN R

In R, the SVD of a matrix Y is given by the command:

```
> svd(Y)
```

Below is an R example of the SVD using the above square matrix Y.

```
> Y <- matrix(c(-149,537,-27,-50,180,-9,-154,546,-25), ncol=3)
> svd(Y)
$d
[1] 8.177597e+02 2.474974e+00 2.964523e-03

$u
             [,1]        [,2]       [,3]
[1,] -0.26906707 -0.6798212 0.6822361
[2,]  0.96200923 -0.1556695 0.2242883
[3,] -0.04627257  0.7166660 0.6958798

$v
            [,1]        [,2]        [,3]
[1,] 0.6822779 -0.6671414  0.29903068
[2,] 0.2287120 -0.1937185 -0.95402513
[3,] 0.6943974  0.7193021  0.02041339
```

Here d is the diagonal matrix D in the SVD of Y. It agrees with the MATLAB version.

Bibliography

1. Gene H Golub and Charles F Van Loan. *Matrix computations*, volume 3. JHU Press, 2012.

Index